Sarbanes-Oxley IT Compliance Using Open Source Tools, 2E

Christian B. Lahti
Roderick Peterson

KEY	SERIAL NUMBER
001	HJIRTCV764
002	PO9873D5FG
003	829KM8NJH2
004	BAL923457U
005	CVPLQ6WQ23
006	VBP965T5T5
007	HJJJ863WD3E
008	2987GVTWMK
009	629MP5SDJT
010	IMWQ295T6T

PUBLISHED BY
Syngress Publishing, Inc.
Elsevier, Inc.
30 Corporate Drive
Burlington, MA 01803

Sarbanes-Oxley IT Compliance Using Open Source Tools, 2E

Printed in the United States of America
1 2 3 4 5 6 7 8 9 0

ISBN 13: 978-1-59749-216-4

Publisher: Amorette Pedersen
Acquisitions Editor: Patrice Rapalus
Project Manager: Greg deZarn-O'Hare
Cover Designer: Michael Kavish
Page Layout and Art: SPi
Copy Editor: Judy Eby
Indexer: SPi

For information on rights, translations, and bulk sales, contact Matt Pedersen, Commercial Sales Director and Rights, at Syngress Publishing; email m.pedersen@elsevier.com.

Lead Authors

Christian B. Lahti is a computer services consultant with more than 18 years experience in the IT industry. He is an expert and evangelist in the field of Open Source technologies in the IT enterprise and has successfully implemented global IT infrastructures. His focus and expertise lies in cross-platform integration and interoperability, security, database, and web development. Christian currently holds the position of Director of IT at a technology startup in Mountain View, CA and is a frequent speaker at both LinuxWorld and O'Reilly's OSCON on a wide variety of topics such as Enterprise authentication and infrastructure monitoring and has contributed to several Open Source projects.

Christian has a degree in Audio Engineering and has several certifications. He is an original co-author of the first edition of this book and served as technical editor and contributing author to Windows to Linux Migration Toolkit (Syngress Publishing, ISBN: 1-931836-39-6).

Roderick Peterson has more than 20 years' experience in the IT industry. He has held various positions with both Fortune 500 public companies and small private companies. Roderick currently holds the position of IT Director at a public technology company in the Silicon Valley. His diverse background includes knowledge of mainframe operations, LAN, Internet, IT infrastructure, business applications, and the integration of emerging technologies. He has successfully led the development and deployment of major applications at several global companies. Roderick also successfully owned and operated his own IT consulting business for more than five years.

Along with being original co-author of the first edition of this book, Roderick has lectured on Sarbanes-Oxley IT Compliance and Governance at the SANS Institute Executive Track.

Contributing Authors

Steve Lanza has more than 20 years of business experience ranging from fortune 500 enterprises to small private and pubic companies. He has held executive positions of Chief Financial Officer at various companies responsible for global business operations, sales, marketing, manufacturing, finance and administration, business development and engineering. His current position is Executive Vice President, Business Development and Chief Financial Officer at a privately held technology company headquartered in Silicon Valley.

Steve has a Bachelors of Science degree in Finance from Cal Poly in San Luis Obispo, an MBA from GGU, and a Certificate of Engineering Management from Cal Tech (IRC). He also holds the title of Certified Management Accountant (CMA).

Bill Haag, William K. Haag (Retired) has over 43 years in Information Technology. During his career he has held various senior management positions, the most recent being the worldwide position of Senior Director of Information Management Services for the Applied Materials Corporation. Previous to Applied Materials he was the CIO of Racal-Datacom, Vice President of Technology and Systems services for the Healthshare Group, and held senior management positions in ATT Paradyne Corporation, Paramount Communication Corporation and Allied Signal Corporation. His accomplishments with these firms include: the development and implementation of both domestic and international information systems to achieve business objectives; significant budget and staff realignments to align MIS with the corporate strategies. His achievements have been recognized in trade and business publications including CIO, CFO, Information Week, LAN World, and Florida Business. He has also been a guest speaker for Bell Atlantic, Information Builders and the Technical Symposium.

Bill received his bachelor's degree in Business Administration from Indiana University and has attended the University of South Florida's Masters program.

Rod Beckström is a serial entrepreneur and catalyst. He is the chairman and chief catalyst at TWIKI.NET, an enterprise Wiki company. He recently co-authored the bestseller "The Starfish and the Spider: The Unstoppable Power of Leaderless Organizations." After working as a trader at Morgan Stanley in London, Rod started his first company when he was 24 and grew it into a global enterprise with offices in New York, London, Tokyo, Geneva, Sydney, Palo Alto, Los Angeles and Hong Kong. That company, CATS Software, went public and was later sold successfully. He has helped start other firms including Mergent Systems and American Legal Net.

He has helped launch more than a half dozen non-profit groups and initiatives including Global Peace Networks which supported the group of CEO's who helped open the border and trade between India and Pakistan, SV2, and the Environmental Markets Network. Rod serves as a Trustee of Environmental Defense and Director of Jamii Bora Africa Ltd., a micro-lending group with 140,000 members. A Stanford BA and MBA, Rod served as President of the graduate/undergraduate student body and was a Fulbright Scholar in Switzerland. His personal website is www.beckstrom.com.

Peter Thoeny is the founder of TWiki and has managed the open-sourced TWiki.org project for the last nine years. Peter invented the concept of structured Wiki's, where free form Wiki content can be structured with tailored Wiki applications. He is now the CTO of TWIKI.NET, a company offering services and support for TWiki. He is a recognized thought-leader in Wiki's and social software, featured in numerous articles and technology conferences including Linux World, Business Week, The Wall Street Journal and more. A software developer with over 20 years experience, Peter specializes in software architecture, user interface design and web technology.

Peter graduated from the Swiss Federal Institute of Technology in Zurich, lived in Japan for 8 years working as an engineering manager for Denso building CASE tools, and managed the Knowledge Engineering group at Wind River for several years. He co-authored the Wiki's for Dummies book, and is currently working on a Wiki's for the Workplace book.

Matt Evans has had a long career in various software development and software quality assurance positions, most of these positions were in early stage startups. Matt graduated from University of Oregon with a Bachelor of Science degree in Computer Science. Currently he holds the position of Senior Director of Engineering Services at a software development startup that specializes in automated test generation tools for the Java Enterprise. Matt has taken advantage of Open Source tools and technologies over the years and is a firm believer in their value and effectiveness for software development and IT infrastructure.

Erik Kennedy has 15 years of experience in the IT industry. His background is in the areas of UNIX/Linux architecture and deployment and IT Security. He has held various positions at Fortune 500 public companies and is currently a Senior Systems Engineer at a public technology company in the Silicon Valley.

John T. Scott has 15 years experience in IT. His background includes end-to-end infrastructure design, implementation and support for PC platforms, IP networks and the security of both for all business models of all sizes. He currently leads an information security incident response team for a global fortune 50 company. He holds CISSP and GIAC certifications and has a bachelor's degree in IT.

Contents

Chapter 1

Overview – The Goals of This Book

Solutions in this chapter:

- IT Manager Bob – The Nightmare
- What This Book Is
- What This Book Is Not
- Why Open Source
- VM Spotlight: CentOS Linux Distribution
- Case Study: NuStuff Electronics, an Introduction

☑ Summary

☑ Solutions Fast Track

☑ Frequently Asked Questions

IT Manager Bob – The Nightmare

"There's no doubt that 404 goes too far, you end up documenting things for the sake of documenting them, even if your judgment says you've gone a bit overboard"."

–Bruce P. Nolop. CFO, Pitney Bowes

The above quote refers to Pitney Bowes's first year audit effort in which they developed testing of 134 processes and more than 2,000 controls in 53 locations and ultimately found no significant weaknesses. We can just imagine the onerous task of managing this huge compliance effort, and can sympathize and agree with Mr. Nolop's final assessment of the outcome. Rather than jump ahead with the language and jargon of compliance, let's step back for a moment and consider a day in the life of Information Technology (IT) Manager, Bob.

It's Monday morning and you have barely had enough time to get your first cup of coffee and log in to check server availability before it starts—your first user call—the Human Resources (HR) Manager system won't boot. After going through the usual—making sure that the correct power button is being pressed, checking to see that it's plugged in, checking the outlet, and so on, you decide, since the HR Manager has a tendency to escalate problems to the Chief Executive Officer (CEO), you will go to the HR Manager's desk to see if you can determine what the problem might be. After querying the HR Manager more intently, you quickly determine the cause of the problem. Apparently, in an attempt to be "Green," the HR Manager turned off the power strip for her PC the Friday before she left work. Well, you guessed it, although she checked to see that everything was plugged in, she never noticed her power strip was off. As you're walking back you think to yourself, well, looks like this Monday is not going to be any different from any other Monday—or so you think.

After returning back from the HR Manager's desk, you take a quick look at your calendar to see what is on your agenda for the day (Figure 1.1). As usual there are more tasks than time to complete them.

Figure 1.1 IT Manager Bob's Calendar

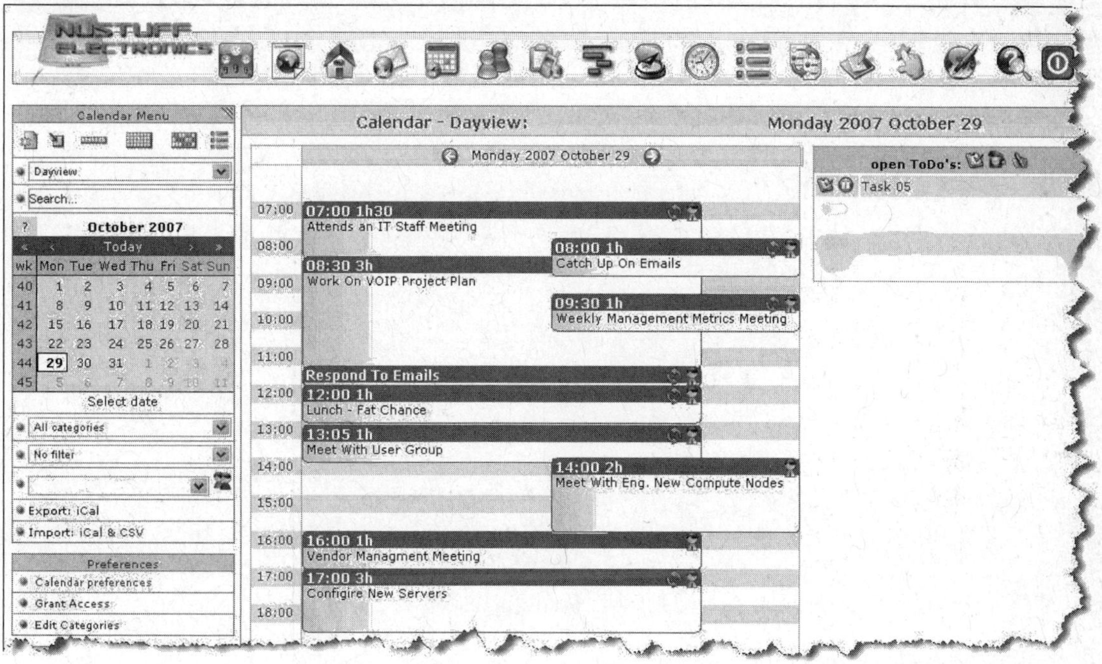

You're halfway through your second meeting when your cell phone rings. You look down at the number and immediately realize it is the CEO's admin. You think about the user this morning, and think, great, she can't switch on a power strip and she still escalates to the CEO. To your surprise, the CEO has asked that you attend a meeting with him, the Chief Information Officer (CIO), and the Controller to discuss this "SOX" thing. You look down to make sure your socks are matching, wondering why on earth they would be concerned with such a nonsensical thing as you enter the meeting. The expected crowd is there as you settle in, along with a couple of those slightly familiar faces you have seen floating about. "Bob, this is Bill and Jane from WeHelpU Consulting, and they have been spending the past couple of months helping us to prepare for our Sarbanes-Oxley compliance audit," says the CEO by way of introduction. The consultants go on to explain that they are there to help finance analyze their business processes and reporting structures for the financial chain. After a few minutes, your eyes begin to glaze over so you decide to read your e-mail. After all, meetings seem like the best time to catch up on this sort of thing. You nod a few times when your name is mentioned, catching phrases here and there like "control objectives" and "material weakness"... say that doesn't sound too good.

Wait a minute! You suddenly realize these people have been here for several months and you are just now getting sucked into something that you instantly know you really don't want any part of, but it is becoming apparent that unfortunately you will have no choice in the matter. To top it off, these people are all acting like you have been clued in from day one!

"Okay, no problem," you say after listening to them intently. "We will just revamp the old audit material from last year and add to it what we need." Everyone agrees that it sounds like a reasonable place to start, and the meeting is adjourned, but somewhere in the back of your mind something tells you this is going to be anything but an ordinary IT audit. In this particular instance, you decide that it would be unwise for you to ignore that feeling, and that you better find out more about this Sarbanes-Oxley thing and PDQ (Pretty Darn Quick). Just then you realize this whole thing seems like a nightmare, and you are right. Whether as a result of your quickened heartbeat, sweating palms, or throbbing headache, you snap out of your Sarbanes-Oxley-induced nightmare back to the realization that you've passed your first year Sarbanes-Oxley compliance audit. You now breathe a sigh of relief as you revel in the knowledge that the worst is over. Or is it? Just as you begin to relax again, you hear the sound of your CEO's voice asking you, "What is the impact of AS5 on our Sarbanes-Oxley compliance? How does our ITIL activities impact Sarbanes-Oxley?" You think to yourself, the nightmare continues.

Whether this story is similar to yours, the simple fact is that as an IT professional, whether you are a system administrator or a CIO, at some point Sarbanes-Oxley compliance should be a major concern if you work for a publicly held company. Therefore, as part of this 2nd edition of Sarbanes-Oxley IT Compliance Using COBIT and Open Source, we will endeavor to provide information that is useful not only for first year Sarbanes-Oxley compliance, but subsequent years' compliance as well.

So, what exactly is this Sarbanes-Oxley, and why do I care? Although we won't delve into this topic in excruciating detail just yet, we will give you some of the highlights. As for what is Sarbanes-Oxley, after various corporate scandals, in order to restore public faith in the U.S. stock market, on July 30, President Bush signed into law the Sarbanes-Oxley Act of 2002 (SOX). The SOX significantly changed the federal regulations for all public companies with respect to corporate governance, financial reporting, and accountability for directors, officers, auditors, securities analysts, and legal counsel.

- The New York Stock Exchange (NYSE) and the National Association of Securities Dealers Automated Quotation (NASDAQ) will not list any public company whose audit committee does not comply with auditor appointment criteria, compensation, and oversight. The audit committee must be comprised of independent directors.
- CEOs and Chief Financial Officers (CFOs) must certify to the validity of their financial reporting and the IT systems that were germane in the process.
- Insiders must report any trading of their companies' securities within two business days after the date of execution for transaction.
- A company must disclose any and all additional information about the company's financial condition or operations that the Securities & Exchange

Commission (SEC) determines is necessary or useful to investors or in the public interest.

- All annual reports filed with the SEC containing financial statements must include all material corrections identified by a public accounting firm.

Now that you have a better idea about what SOX is and how it has and/or will change life in publicly traded companies, we will now touch on the financial impact:

- According to Warren Buffett, the CEO of Berkshire-Hathaway spent $24 million on auditing this year; a figure he says would have been closer to $10 million without SOX. (DealBreaker – A Wallstreet Tabloid, March 2007)
- Investors are taking companies private at a record pace. On Monday, it was Sallie Mae, the mammoth school-loan company, in a $25 billion deal. Do private equity firms know something the rest of us don't? (Investor's Business Daily, April 2007)
- 100,000 fans flock to Shelbourne, Vermont, each year to tour the factory of the Vermont Teddy Bear Company. Although they can buy the bears, they can no longer buy the firm's shares. That's because Vermont Teddy Bear went private in September 2005, after 12 years as a public company. The company's CEO, Elisabeth Robert, says a major reason was the SOX. Had the firm remained public, she estimates the cost of complying with the law would have doubled to about $600,000 a year. (Nightly Business Report, April 2007)
- Financial Executives International, a professional association, suggested that the cost of complying with Section 404 has been falling as companies become more efficient, but is still substantial. The survey showed that companies with a market capitalization greater than $75 million spent an average of $2.9 million in fiscal 2006 to comply. That was a 23 percent decrease from the 2005 figure.—Michael Hardy (Quote.com, July 2007)

So what does this mean? You might surmise from the figures above that SOX compliance is proving to be an expensive, resource-intensive undertaking, and that IT plays an integral role in that process.

NOTE

Although compliance methodologies and requirements other then SOX will be presented in this 2nd edition of "Sarbanes-Oxley IT Compliance Using COBIT and Open Source," in keeping with the previous book, SOX will be used as the basis for compliance.

What This Book Is

In reading the next few chapters, you might get the feeling that this book has very little to do with implementing open source, since the subject matter seems very geared toward explaining the business aspect of SOX compliance. However, due to the inevitability that SOX compliance will permeate your organization, this fact makes it a requirement that IT staff, from the CIO down, have a certain level of understanding of what SOX compliance means, some of the how's and why's of business processes, and the impact this will have in their daily jobs. In fact SOX is so far reaching, that virtually every person in your organization will be affected to some degree. So as a reader, one could almost view this as two books in one. On one hand we delve into the business processes and organizational considerations surrounding SOX compliance, and in the next breath we talk about specific open source tools and implementation strategies on how best to exploit the applicable open source technologies.

By way of analogy, we can compare the SOX compliance audit experience with training for a marathon. During the months preceding the race, you can choose not to change your daily routine, ignore your coaches by eating the wrong foods, and not exercising. That is certainly your right; however, once race day comes, those extra 20 pounds and the shortness of breath after ten minutes of effort are going to make for a very long and unpleasant uphill climb. Or you could do the opposite and prepare yourself as much as possible by eating healthy, performing weight training, and running several miles daily. As with anything in life, these activities are no guarantee that you will have an easy and cheery marathon or even win the race. However, you are certainly guaranteeing an unpleasant, if not terrible, experience if you do not adequately prepare. The point is that you at least want to finish without having a heart attack in the process. We hope this book serves as a guide for your SOX compliance, by illustrating open source technologies and demonstrating concepts to help you survive compliance activities with your sanity, and enable you to better manage compliance costs.

What This Book Is Not

Honestly, it would be impossible to write a book on how to pass your SOX audit. Every business is different in operation and philosophical approach, and we could not begin to write a do-this, do-that, and voila, somehow the auditor's magically accept your IT infrastructure at face value and give you three gold stars. Speaking of IT, if you are looking for advice on anything remotely related to your finances, this is also not the book for you.

Disclaimer

The authors of this book and its publisher, Syngress/Elsevier, do not assert that the use of this book or technologies presented within it will affect your compliance efforts positively or negatively, and the contributors make no representation or warranties that the use of principles provided by this body of work will, by its nature, influence the outcome of an audit. Although many examples of IT controls, policies, procedures, and tests have been presented, these are merely examples of what could be utilized as part of a compliance effort. Readers should apply appropriate judgment to the specific control circumstances presented by their unique environment. This book has not received any endorsement from the SEC or any other standards-setting organization; companies should seek specific advice regarding their compliance from their respective auditors.

This book is intended to give the reader an understanding of how open source technology and tools might be applied to their individual requirements. Without specific knowledge of your environment and business practices, it would be impossible for the authors to make specific recommendations in a work intended for general consumption.

Conventions Used in this Book

In every chapter we will be introducing sections to accomplish the goals of the book, namely highlighting the use of open source technology in IT organizations that enable them to deliver quality services that naturally avail themselves toward compliance. In doing so, there are a few conventions we use throughout the book, which we would like to introduce.

The Transparency Test

In the course of writing this book, we have tried to expand our discussion to include the perspective of a wide range of people who have a stake in the compliance process. In each of these sidebars, we hear from executives and stakeholders in the compliance process on how compliance impacts their daily activities, or has changed how they approach a particular task due to the need for compliance.

Lessons Learned

These sidebars provide narratives on actual in-the-trenches experience we have had in dealing with real-world IT issues, and how compliance activities ultimately changed the way we thought of the problems to be solved. Here we attempt to impart some wisdom and commentary on the benefits (or detriments) of deploying open source solutions as the genuine article. Additionally, in some of these sidebars, we hear the voices and concerns of other frontline managers and administrators in relation to compliance issues.

Tips and Notes

Here you will find notes, exceptions, pitfalls, warnings, and pointers that relate to the subject matter being discussed. We try to include information in these sections to arm you with information that might save you time and effort.

VM Spotlight

Here we focus on a specific open source technology that is available and/or has been implemented on the VMware virtual machine provided on the companion DVD. We showcase the technology in detail, running as a real-live example and give the reader an opportunity to actually sample and use the software, hopefully giving a broader sense of what open source has to offer by specifically highlighting the capabilities of open source applications in real time, and the configuration and operational considerations for actual deployment. Most important we try to show how they either satisfy compliance requirements specifically, or how they can assist in the actual process. This is by no means an exhaustive discussion or how-to on each application; however, we have attempted to provide further reading and reference pointers so that you can learn more about each technology discussed. We also list competing or similar open source projects so you may compare and contrast the relative merits of each.

Case Study

This is the section where our sample company, NuStuff Electronics, becomes the center of attention. We try to demonstrate by example, the concepts outlined in each chapter with a hypothetical use case as we build upon the material of each proceeding chapter to walk through what you might expect when partaking in the compliance journey from start to finish.

The Transparency Test

The CFO Perspective

"Today's managers have a tremendous number of areas clamoring for their attention. Unfortunately to remain a public company, or become one if you are private, SOX is dominating the priorities. While there is no debating the detrimental impact the Enron's and TYCO's have had on the investor community, and that corporate governance and control did need to increase; it is not at all clear that the monies and time spent on SOX are merited. Hopefully approaches such as those included here, will begin to streamline the process and thus the time and cost involved with being certified and thusly allow top management to return their focus to market share, profitability and growth."

–Steve Lanza

Why Open Source?

In order to answer the "Why Open Source" question, we initially take a brief departure from discussing SOX to discuss open source software, its developmental methodology, and some of the benefits that can be realized by its implementation into your organization. Undoubtedly, you have read about open source in trade periodicals, news publications, or other sources, or you have had some exposure to the phenomenon in the actual deployment of a project. The purpose of this book is not necessarily to educate you on the philosophy of open source per se, but rather to provide an understanding of the underlying concepts and correct possible misconceptions concerning open source to better enable you to gain the most benefit from the technologies presented here. Before we discuss the pros and cons of the open source model, we should spend a few minutes discussing how software is developed in general, and highlight the differences between this and closed-source methodology.

Open Source Licensing: A Brief Look

When most people talk about an open source-compatible license, they are usually referring to a license that has been reviewed and certified by the Open Source Initiative (OSI) (www.opensource.org), a nonprofit organization whose sole purpose is to promote the idea of Free/Libre/Open Source Software (FLOSS). At last count, there were 58 distinct OSI-approved licenses for open source. In fact, the OSI has an ongoing project aptly named "The License Proliferation Project," in an attempt to reduce the number of open source licenses to simplify and streamline their application and selection based on the principle that sometimes less (or fewer) is more. Below is a brief look at a few examples of OSI-certified licenses and how they differ. A full listing of these is provided in the index at the end of this book.

TIP

Any open source licensing restriction actually applies to only the licensees of a project's source. The original developer(s) of an application can do what they like with their source, including selling a proprietary version if they so desire. Only derivative works or improvements to a version the developer may choose to release under an open source license are affected.

GNU General Public License

The General Public License (GPL) is what is termed a "strong" license, because it is completely incompatible with proprietary software. The main reason is that the GPL compels a user to make the source code available when distributing any copies of the software, and that all modifications to the original source are also licensed under the GPL. In addition, if any GPL-licensed source code is incorporated into another project (known as a "derivative work"), the entire project would be required to also be released under the GPL. For this reason, GPL-licensed software cannot be mixed with proprietary offerings, because it inherently would render the proprietary source GPL licensed as well. Users are free to make copies and changes, redistribute, and charge money for derivative works as long as the source code is available and a copyright notice is attached. The GPL has currently undergone a controversial revision from version 2 to version 3, which was just released after more than a year of public request for comments.

GNU Library or "Lesser" General Public License

The "Lesser" General Public License (LPGL) is essentially the same as the GPL, with a notable exception. Unlike the GPL, which requires the source code for the "derivative work" to be licensed under the GPL and the source be made available, the LPGL allows binary-only linking of applications, typically libraries, with any other application, including proprietary software. Thus, under the terms of the LPGL, the original source and any changes made to it must be made available along with a copyright notice. However, if a binary version of it is used by a non-free application, the source of that application is not required to be released under the LPGL.

The New Berkeley Software Distribution License

By contrast, the original Berkeley Software Distribution (BSD) license and the more recently modified version of it are the most permissive in nature. These basically say that users are free do to with the software whatever they like, including modify the original source or incorporate it into another project. Users are free to redistribute their derivative works without any requirement to make the source code available or any of their modifications. The only requirement is that the original authors be acknowledged in the license that does accompany the released application, whatever it may be. The only difference between the new BSD License and the original BSD License is that the advertising clause in the license appearing on BSD UNIX files was officially rescinded by the Director of the Office of Technology Licensing of the University of California in 1999, which states that the applicable clause is "hereby deleted in its entirety."

Lessons Learned

Deja' Vu All Over Again

Back in 2000 when we were re-architecting all of our enterprise data storage, we interviewed each department to find out how they currently stored their files and what the typical usage model for access was. As we defined what was currently out there, we attempted to reorganize and restructure most of this data to fit into our project goals as an IT organization, which were manageability, security, and disaster recovery. During this process, we received considerable bottom-up resistance to change, especially in the areas of file access and permissions. We ended up with a few compromises that *we* considered less than optimal, however, management did not at the time provide top-down support for the changes we felt were necessary.

Fast forward to 2004, when we were going through our discovery phase for the SOX compliance audit. We revisited many of the same topics we had previously covered with varying levels of success, including our storage footprint. We again identified requirements and changes that needed to be made for manageability, security, and disaster recovery, and again experienced the same resistance to change from the general users. The main difference this time was the top-down support we received from management to make the necessary changes in order to meet our goals.

As a final chapter to this mini-story, the re-architecture of the storage systems was ultimately beneficial to the business. By 2007, not only did this satisfy ongoing compliance requirements, it also made the administration of storage vastly simpler. Clearly defined processes made it much simpler for access controls, backups, and identifying ownership of data, which translated into less time spent by administrators trying to "figure out" what to do, and more time doing what needed to be done accurately and reliably. Although from an IT perspective the aforementioned benefits were substantial, the biggest benefit derived from this process was that the company had greater security of critical data and more timely access to this critical data than existed previously.

Open and Closed Source in Contrast

One of the easiest ways to compare and contrast open source and proprietary software is to point out some of the differences in the development cycle. A different set of motivators exist for each, so the next section attempts to illustrate this to give you an idea of each approach. Generally speaking, the term "open source" refers to a method of software development where volunteer developers contribute to a particular project and donate all of their source code and documentation efforts to the public for the benefit of all. Altruistic as this may sound, most people who get involved with open source at the coding level do so for several reasons. Some developers may join to avail themselves of the expertise of other developers on the project and benefit from their work, some desire peer recognition, and some simply may be paid by a company to develop software for a need the company has and the resulting application is released to open source. The salient point is that very often a group of interested individuals both drive the requirements of the software project and directly develop the end result to their own satisfaction. Figure 1.2 diagrams a typical closed source development model compared to Figure 1.3, which is a typical open source development model.

Figure 1.2 Proprietary Software Development Model

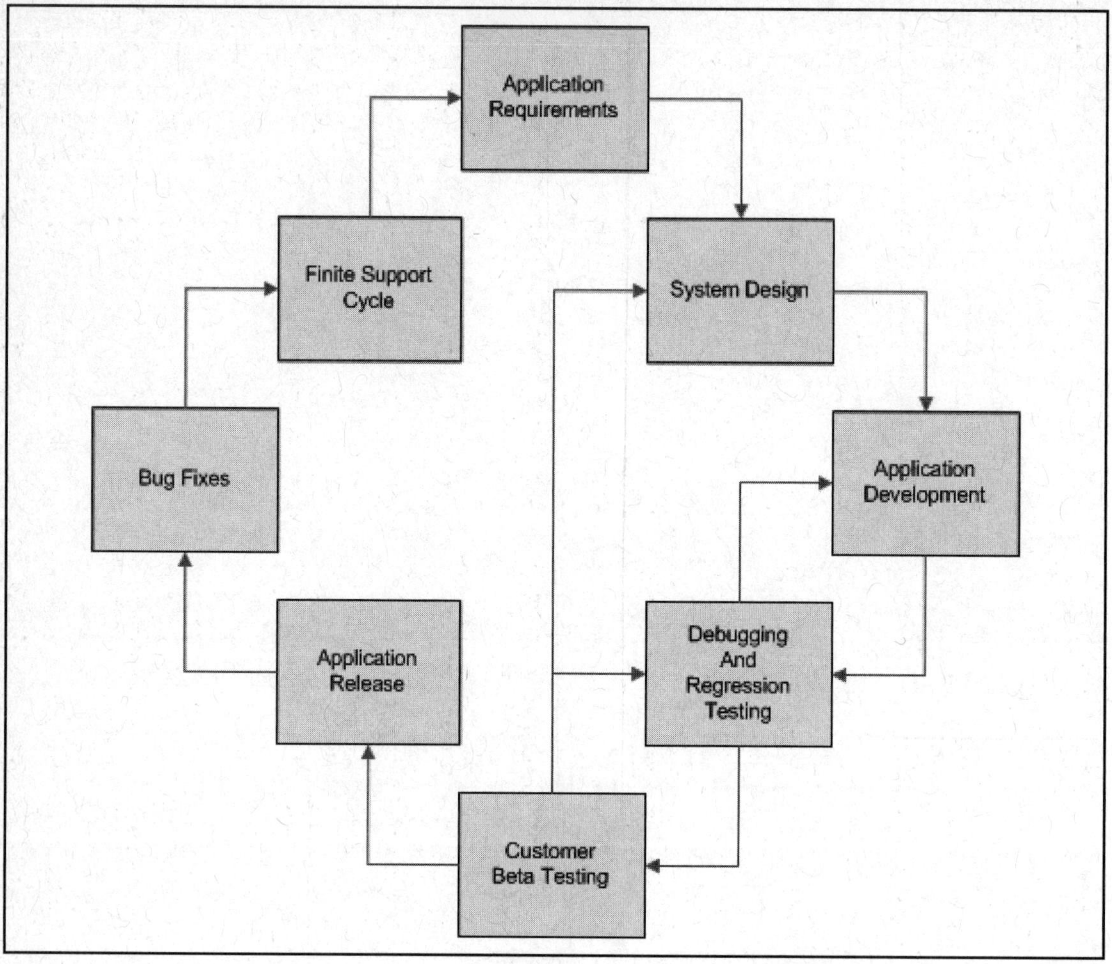

Figure 1.3 Open Source Software Development Model

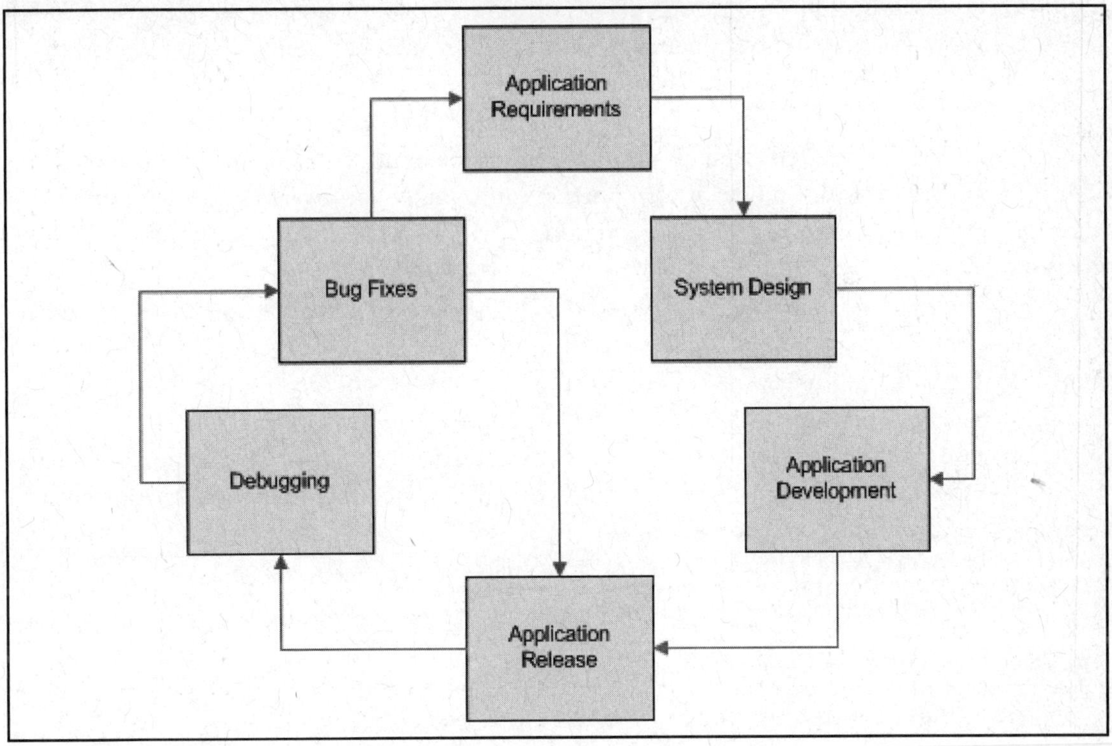

NOTE

Although these diagrams are useful for our discussion, there are many aspects of the software development model that may not be represented in detail. The goal of this section is to give you a flavor of the typical development considerations and major phases in both open and closed source projects, as well as some of the fundamental differences that drive each one.

When setting out to develop a software product, the first step is to decide what need this will satisfy and what functionality should be provided. This set of requirements can come from many sources such as customer feedback, expertise of those involved in the venture, or your basic "light bulb" type of idea. In closed source development, this is often a formal process requiring significant time, energy, and financial resources, with market competition and time to market considerations thrown into the mix. Open Source, on the other hand, is usually approached as an "itch that needs to be scratched."

Because a significant portion of a closed source company's revenue stream comes from the prospective sales of the application, most software companies attempt to select projects that will maximize their ability to generate profits, either by identifying a "vertical market" in which they can write a specialized application and charge a comparatively large sum for each software license, or by developing an application that has mass appeal where the company might not necessarily charge as much per license, but make up the difference with volume sales. As with any for-profit model, it is important to note that a software company's motivation must be for the maximum salability of its product for the least amount of research and development costs accrued.

Comparing a programmer who is paid to develop a piece of software that he or she neither uses nor particularly cares about, and the community that springs up around an open source project that consists of people interested in using the software and adding features they find useful, may explain the reason the quality of code for the latter is so characteristically high. The Internet is what makes volunteer involvement via distributed and parallel hacking eminently possible, and as we will see in the release and debugging stages, this is crucially important to the success of the FLOSS phenomenon.

NOTE

Sometimes companies will decide to release a software application that they previously developed as closed source for many reasons. Examples may be that the software is in danger of becoming obsolete, but the company does not have the developer resources to continue to innovate and maintain the software, or a company may be transitioning from software sales to being service providers and opening its software would generate or expand its customer base. Netscape Communications was one of the notable firsts with its release of its well-known Communicator Web browser suite, which later became the Mozilla Foundation. Others include Borland's Interbase (now know as the Firebird project), and more recently, Sun Microsystems has released the Solaris 10 operating system under the Common Development and Distribution License (CDDL). The CDDL is one of many licenses approved by the Open Source initiative (OSI). A full listing can be seen at www.opensource.org/.

As Eric Raymond, a famous proponent of Open Source once famously stated and is often quoted: *"Release early. Release often, and listen to your customers."* In contrast to a closed source project, where release happens when the company is convinced it is of sufficient quality to be able to charge money for and not completely upset their customer base, most open source projects release their code as often as possible. Although it may seem that users would choose not to muck about with something that might be buggy and wait for a "stable" release (although there are plenty that fall into this category), users will often embrace early releases for several reasons. First, they are regularly stimulated and rewarded with new features, and this fosters a constant flow of communication in the form of bugs and feature requests/ refinements, particularly when they see one of their own requests quickly incorporated into the application. Second, it contributes to the rapid stability of an application, because debugging happens in consort between the developers and the actual users of the project rather than by the developers of a closed source project who attempt to envision how their user base might use the product.

Many small incremental releases are also rewarding to the developers, who see their work being used and problems being fixed continuously, which gives them a sense of accomplishment very early in the process. This is sufficient to keep most developers interested in continuing to develop, while recognizing the contribution of others to the project. It is a truly win-win situation.

Another famous quote often cited comes from Linux Torvalds, the creator of the Linux operating system: *"Given enough eyeballs, all bugs are shallow."* This expression means the odds that someone will find a bug in a piece of software is the greatest when many people are using the software. Having access to the source code means that someone, somewhere is very likely to see a solution that the code's original developer may not be able to find so quickly, much less have a fix for it. In fact, many users are also hackers themselves, and will often report a bug and submit a patch to fix the bug they found all in one go. This is clearly not an option any user of a closed source application could possibly do. Thus, the quality of open source tends to be very high because of the constant peer review of developers, users, and hackers that make up the project community.

The Business Case for Open Source

As we have seen, there are many compelling reasons to consider the use of open source in an organization. When examining the pros and cons, it is important to understand the factors that will ultimately make the decision a good or bad one. Now that you have a better understanding of how open source is developed, the next logical question one might ask is "What's the catch?" Here we discuss some of the more practical considerations in introducing open source into your environment.

Free != No Cost

While open source software is freely available, and you could theoretically run any one of thousands of projects available without spending any money, therein lays a problem. Because there is so much choice (a good thing), users could (and would) spend significant time finding software suitable for their needs that also plays well with their OSS brethren (maybe a bad thing). This being the case, we will examine three ways in which open source can, and maybe should, cost you money.

- **Distribution Vendors** Linux is a shining example of the power and success that open source software can achieve. One often overlooked fact is that Linux is actually only the kernel of the operating system, not the thousands of applications that run on top of it. Because this is the case, there are many collections of software known as "distributions" put together by various people and companies. One example is the Debian project, which is a Linux distribution maintained by thousands of volunteers all over the globe. At last count there are more than 16,000 distinct packages in the Debian distribution. Many businesses that deploy Linux, however, choose to use a distribution that is tested and supported by a company. Red Hat's Enterprise Server and Novell's SUSE are two example Linux distributions provided by companies that charge for the regression and integration testing they perform on the packages offered in the distribution and the after-sale support of these products.
- **Project Developers** In addition to distribution vendors, some project developers also provide per-incident and/or support contracts for the software they help develop, as a means to make a living while donating their development efforts back to the project. Charging for deployment assistance or one-off custom integration tasks is also very common.
- **In-House** Companies without in-house development expertise may choose to sponsor development, either to ensure that a project continues in a healthy manner, or for a specific set of features the company needs. This allows a company to lower costs by embracing open source, while ensuring that they get the features they need to meet their business goals and mitigate the risk that its implementations might otherwise pose. Sometimes it is as simple as a project donation without any particular goal in mind other than to reward the developers or give them equipment or Internet bandwidth to ensure the project has the ability to continue uninterrupted.

Lastly, a company may employ in-house developers to steer an open source project in the direction they wish it to go, while leveraging the benefits of outside resources for a myriad of details such as development, testing, and documentation.

Does It Really Save Money?

We have discussed how the implementation of open source software in your IT environment is not necessarily free of cost. While there are ample opportunities to donate time, human resources, and money to your distribution vendors and favorite projects, the simple fact is that there are no software licensing fees associated with an open source project. This means that once you decide to deploy an open source undertaking for the life of that project, you will not be burdened with initial or subsequent fees for the use of the software unless you elect to pay for some form of support. Because this makes sound business sense and is an ongoing cost even with proprietary software, you actually pay just for support when you use open source software.

Another consideration is that many large, over the counter (OTC) software applications are of sufficient size and complexity that significant money is spent in customizing it into your business to work the way you want it. Because you are paying for deployment in this regard, a one-time sponsored development can be significantly less expensive than paying a company to customize the OTC package. Because you also have access to the source code, you mitigate any risk of the company that performed the customizations getting hit by the proverbial truck and going out of business. You have the option of continuing support of your customizations with in-house developers, or finding another source that can continue the maintenance of the code. Another tangible cost-saving feature lends itself to the quality of open source, which tends to be high for reasons previously discussed, contributing to its efficiency, performance, and ability to run on older hardware that can save companies' significant money by extending the usable life of the hardware and older PCs.

There are other intangible benefits that can reduce a company's overall IT expenditures, such as better security. For example, it is now a widely accepted fact that open source software tends to be more secure than closed source offerings. Wikipedia (http://en.wikipedia.org/wiki/Security_by_obscurity) defines "Security through obscurity" as the attempt to use secrecy (of design, implementation, and so on) to ensure security. The premise of this is that if the methodology of an implementation is not revealed, then an attacker is not likely to discover any vulnerabilities, because he or she does not have access to the implementation details. In practice, this is far from the truth. One look at the pace with which Microsoft releases security updates for its products supports this position. Once a security flaw is revealed, it is the sole responsibility of the company to provide a patch to deal with the breach. In the open source world, however, if a vulnerability is discovered, it reverts to the "many eyeballs" approach and typically severe security weaknesses are patched far faster than their non-free counterparts.

Platform-agnostic Architecture

Microsoft can and should take credit for being one of the main contributors to the wide acceptance of personal computing over the last couple of decades. However, it is no secret that Microsoft has become a company with more than $35 billion in annual revenue as a result of what some might consider predatory business practices. Limiting choice and dictating pricing models ultimately resulted in an indictment and conviction of this behavior by the Department of Justice in 2000. In the end, open source software offers freedom, not necessarily in cost (although this is a demonstrable benefit), but in choice.

Open source software is more than Linux. Linux has gained much fame and fortune over the past ten years, but there are many other wildly successful examples such as the Apache Web server and Samba. One of the advantages that open source enjoys is having a passionate yet diverse support organization made up of people with all sorts of needs, desires, and agendas. This is a good thing, because anyone who wants to can port the project to run on his or her own favorite hardware or operating system. Thus, open source projects tend to have broad support across multiple architectures, particularly prominent projects. From a SOX point of view, this feature is important, because many companies run their IT systems on a multitude of different platforms and technologies, and it is important to understand where free software fits into the equation.

Open Source and Windows

Most of the major open source projects (other than operating systems such as Linux, to state the obvious) run on a Windows platform. As we will see a bit later in this chapter, this can be helpful when assessing your infrastructure, as most companies have some type of Windows software deployed in their organizations. In fact, open source software is useful even if your IT infrastructure is completely Windows-based using non-free software. For the purposes of SOX, the main goal is to avoid any deficiencies that could lead toward a material weakness. If a risk is identified and no in-house, closed-source solution lends itself to immediate remediation, there is most likely an open source solution that can be utilized or modified to mitigate the risk and satisfy the auditors. We will examine these projects in detail in the remaining chapters of the book, but for now suffice it to say that a primarily Windows-based platform does not prevent the use of open source software to assist in your compliance requirements.

TIP

If you are interested in investigating the many open source projects that are available on the Windows platform, you should visit The OSSwin project (http://osswin.sourceforge.net). This site contains information about most OSS applications that can be run natively in a Windows environment, including many applications outside the scope of this book that we will not cover.

Mixed Platforms

In today's business IT environments, there is a good chance that you are using a mix of technologies, particularly if your business is related to technology, health care, research, or manufacturing, to name a few. Not too long ago if you needed to run a UNIX and Windows environment side by side, your IT infrastructure may have looked something like this:

- Windows and UNIX network segments physically separate
- Windows using domain or Active Directory, and UNIX using Network Information Systems for authentication
- Engineers with UNIX workstations also have a Windows box for e-mail, Web browsing, and Microsoft Office applications for documentation.

This is not necessarily a bad setup, and some environments may still be similarly laid out; however, when considering IT controls for SOX, it is best to approach your environment in the simplest terms possible. The more complex the environment, the more work will be needed in all phases of the compliance process in order to get through your audit. Even if you survive the audit process with a few strands of hair in tact, keep in mind that compliance is an ongoing requirement and complexity breeds over time. The good news is that open source software fits well into a mixed environment; you will be able to use this to help you wrap your arms around compliance, and the examples in the book will aid you in the use of open source tools.

Migration: a Work in Progress

If you are already migrating some or your entire IT infrastructure away from proprietary systems, you'll need to consider a few things about SOX compliance. You must keep in mind that section 404 of Management Assessment of Internal Controls, means that every IT system that touches, contributes to, or in any way supports the financials of the company and thus the reporting thereof, is affected by the act. Any changes as a work in progress must be stringently documented, that may make up the support infrastructure for the financial controls identified for your business processes. Subsequently, deviations must be subjected to the appropriate approval chain and documented as well.

VM Spotlight: CentOS GNU/Linux Distribution

CentOS http://centos.org

For our first VM Spotlight, we will cover the operating system used to develop the ITSox2 VM Toolkit, which is a Linux distribution called CentOS. The CentOS Web site coyly states: "CentOS is an Enterprise-class Linux Distribution derived from sources freely provided to the public by a prominent North American Enterprise Linux vendor. CentOS conforms fully to the upstream vendors redistribution policy and aims to be 100% binary compatible." CentOS is basically a repackaged distribution of the most popular paid-for vendor distributor in the United States (think RedHat). The repackaging effort only modifies the official distributed packages to remove the upstream vendor branding and artwork. CentOS is free to download and use without cost, however, they do accept and appreciate donations via their Web site at http://www.centos.org.

CentOS is developed by a small but growing team of core developers. These developers are supported by an active user community that includes system administrators, network administrators, enterprise users, managers, core Linux contributors, and Linux enthusiasts from around the world. CentOS is not the first or only repackaging of a Linux distribution, but currently is the most popular and has numerous advantages over some of the other clone projects including:

- A fairly large, active and growing user community
- Quickly rebuilt, tested, and quality assured errata packages
- An extensive mirror network
- Developers who are contactable and responsive
- Multiple free support avenues
- Commercial support offerings via a number of vendors

A Word on Linux Distributions in General

Something to keep in mind when evaluating any software for your enterprise in regards to SOX is whether the software meets your needs on two key aspects, whether the software is open source or not. The two key aspects are:

- **Functionality** Can the software do what I need it to do? Can it be modified to suit our business processes? If so, how do these changes affect other dependent systems? Who else is using this software that can contribute mind share back to the developers to make it a better, and important to our discussion, compliant offering?
- **Support** Who will answer my auditor's questions regarding the technical bits of the application that I do not already know the answer to? From a change management perspective, how do I know whether application ABC plays nice with XYZ?

If we use Linux as an example, your selection criteria will depend in part on your goals for open source, whether it is to serve as a one-off application or two, or you will be basing your entire core infrastructure on open source. The more you consider the latter, the more valuable a distribution becomes. In the writing of this book, we have been conscious not to advocate any particular Linux vendor, as the concepts and principles are truly neutral. That being said, you must consider that an auditor's experience with open source may be limited. Using a recognized distribution, such as Red Hat's Advanced Server or Novell's SUSE Enterprise may save you a fair amount of time and energy in documenting and validating the origin of your open source software.

Now, for the exception that proves the rule. We have seen the successful use of community driven distributions such as CentOS used in roles throughout the enterprise as the basis for core IT infrastructure. The salient point is these distributions are based on vendor-supplied source Red Hat Package Management (RPM) in accordance with the GPL, and recompiled unmodified and in most cases are the binary equivalents of their commercially distributed brethren. If there are any differences due to compile time environmental variances, these are usually very minor and documented.

NOTE

RPM refers to Red Hat Package Management, the mechanism used to package open source software into an installable format. RPM supports the requirement for dependencies in both the source RPM that needs to be compiled, and the binary RPM that may require other applications and libraries to compile/run properly. RPMs are usually digitally signed by the vendor to ensure that you are getting the genuine tested package. Novell SUSE also uses the RPM format. Ubuntu, Debian, and their derivatives use a similar but not interchangeable package management system called APT, and have an extension *.deb.*

You might choose to use Debian, Gentoo, or one of the many other fine Linux distributions out there in the wild. Many of these distributions offer compelling features in the areas of performance, security, and stability. In fact, the Live CD used as a companion to the first edition of this book is based on the XFLD distribution, which is based on the original Knoppix distribution, which in turn uses Debian as its core. The new CentOS-based virtual machine used in this second edition, together with the example of the first, illustrates a wonderful example of the power of open source and the ability to adapt to a certain need or set of requirements. It is, however, very important to communicate with your auditors to make sure they have an understanding of your environment. Sometimes that involves using a toolset both of you can come together on, as a basis to move forward with your compliance audit. Ultimately the choice is yours, but this should be considered.

Linux Distributions and References

- **Red Hat Linux and the Fedora Project** Red Hat is the leader in development, deployment, and management of Linux and open source solutions for Internet infrastructure, ranging from embedded devices to secure Web servers. Red Hat was founded in 1994 using open source as the foundation of their business model, which represented a revolutionary and fundamental shift in how software was created. The code that makes up the distribution is available to anyone, and developers who use the software are free to improve upon it. The net result is rapid innovation. Red Hat markets solutions such as paid-for updates for Red Hat Linux, training, management services, and technical support. The Fedora Project is an openly developed project designed by Red Hat, that is open for general participation and is led by a meritocracy following a set of project objectives. The goal of The Fedora Project is to work with the Linux community to build a complete, general-purpose operating system exclusively from open source software. The project produces releases of Fedora two to three times a year, and the Red Hat engineering team participates actively in building Fedora along with outside participation. The main reason for

Fedora's existence is to develop new and improved existing open source technologies that ultimately find their way into the official Red Hat Linux product line. (http://www.redhat.com and http://fedoraproject.org).

- **Novell Linux Desktop and OpenSUSE** Novell entered the Linux arena by acquiring SUSE Linux AG in 2003, which at the time was a German-based distribution that enjoyed popularity in much of Europe. Novell's goal is to provide a leading end-user productivity environment designed specifically to empower businesses to leverage Linux and open source with confidence. It can be deployed as a general-purpose desktop platform, or tailored for use in information kiosks, call-centers, or stations for infrequent PC users. Novell Linux Desktop also provides an alternative (as do virtually all Linux vendors) to high-cost UNIX-based engineering workstations. Novell Linux Desktop is backed by Novell support, training, and partners, aiming to allow businesses to deploy Linux systems with confidence. The openSUSE project is a community program sponsored by Novell, promoting the use of Linux by providing free, easy access to a complete Linux distribution with three main goals: make openSUSE the easiest Linux for anyone to obtain and the most widely used Linux distribution; leverage open source collaboration to make openSUSE the world's most usable Linux distribution and desktop environment for new and experienced Linux users; and dramatically simplify and open the development and packaging processes to make openSUSE the platform of choice for Linux developers and software vendors. (http://www.novell.com/linux and http://www.opensuse.org).
- **Ubuntu and Debian Linux** Ubuntu (from an ancient African word meaning "humanity to others") is a complete desktop Linux operating system, freely available with both community and professional support. The Ubuntu community is built on the ideas enshrined in the Ubuntu Manifesto: that software should be available free of charge, that software tools should be usable by people in their local language and despite any disabilities, and that people should have the freedom to customize and alter their software in whatever way they see fit. The Ubuntu distribution brings the spirit of Ubuntu to the software world. Ubuntu is actually a derivative of The Debian Project, which is an association of individuals who have made common cause to create a free operating system. This operating system is called Debian GNU/Linux, or simply Debian, which comes with over 20,000 packages of precompiled software that is bundled up in a nice format for easy installation and maintenance. (http://www.ubuntu.com) and (http://www.us.debian.org).

CentOS in Detail

We chose to use Community ENTerprise Operating System (CentOS) as our platform for all of the open source technology demonstrated throughout this book for several reasons. Whereas in the first edition the Knoppix Live CD did a fine job of providing a useful platform for our needs, CentOS is a clone of the widely deployed RedHat Enterprise Linux product line, and is very likely a Linux distribution you will have encountered or may have already deployed to some degree in your environment. We could have also easily substituted any one of the fine distributions outline above, however, we felt it was important for the open source stack discussed throughout the book to be representative of production systems as closely as possible. Some of the other features that make this a great distribution are:

- **Rock-solid Reliability** RedHat puts considerable research and development into every package in its Enterprise line, and CentOS represents the binary equivalent of this effort. Of course you do not receive formal support when running CentOS, however, it is completely free to run and distribute.
- **Freely Available Support** There are many avenues of assistance available to you when running CentOS in the form of mailing lists, robust user documentation, and user forums. We provide a short list of resources for your reference in the "Frequently Asked Questions" section at the end of this chapter to get you started.
- **Long Product Lifecycle** RedHat officially supports their Enterprise products for seven years, providing bug and security updates. In compliance with the GPL, these updates are made available in source form, which the CentOS team recompiles and redistributes as updates to the CentOS distributions. Currently as of this writing, we are at version 5 and the CentOS team will provide maintenance updates until at least 2014.
- **Add-on Repositories** Although RedHat provides a very complete distribution with many open source packages, they do not include everything one might desire to run. The active CentOS community provides additional software not available in the original distribution in the form of the "plus" and "extras" yum repositories. We take a closer look at yum in the next chapter when we explore the CD in more detail.
- **Simplified Packaging** One difference between Red Hat Enterprise Linux 5 and CentOS 5 is that CentOS 5 includes packages from both the server and client distributions. Both the Red Hat enterprise server and the workstation flavors have been combined into a single distribution for easier manageability and maintenance, thus eliminating any guesswork on the end user's part.

- **Xen Virtualization** We will discuss virtualization in detail in later chapters. One of the new features to CentOS 5 is Xen virtualization which provides built-in support to run multiple operating systems on a single server simultaneously. Prior to version 5, this was an add-on from the "plus" repository.
- **Clustering** CentOS provides out-of-the-box support for clustering technologies, both for high availability and load balancing. Storage clustering is provided via the Global File System (GFS), which allows a cluster of Linux servers to share data in a common pool of storage.

Case Study: NuStuff Electronics, an Introduction

NuStuff Electronics is a successful semiconductor designer of baseband communication chips for original equipment manufacturers (OEM) of digital telephones. Operations span the globe with offices in, India, Japan, Singapore, the United Kingdom, and two offices in the United States. The majority of the design work is done in India, and research and development on new products is primarily done in the UK branch, with corporate headquarters in the US and the remaining offices performing sales and customer support. NuStuff out-sources its manufacturing needs to contract electronics fabrication firms, and has approximately 800 employees worldwide. NuStuff has $60 million in assets and quarterly revenues averaging $20 million.

IT Infrastructure

Because electronic design automation (EDA) tools have strong historical roots in, NuStuff has already embraced open source and Linux technologies to a great extent. NuStuff recognized early on the cost-saving benefits of migrating away from proprietary and Windows systems on both the client and server sides for engineering, while concurrently maintaining mostly Windows-centric clients for non-engineering and support personnel. To consolidate its IT infrastructure as much as possible, the company has standardized on Linux in the server room, and eliminated as many Windows servers as possible, although it does have a few proprietary and legacy applications that run in only a Windows environment. Figure 1.4 illustrates the interoffice topology for NuStuff's global operations.

Figure 1.4 NuStuff Electronics Interoffice Network (Global Operations)

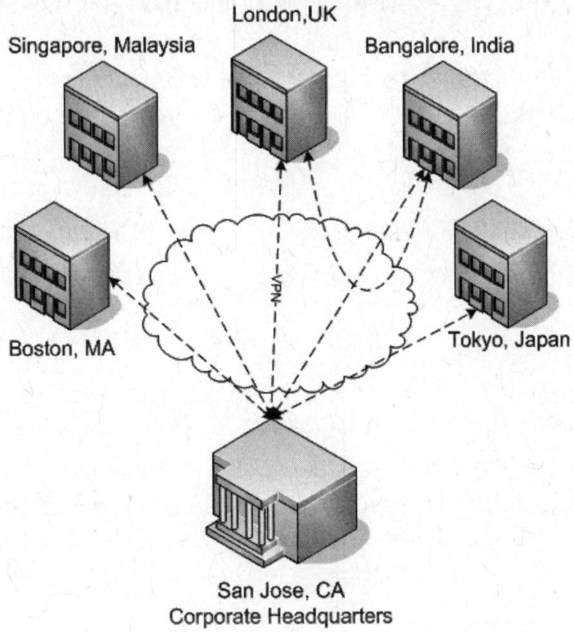

Server Room (General, Sales, Support, and Executive)

Although not a particularly large environment, NuStuff's servers support diverse in the technology and has all of the functions and requirements of a much larger company. Therefore, they need to address the same audit concerns as a larger company.

- SAN storage for network services and departmental file services
- Red Hat Advanced Server Linux and CentOS servers in a high-availability cluster for network services such as Domain name system (DNS), File Transfer Protocol (FTP), and Hypertext Transfer Protocol (HTTP)
- Oracle Financials managed by an outsourced provider and a financial analyst on staff
- Fedora Directory Server Lightweight Directory Access Protocol (LDAP) and Samba for cross platform, single-sign-on authentication services
- Scalix for groupware/messaging services
- Astaro Firewall/Virtual Private Network (VPN) with dedicated interoffice Internet Protocol Security (IPSEC) tunnels

Server Room (Engineering and Design)

NuStuff Electronics' Engineering Department is the lifeblood of the company. There is little to no tolerance for down time in their Engineering environment. Therefore, the IT department has endeavored to provide them with a highly available network utilizing SAN storage, Virtual Local Area Networks (VLANS), blade servers, and virtualization.

- Linux engineering compute farm
- SAN storage for engineering data
- Separate VLAN for engineering traffic

Desktops (Sales, Support, Executive, Finance, and HR)

NuStuff Electronics' desktops for General and Administrative support applications and functionality inline any regular company. For the most part, all non-engineering staff falls into this category from an IT desktop standpoint. NuStuff has separated their Finance and HR staff onto a separate subnet for security reasons as these workstations must make a VPN connection to Trusty database administrator (DBA) services for their outsourced financials system. The standard suite of applications are:

- Windows/XP desktops for general support staff, XP laptops for field sales
- Microsoft Office and Open Office for desktop applications
- Mozilla Firefox Web browser for Internet/intranet access
- Microsoft Outlook, Mozilla Thunderbird for e-mail clients
- TrendMicro AntiVirus for virus and spyware prevention and detection
- Microsoft Visio for diagrams

Desktops (Engineering and Design)

Because the engineering team already has Linux workstations for EDA design work, the NuStuff IT department has strived to consolidate the engineering footprint to one desktop per user. To achieve this goal, it has deployed the following

- CentOS Linux workstations
- Open Office for desktop applications

- Mozilla Firefox Web browser for Internet/intranet access
- Mozilla Thunderbird for e-mail client

Network Topology

Figure 1.5 diagrams NuStuff's corporate headquarters' IT landscape, followed by a detailed diagram in Figure 1.6 of their core network services, which are provided with a high-availability cluster solution. Each of these topics will be explored in more detail in subsequent chapters.

Figure 1.5 NuStuff Electronics Network (Corporate Office)

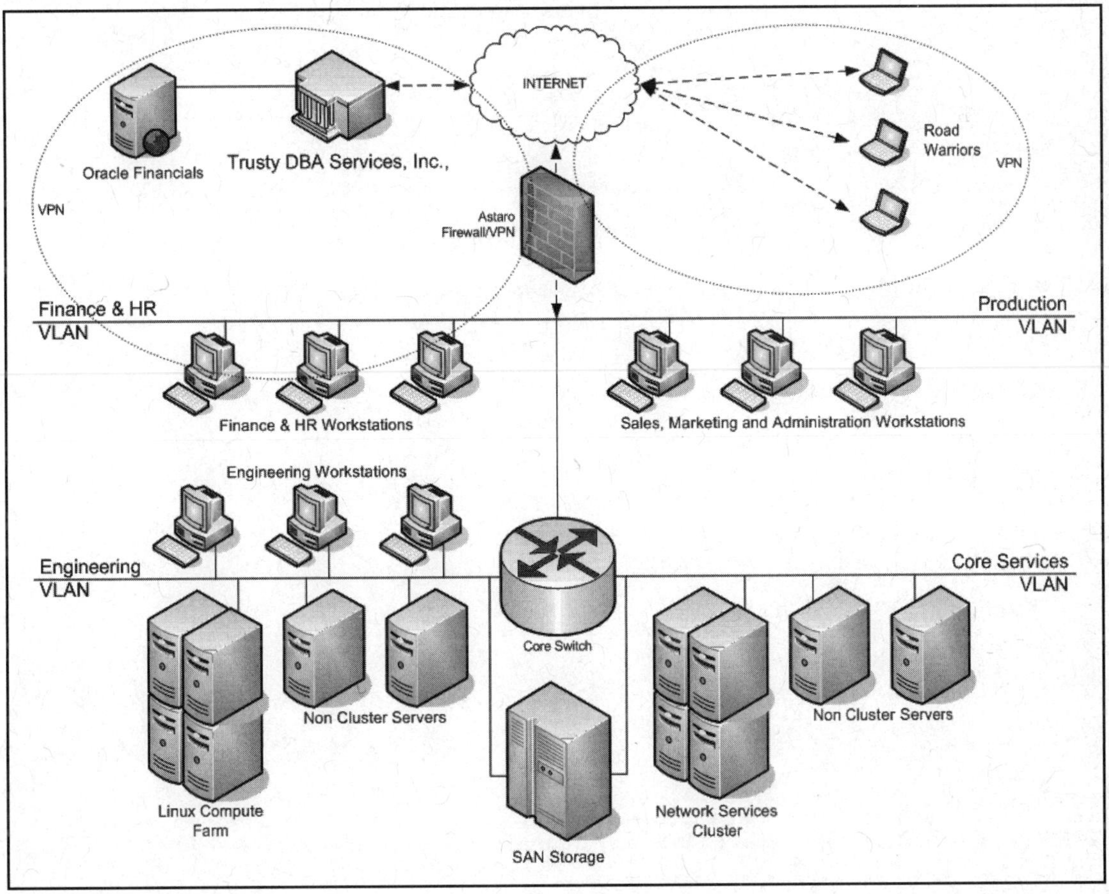

Figure 1.6 NuStuff Electronics Network Services Cluster Detail)

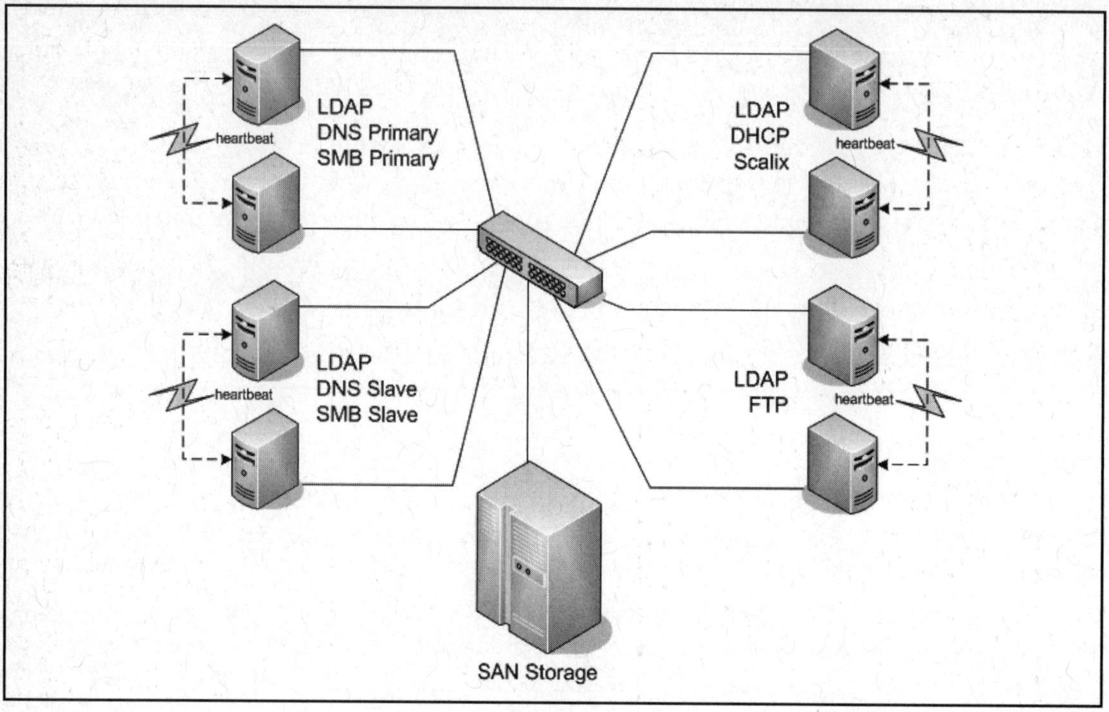

Summary

We discussed how all publicly held companies are required to prove compliance with SOX and how it will impact virtually every person in your organization. This impact promises to be quite different in terms of depth, cost, and resources than your normal year-end audit. In an attempt to assist you in mitigating the impact of SOX compliance as much as possible, we talk about the two main focuses of this chapter. The first is to better educate those of you who have or are considering deploying open source in your IT organization, and the second is to illustrate opportunities to reduce the amount of your audit cost in terms of budgetary and personnel resources by employing open source technologies to help in the monitoring, process, and documentation aspects regardless of your current IT landscape.

We also discussed how IT Management will see how open source can assist with and automate the task of documenting and tracking compliance and internal controls, independent of whether they are derived from proprietary or open source systems, and understand benefits derived from such an environment. IT/Financial Consultants will be able to use the tools and technologies on the ITSOX VM as a valuable toolset to improve their client's IT processes, and hopefully enable their SOX compliance to be a less painful and costly experience. Finally, principals of non-public companies who might be considering an IPO, can better understand some of the implications SOX brings to the table together with an idea of how open source can offset some of these requirements.

We finalize the chapter with a spotlight on the CentOS Enterprise Linux distribution. Here we discuss a real-world open source application as well as an overview of Linux in general and its use in the IT landscape.

Solutions Fast Track

IT Manager Bob – The Nightmare

☑ The SOX audit is much more in depth, costly, and resource intensive than any other audit you might have experienced before.

☑ SOX will affect virtually every person in your organization.

☑ Companies are compelled to comply if they are publicly held, regardless of size or revenue.

What This Book Is and Is Not

☑ This book is a technical book at heart; however, much material on the business side is presented to place the technology into a frame of reference for SOX compliance.

☑ The examples and technologies presented here are based on open source technologies to help you save time, resources, and money.

- ☑ This book is not a road map on how to comply; that would be impossible, since every business is unique.
- ☑ This book is not about financial compliance; it is strictly focused on the IT considerations for SOX. There are many other references for that particular aspect.

Who Should Read This Book

- ☑ **Non-IT Management** Even though you may not be in the IT department directly, you will have an over-arching understanding of how open source can help.
- ☑ **IT Management** This book demonstrates open source to assist and automate the task of documenting and tracking compliance and internal controls, independent of whether they are derived from proprietary or open source systems, and outlines the business reasons and benefits derived from such an environment.
- ☑ **IT/Financial Consultants** The live CD provides a valuable toolset one can use to improve their client's IT processes, and hopefully lead their SOX compliance to be a less painful and costly experience.
- ☑ **Principals of Non-public Companies** If you are considering an IPO, you should read this book to understand some of the implications SOX will bring to the table, together with an idea of how open source can offset some of the requirements.

The Open Source Model

- ☑ The quality of open source tends to be very high because of the constant peer review of developers, users, and hackers that make up the project community.
- ☑ The GPL is the most common open source license in use today, but there are many others.
- ☑ The GPL's main features are that it compels one to make the source code and a copy of the license available when distributing copies of the software, and that all modifications to the original source and any source it is tied to be also be licensed under the GPL.
- ☑ The LPGL requires the source code available for the LGPL-licensed project, but allows for binary-only linking to other applications and does not require that application to be licensed in any particular way.
- ☑ The BSD license allows users to do whatever they like to the software with very little or no restrictions.

The Business Case for Open Source

- ☑ Deploying open source can have costs associated, such as purchasing a support contract from distribution vendors, paying project developers for specific functionality that your organization requires, hiring in-house developers to work on a project, and leveraging community support. In addition, you might consider a donation to help defray project costs.

- ☑ Open source can save money in many ways. There are no initial license fees to pay, deployment costs are not usually more expensive, and often are less since closed source deployments may require customizations. High reliability and security reduce costly down time.
- ☑ Intangible benefits include freedom of choice and risk mitigation, since having access to the source ensures the continuation of a project regardless of whether a company goes out of business or decides to end-of-life a product.

VM Spotlight: CentOS GNU/Linux Distribution

- ☑ CentOS is an Enterprise-class Linux Distribution derived from sources freely provided to the public by a prominent North American Enterprise Linux vendor. CentOS is developed by a team of core developers, which are backed by a large, active, and growing user community.
- ☑ The two main criteria in choosing a Linux distribution are functionality and support. CentOS provides rock-solid reliability via the repackaging of the Red Hat Enterprise product line, including Xen virtualization and out-of-the-box clustering support. There are many avenues of freely available support in the form of mailing lists, robust user documentation, and user forums.

Case Study – NuStuff Electronics Inc.

- ☑ NuStuff Electronics is our example company for the case study used throughout this book. Our fictional company is a successful semiconductor designer of baseband communication chips for OEM of digital telephones. Operations span the globe with offices in, India, Japan, Singapore, the United Kingdom, and two offices in the United States. NuStuff out-sources its manufacturing needs to contract electronics fabrication firms, and has approximately 800 employees worldwide. NuStuff has 60 million in assets and quarterly revenues averaging $20 million.
- ☑ NuStuff has already embraced open source and Linux technologies to a great extent, and has consolidated its IT infrastructure by standardizing on Linux in the server room and eliminating as many Windows servers as possible, although it does have a few proprietary and legacy applications that run in only a Windows environment. It's desktop topology is a mixture of Linux and Windows clients.

Frequently Asked Questions

Q: You mentioned other resources that cover the financial aspects of SOX compliance. Can you name a few?

A: Here is a short list:

Web Sites

- SEC Spotlight on Sarbanes-Oxley Rulemaking and Reports http://www.sec.gov/spotlight/sarbanes-oxley.htm
- The Sarbanes-Oxley Act Forum http://www.sarbanes-oxley-forum.com/
- American Institute of Certified Public Accountants (AICPA) http://www.aicpa.org/index.htm
- Sarbanes-Oxley Disclosure Information http://www.sarbanes-oxley.com/

Books

- What Is Sarbanes-Oxley? (ISBN: 0071437967)
- Manager's Guide to the Sarbanes-Oxley Act : Improving Internal Controls to Prevent Fraud (ISBN: 0471569755)
- Sarbanes-Oxley and the New Internal Auditing Rules (ISBN: 0471483060)

Q: Under what license are the applications and examples on the CD being released?

A: All of the software available on the CD, including any customizations, are released under the GPL. The example policies and procedures are copyright 2005–2007, Syngress Publishing Inc. This being said, all of your policies and controls should be original and applicable to your own organization, but you may use those provided on the CD as examples to get you started. A copy of the GPL can be found in the Appendices at the end of this book.

Q: I would like to know more about open source. Are there any other books or Web sites where I can get more information?

A: The short answer is: plenty! A Google search of the term open source reveals some 199,000,000 results, so although there are far too many to catalog in one list, here is a short list to get you started:

Web Sites

- Open Source Initiative http://www.opensource.org/
- Free Software Foundation http://www.fsf.org/

- Open Source News http://osdir.com/
- Linux Online http://linux.org

Books

- The Cathedral & the Bazaar, Musings on Linux and Open Source by an Accidental Revolutionary (ISBN: 0-596-00108-8)
- Open Sources, Voices from the Open Source Revolution (ISBN: 1-56592-582-3)
- The Success of Open Source (ISBN: 0674012925)
- Understanding Open Source and Free Software Licensing (ISBN: 0596005814)
- Succeeding with Open Source (ISBN: 0321268539)

FAQ

Q: Why does some open source software run on Windows? Isn't that contrary to the mind-set of the open source movement?

A: Although Linux and open source may be synonymous in many people's minds, the fact is that the open source development model existed for many years before Linus released their university project into the wild. The motivations that drive open source projects, such as freedom of choice, lower cost, increased security, and better quality, are not a function of platform, but rather a philosophy of the individual contributors of each project. If the infrastructure happens to be Apple, Solaris, or Windows, this does not preclude or prevent anyone from taking advantage of the open source paradigm.

Q: If open source software is frequently updated after bugs are found and the code is always changing, does that mean that an attacker would have more difficulty trying to exploit systems based on open source software than Windows-based systems?

A: The fact that the code may be on a rapid development cycle might deter a lazy attacker, but certainly not a determined one. In addition, as many open source projects mature, they tend to have maintenance releases for those users who wish to deploy a "stable" version, so the rapid-development cycle tends to not be applicable in these cases; however, the security needs remain the same. The real security advantage is the rapidity in which exploits that are discovered are fixed in the code. Anyone can develop an exploit to either open or closed source; however, the flip-side cannot be said. Whereas anyone can develop a fix for the open source exploit, in closed source you are betting the farm on the closed source provider having the expertise and priorities to close the hole. Transparency in this regard, not the changing code base, is what differentiates open and closed source for security.

Q: How complicated is using Linux or open source?

A: If you haven't seen a modern distribution of Linux, I highly encourage you to do so; you may be pleasantly surprised. Linux has come a very long way in the past 10+ years and we think it rivals any available operating system out there, for both corporate and personal use. In the past, one of the main problems with Linux was the distinct lack of software that was not specialized for some obscure IT or scientific need. That has now changed with the advent of contemporary software such as KDE and Gnome, Open Office, Evolution, and Firefox. The main point is that there is sufficient high quality and mature open-source software that combines with Linux to make a compelling choice. If you want to try Linux without installing it, we recommend downloading Knoppix (www.knoppix.org), which is a full desktop Linux distribution that runs entirely on CD and memory, without disturbing your computer's existing OS. Or you could simply run the live CD included with this book, which makes use of virtualization (covered in detail later in this book). You be the judge.

Q: Where can I get more information on the CentOS Linux distribution, including support?

A: Although this is not an exhaustive list, here are some to get you started:

- Main Web site http://www.centos.org
- The FAQ http://www.centos.org/modules/smartfaq/
- User forums http://www.centos.org/modules/newbb/
- Public mailing lists http://www.centos.org/modules/tinycontent/index.php?id=16
- Internet Relay Chat (IRC) information http://www.centos.org/modules/tinycontent/index.php?id=8

Chapter 2

Introduction to the Companion DVD

Solutions in this chapter:

- Installing the ITSox2 Toolkit VM
- Overview of the CentOS desktop and selecting your desired window manager
- Overview of eGroupware and applications
- Case Study: NuStuff Electronics SOX portal and cast of characters

☑ Summary

☑ Solutions Fast Track

☑ Frequently Asked Questions

The DVD Redux

"I'm doing a (free) operating system (just a hobby, won't be big and professional like gnu) for 386(486) AT clones."

–Linus Torvalds, creator of the Linux kernel at its initial release in 1991

In the first edition of the book we provided a Live CD based on the Knoppix Linux distribution. In the second edition we have improved the Live CD by reworked the entire toolkit utilize a VMware virtual machine in order to provide a robust platform in which to present the examples of Open Source and SOX concepts booking the second edition. While the Live CD was in many ways an appropriate stage for demonstrating a myriad of Open Source technologies our decision to move to a virtual environment was made for several reasons. First and foremost we recognize that there are a multitude of computer systems and software, both Windows and Linux, which might serve as the underlying host system for our toolkit. VMware runs on most computer systems and presents a stable but more importantly a generic (i.e. predictable) platform to take advantage to ensure the toolkit as robust and feature rich as possible. Second, while the Live CD did have methods for installing to a hard drive, this process required the host system to be able to dual boot with an existing OS or take over the target system entirely. On the original Live CD installation of the toolkit is necessary in order to save any changes made to the system, as a result of working through the book examples, and we felt this may have been problematic for some people.

Another issue was the Live CD was confined to the limited space available on the CD-Rom as well as the performance restrictions of a read-only CD medium. With our VM approach we have the best of all worlds; broad system support, performance, and the ability to use the system as a normal Linux computer. Last but not least, this also provides a handy segue into discussing the importance of virtualization and what it means for SOX compliance! We will discuss virtualization as a concept in detail later in the book, but for now here is brief synopsis of what virtualization actually is: a technology that allows multiple "computers" (normally referred to as guests) with different operating systems to run in isolation, side-by-side on the same physical machine. Each virtual machine has its own hardware such as RAM, CPU, networking components, etc. which are emulated as a generic set of hardware components that the VM sees as normal devices and can use like any normal computer regardless of the actual underlying physical hardware components. Virtual machines reside in a completely self-contained set of files, and these files which represent full "computers" can be moved from system to another, and assuming that the hardware minimum requirements are met, should continue to run without modification.

Installing the ITSox2 Toolkit VM

Host System Requirements

For a typical host system, we recommend that you have a 750MHz or faster processor (1GHz recommended) and 512MB RAM minimum (1GB RAM recommended). You must have enough memory to run the host operating system, plus the memory required for the guest VM operating system with its applications on the host machine. The VM provided on the DVD is set to require 256MB of memory however this can be reduced if necessary by editing setting the Ram threshold in VMware player or editing the ITSox2-Toolkit.vmx file (see below). The VMware Player requires approximately 150MB of disk space to install and the guest CentOS distribution will need a minimum of 3.5GB and depending on the amount of stuff *you* add may consume up to 8 GB of total disk space. If you do not plan to add to the VM content in any significant way then the disk requirements should stay closer to 4GB.

Installing the VMware Player

VMware Player is free software that enables users to easily run any virtual machine on a Windows or Linux PC. VMware Player runs virtual machines created by VMware Workstation, VMware Server, or VMware ESX Server. The player for both Windows and Linux has been provided on the DVD for your convenience, or you may alternatively download the latest version from the VMware website at http://vmware.com/download/player/.

Windows Installation

VMware player version 2.x runs on the following Windows platforms: Windows Vista, XP, Server 2003, and Server/Workstation 2000. To begin the installation you should navigate to where the VMware player installer is and double-click the setup file (ex. VMware-player-2.0.0-45731.exe). Figures 2.1 through 2.6 walk you through the process.

Figure 2.1 VMware Player Welcome (select Next)

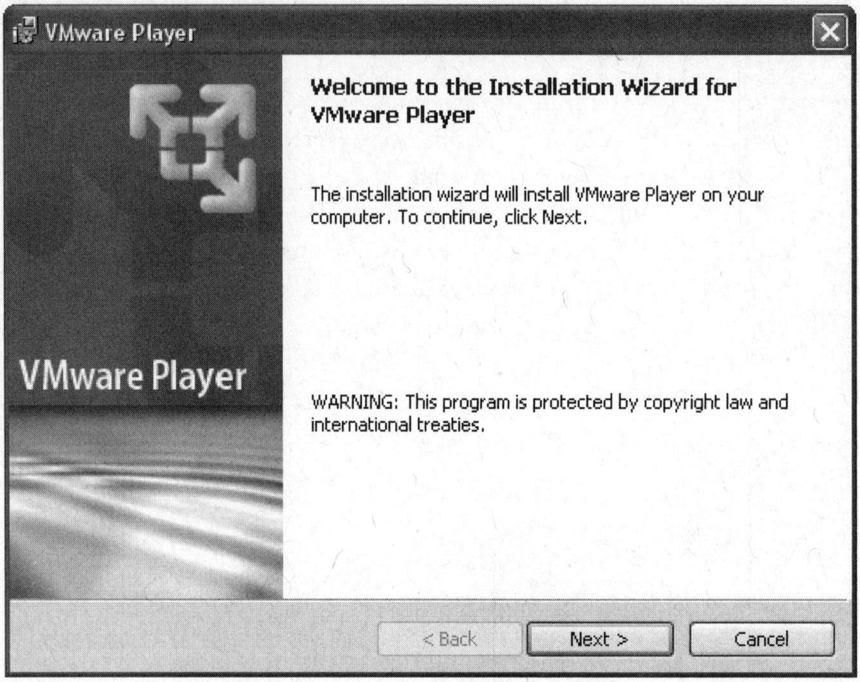

Figure 2.2 Destination folder (make any desired changes, select Next)

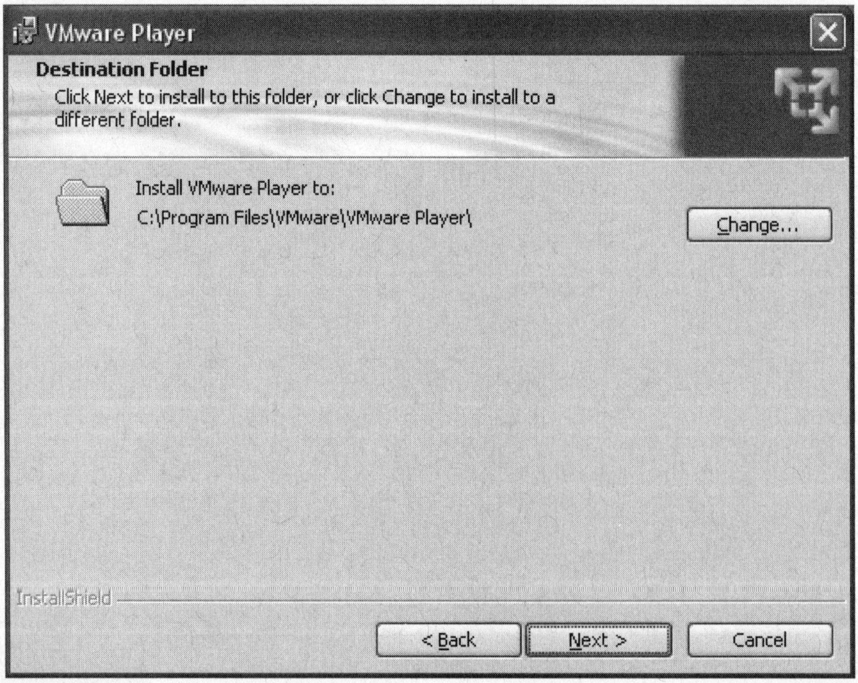

Figure 2.3 Shortcuts (make any desired selections, select Next)

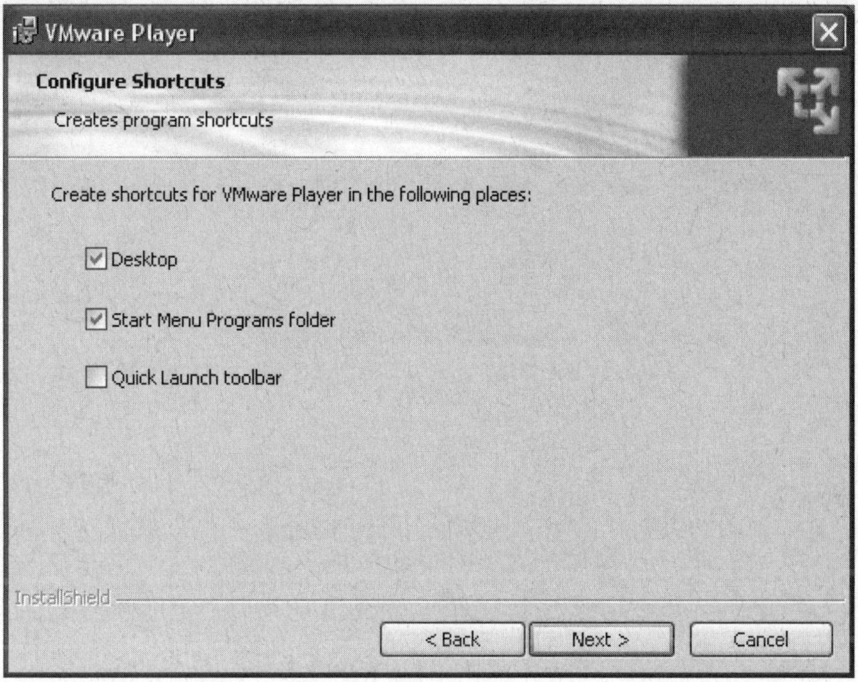

Figure 2.4 Install confirmation (select Next)

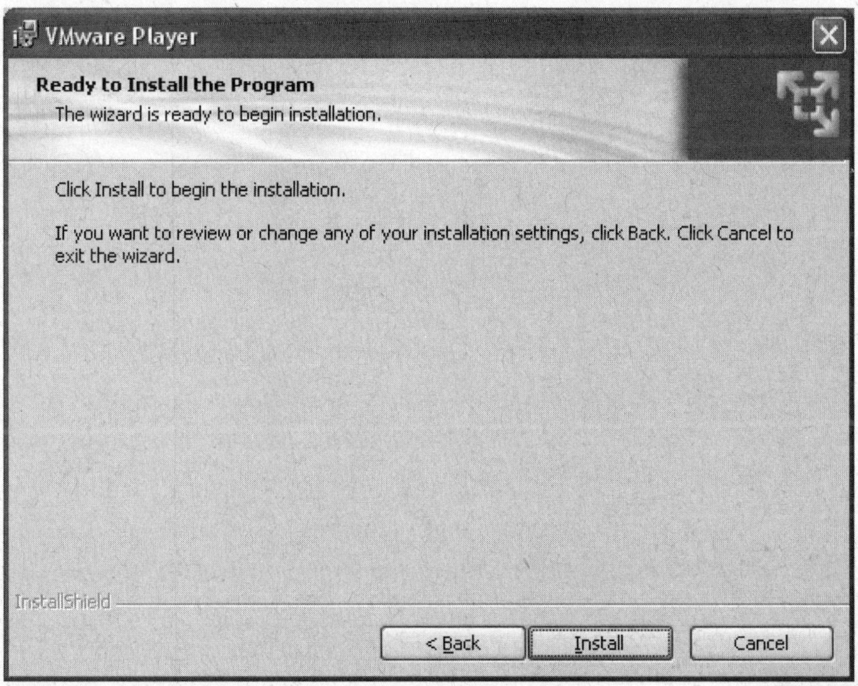

Figure 2.5 Installation Progress

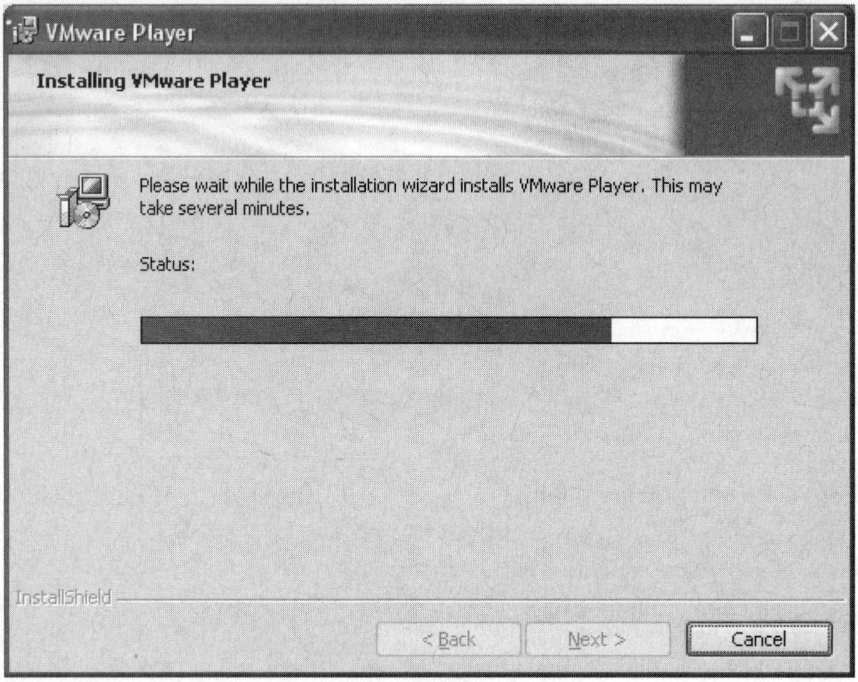

Figure 2.6 Installation Complete (select Finish)

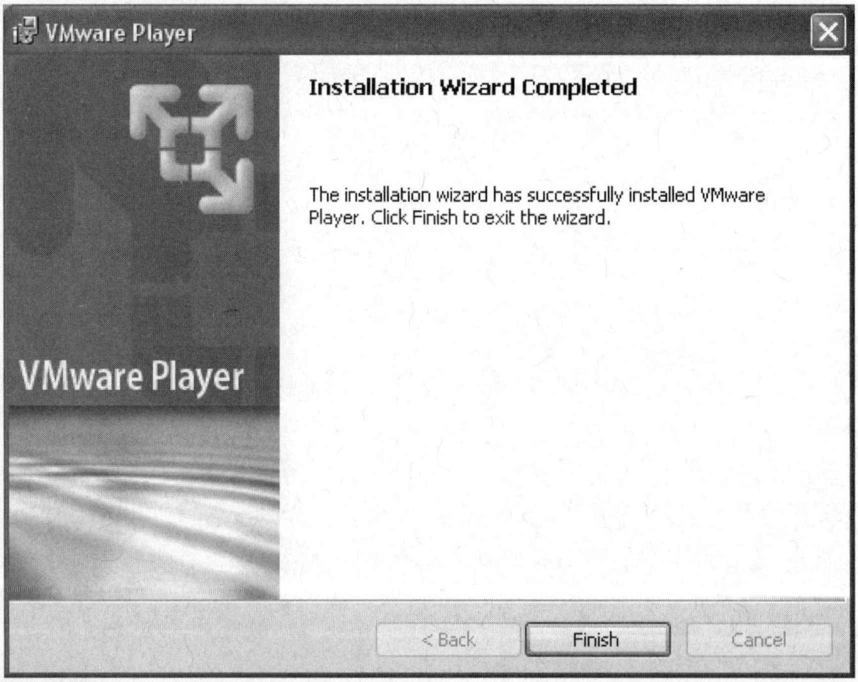

Linux Installation

For a Linux host you will need a running X-Windows desktop such as Gnome, KDE or XFLD. For Red Hat based host systems such as Enterprise Linux versions 3–5 or Fedora Core version 3–6 you may install the player via RPM package. In a root shell navigate to the folder containing VMware-player-2.0.0-45731.i386.rpm and type:

```
Rpm -Uvh VMware-player-2.0.0-45731.i386.rpm
```

For other Linux distributions such as Ubuntu or Novell/SUSE Linux you will have to install from tarball. In a root shell navigate to where the VMware player tarball is located and type:

```
tar zxvf VMware-player-2.0.0-45731.i386.tar.gz
cd vmware-player-distrib/
./vmware-install.pl
In which directory do you want to install the binary files?
[/usr/bin]
What is the directory that contains the init directories (rc0.d/ to rc6.d/)?
[/etc/rc.d]
What is the directory that contains the init scripts?
[/etc/rc.d/init.d]
In which directory do you want to install the daemon files?
[/usr/sbin]
In which directory do you want to install the library files?
[/usr/lib/vmware]
The path "/usr/lib/vmware" does not exist currently. This program is going
to create it, including needed parent directories. Is this what you want?
[yes]
In which directory do you want to install the documentation files?
[/usr/share/doc/vmware]
The path "/usr/share/doc/vmware" does not exist currently. This program is going
to create it, including needed parent directories. Is this what you want?
[yes]
Before running VMware Player for the first time, you need to configure it by
invoking the following command: "/usr/bin/vmware-config.pl". Do you want this
program to invoke the command for you now?
[no]
```

Once you have installed the Linux VMware player this requires some additional setup after installing the package so please use this procedure to setup bridged networking for the host system, NAT and host-only networking are not required for running the ITSox2 Toolkit VM:

```
/usr/bin/vmware-config.pl
In which directory do you want to install the theme icons?
[/usr/share/icons]
What directory contains your desktop menu entry files? These files have a
.desktop file extension. [/usr/share/applications]
In which directory do you want to install the application's icon?
[/usr/share/pixmaps]
Do you want networking for your virtual machines? (yes/no/help)
[yes]
Do you want to be able to use bridged networking in your virtual machines?
(yes/no)
[yes]
Do you want to be able to use NAT networking in your virtual machines? (yes/no)
[yes]no
Do you want to be able to use host-only networking in your virtual machines?
[no]
chkconfig vmware on
service vmware restart
```

Installing the ITSox2 Toolkit VM

The procedure for installing the toolkit is essentially the same for both Windows and Linux hosts. Navigate to where the root of the DVD and double-click the appropriate (ex. ITSox2-Toolkit-VM-w32-x86-setup.exe for Windows or ITSox2-Toolkit-VM-Linux-x86-install for Linux) to begin the installation.

Figure 2.7 ITSox2 Toolkit Confirmation (select Yes)

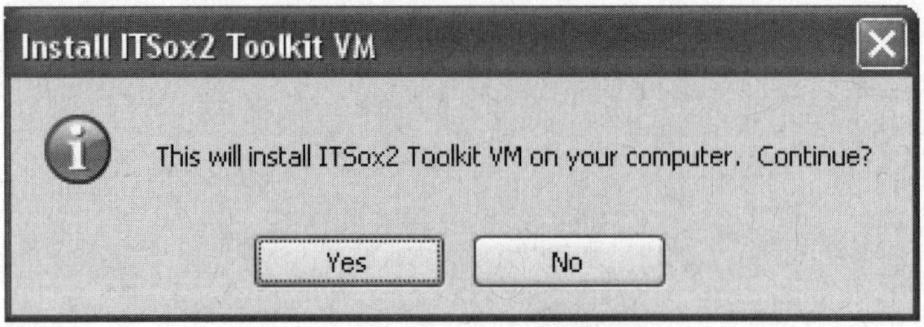

Figure 2.8 Welcome screen (select Next)

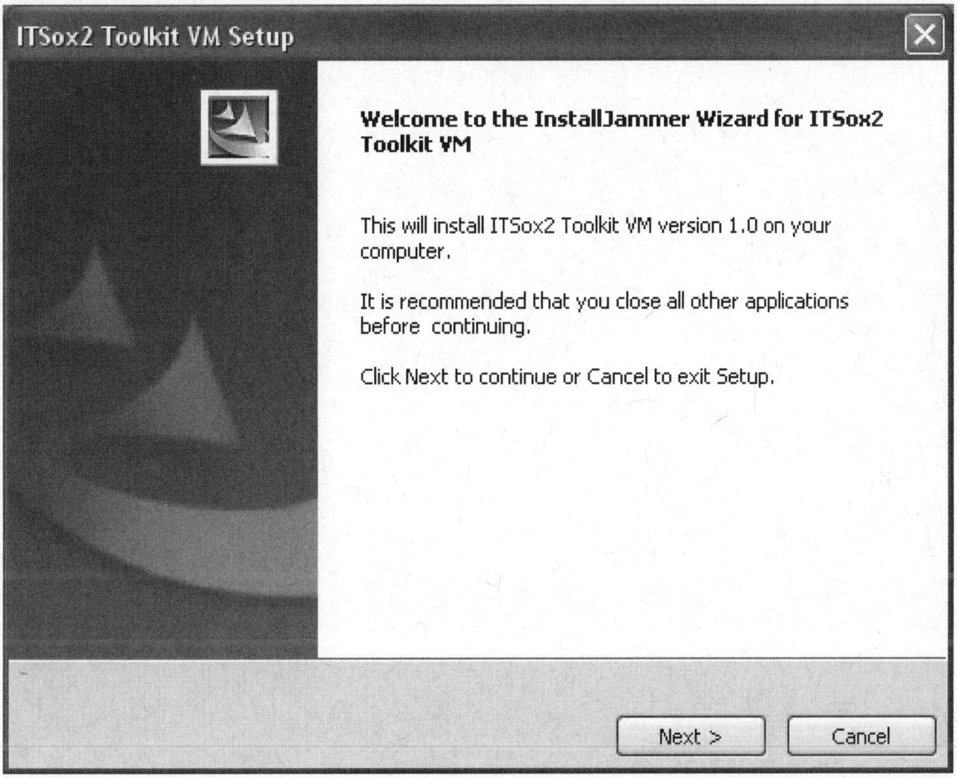

Figure 2.9 Default Location for Windows and Linux (make any desired changes and select Next)

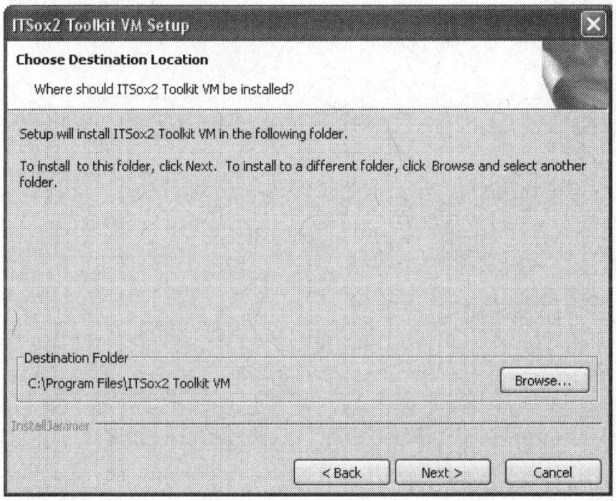

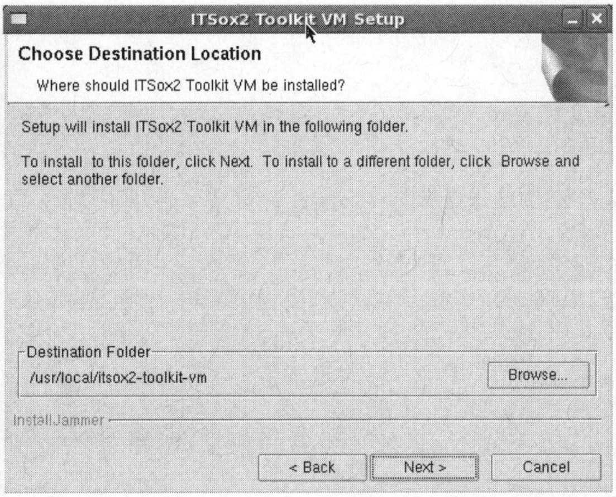

Figure 2.10 Settings confirmation for Windows and Linux (select Next)

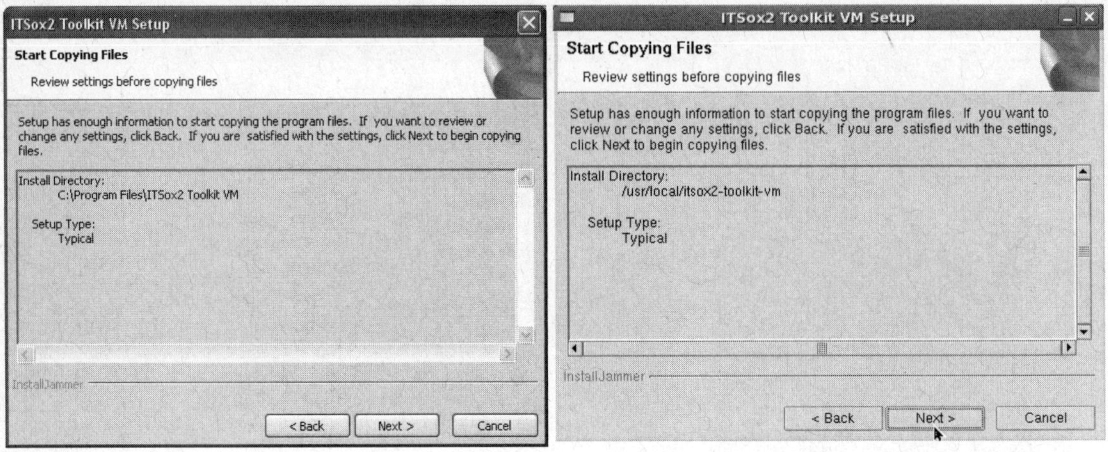

Figure 2.11 Installation progress

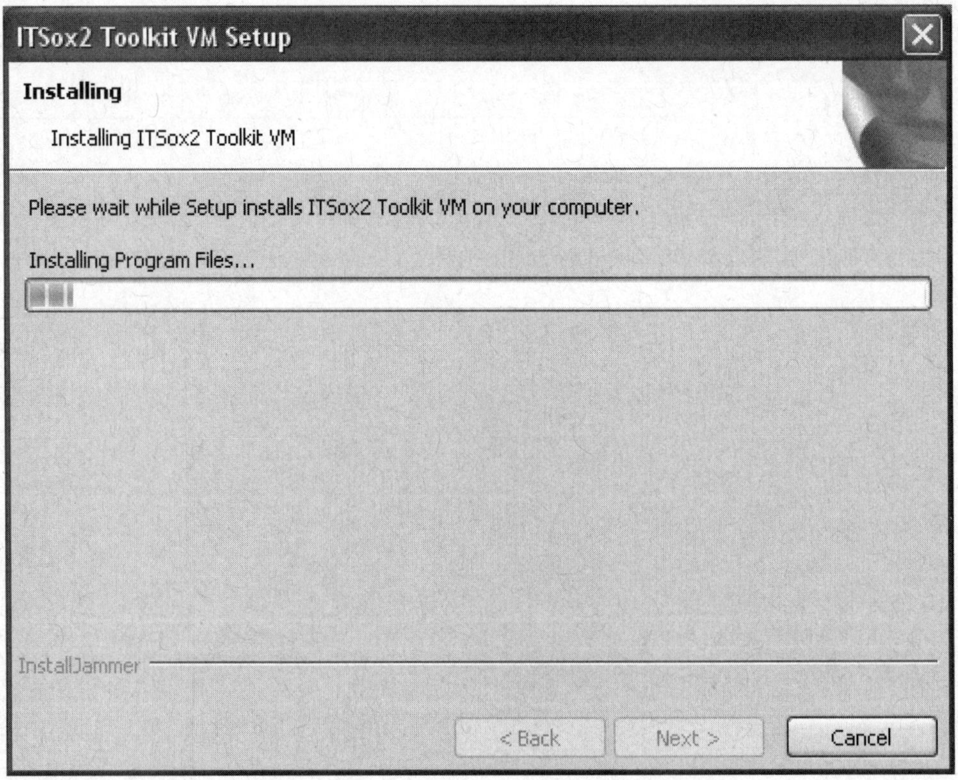

Figure 2.12 Installation complete (select Finish)

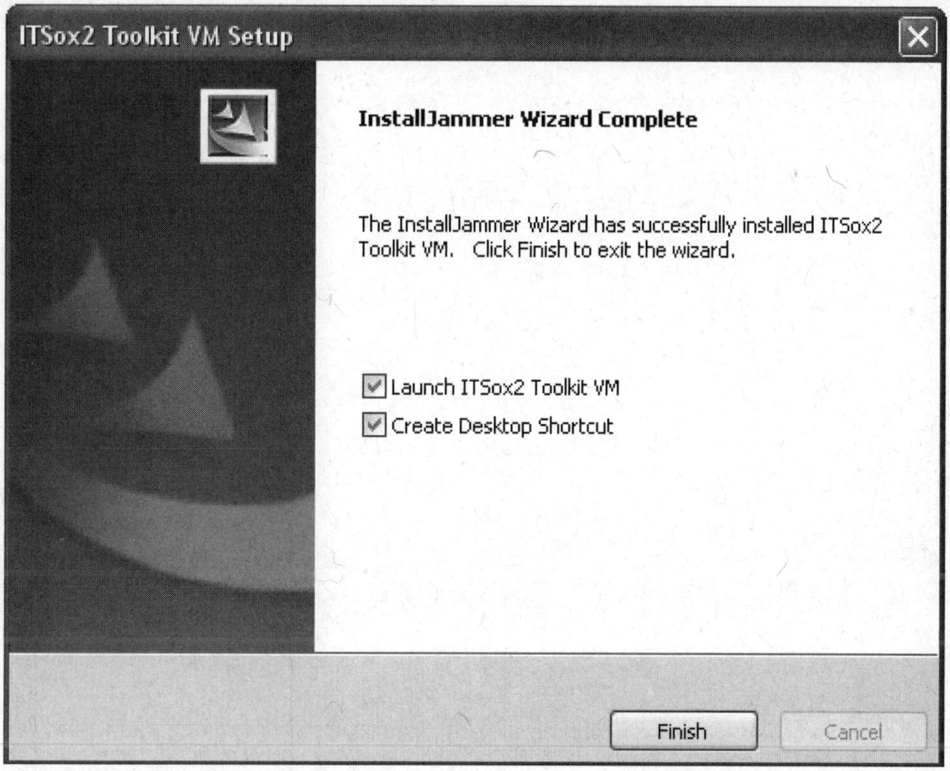

Launching the ITSox2 Toolkit VM

On windows you can choose the desktop icon or load the VM from Start Menu. On Linux you will need to launch the Player by either typing `vmware-player` from a shell or double click the icon if one was created by the installer. If necessary you can then open the VM by navigating to the installation directory (which is `/usr/local/itsox2-toolkit-vm` on Linux and `C:\Program Files\ITSox2 Toolkit VM` on Windows by default) and select the itsox2-toolkit-vmx file. Either way you will be prompted the first time you run the VM on either platform as illustrated in Figure 2.13 to create a unique identifier for this VM instance. This will setup a unique MAC address for the virtual network card so you don't have any collisions on the network if you run multiple instances of this VM either on the same host machine or elsewhere on the network. You will only have to do this once:

Figure 2.13 UUID Option (select Create)

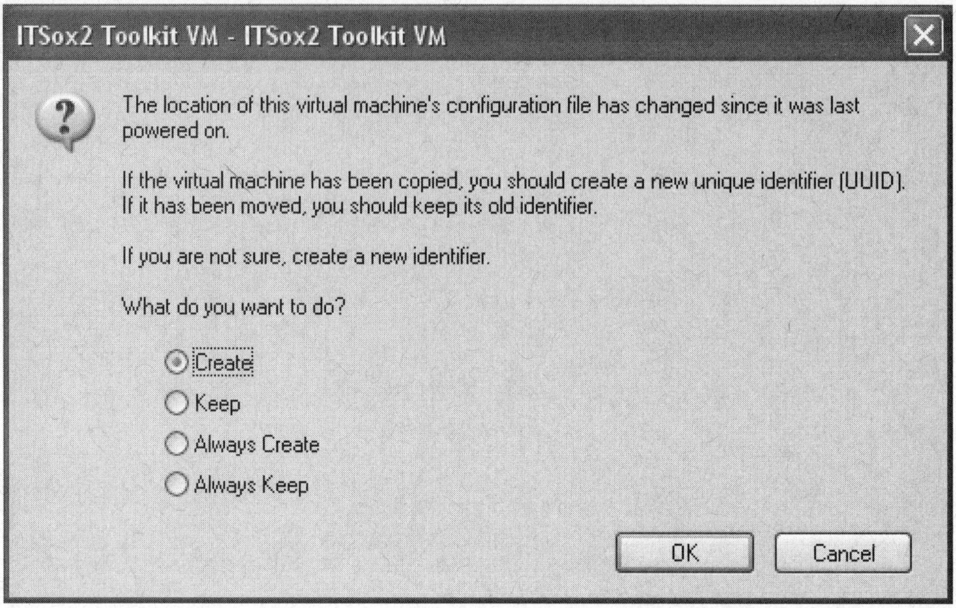

Uninstalling the ITSox2 Toolkit VM

To uninstall the toolkit simply remove the folder containing the VMware files. Since this is self contained this will completely remove the VM from your system, there are no registry entries or configuration files to have to deal with. Please use the normal Windows or Linux procedure if you wish to uninstall the VMware player.

The Transparency Test

The CEO Perspective

In today's information age, many collaboration tools are taken for granted, such as e-mail, calendaring, instant messaging, RSS feeds, discussion forums, social networks, blogs and Wikis. Over time we have seen collaboration move away from paper and pens into online systems that are available using a web browser, cell phone, or a PDA. Although these collaboration technologies have brought us significant benefits, they have also created challenges in information management, privacy, audits, security, data retention, and legal compliance. For example, the ubiquitous e-mail is a great communication tool, but it has also drawbacks: E-mails get lost over time; too many people in to loop; not reaching all interested parties; overflowing inboxes.

With so many tools at hand it is not easy to find a balance on what tool to use for what purpose. Often, a tool a person feels most comfortable with gets used most. But is e-mail the best tool to review a system requirements document? What is a suitable tool for my task at hand today, or tomorrow? Tools governance tries to address these questions, alongside with user training. For each collaboration tool, who is responsible for the content? Who manages access rights? Who monitors security? Who defines/refines processes and policies? Who enforces policies and legal requirements? Data governance tries to address these questions.

Wikis are very well suited for collectively sharing and organizing content. Many people learn to appreciate Wikipedia, which has become one of the most popular reference sites on the internet. Wiki is no longer a word just known by geeks. Wikis proliferate into the workplace, typically in grass roots. The younger generation wants to bring web 2.0 into the enterprise, IT is pushing back. Before long, more and more mission critical data gets stored in the grassroots Wikis. At some point, these Wiki get at the radar screen of CTOs and CIOs. They ask questions. What about backup? What about protecting our intellectual property? This is chaos, where is the data governance? With this, there is a push to consolidate the grassroots Wikis into a corporate Wiki maintained by IT.

Are Wikis suitable collaboration tools for the business environment? That is an affirmative yes for users who would like to simply share content with their colleagues. But most Wikis built for internet communities (so called publishing Wikis) lack features needed at work, such as fine grained access control, version controlled page data, attachments and meta data (e.g. a complete audit trail). These items are essential for SOX compliance as you need to know who the users are, what content they changed and looked at, what content they are supposed to see, and what content they are not supposed to touch. Whether it be Wiki or any other collaboration tool, it is essential to carefully select the right tool with the right features to ensure you have no problems with compliance down the road.

–Rod Beckström

Exploring the CentOS Linux Desktop

Once you have loaded the VM in the player the CentOS operating system should begin booting and you will eventually be presented with the initial GDM login screen shown in Figure 2.14. You will need to login to continue exploring the VM, so for now use the following credentials to gain access to the system. A complete discussion of the user roles is provided in this chapter in the case study section but for the impatient we are logging in as Biff Johnson.

NOTE

Normal Linux system users do not have by default administrative privileges. This is a great design of Linux and Unix in general; however you will need to be able to perform some tasks which require this type of access. Linux/Unix has a user called "root" which is the deity of all users for that system, and there are two ways of gaining administrative access via this account:

Login directly in a terminal or shell and change to the root account: `su - root`

Use sudo: `sudo <the-thing-you-want-to-do>`

We like sudo because access to administrative tasks is limited to a configuration file stored in /etc so you can grant fine-grained administrative access for ordinary users. A more complete discussion of sudo is provided in chapter 3 and 4. In any case the password for all accounts on the ITSox2 Toolkit VM is the same, for example:

Username: `bjohnson (or root)`

Password `letmein!1`

Figure 2.14 CentOS GDM Login Screen

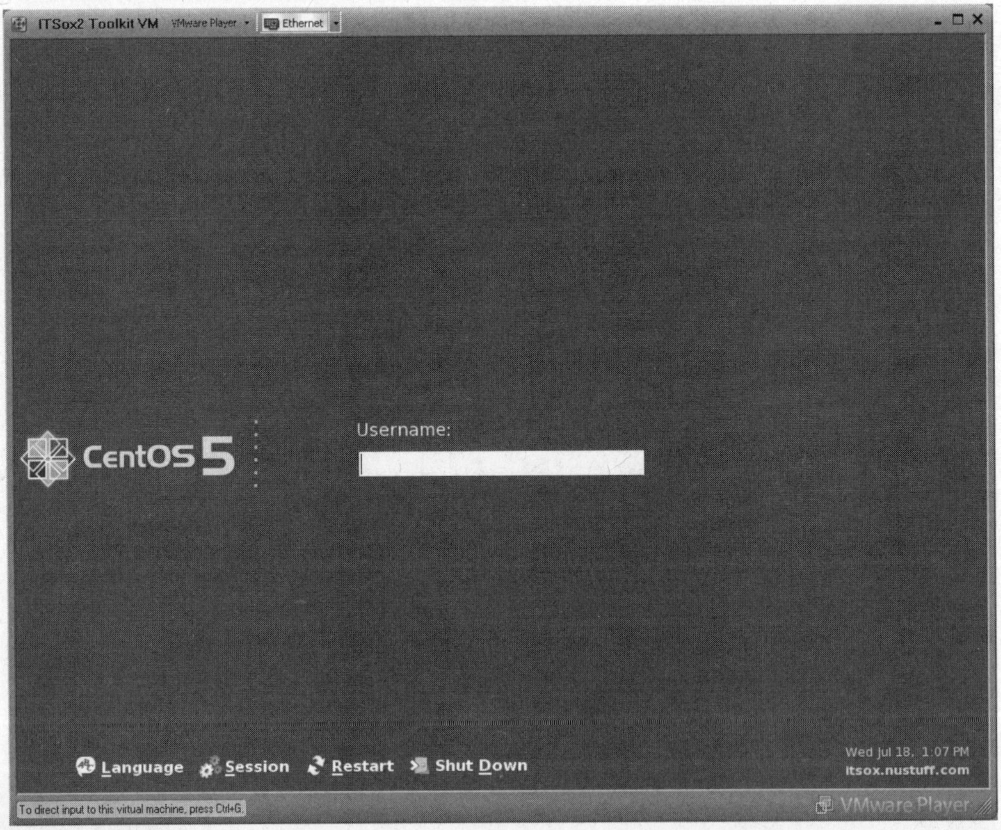

Selecting your Window Manager

If you are at all familiar with CentOS or any recent Red Hat distribution the first thing you will notice is the window manager in our VM is not what you might expect. We have customized the X-Windows environment for the toolkit to use IceWM as the default for all users which is an extremely light-weight window manager that does everything we need it to do and absolutely nothing else. It is not quite as no-frills as TWM (the fallback window manager that comes with Red Hat's X11 implementation) however if you are used to Gnome or KDE desktop environments this might seem a bit spartan, shown in Figure 2.15.

Figure 2.15 CentOS running the IceWM Window Manager

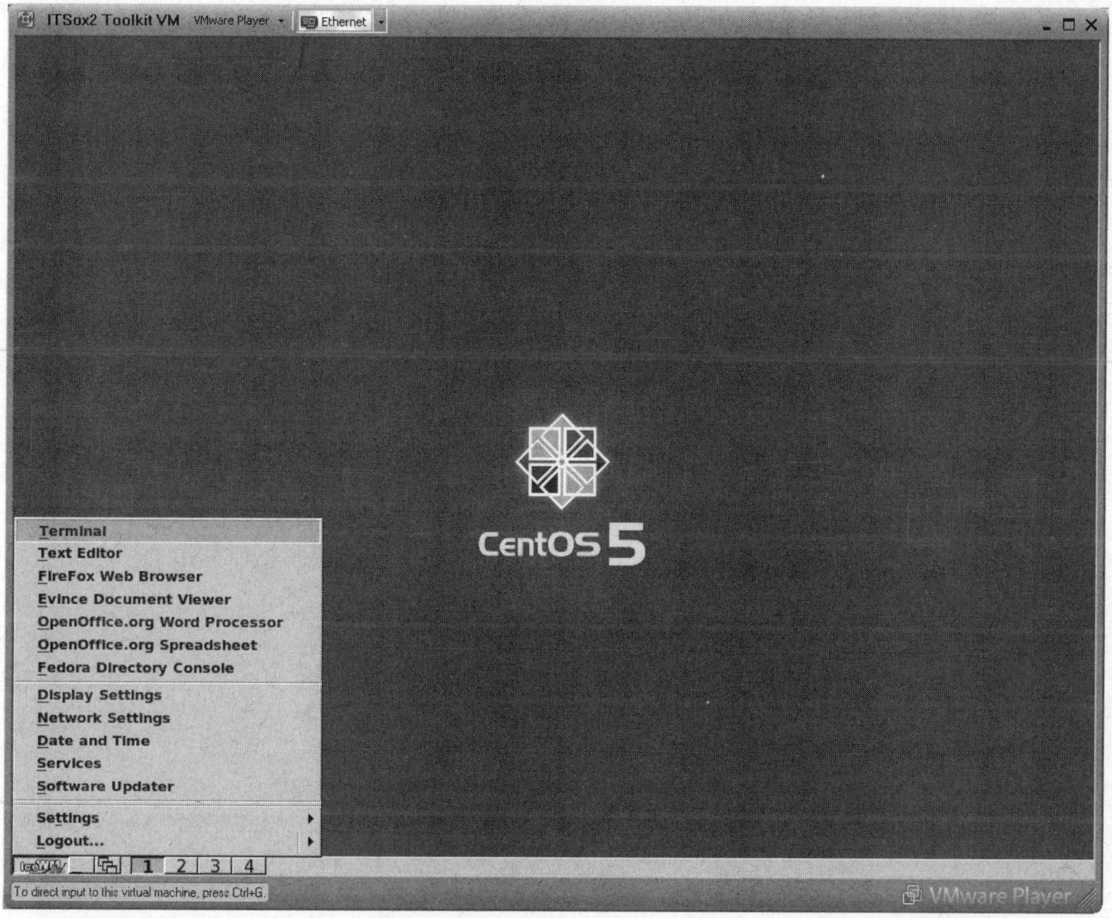

We chose this window manager to drastically reduce the memory requirements of running this virtual machine for those readers who did not have more than 256MB to dedicate to the system but for those of you who do have 1GB of memory or more and wish to run your favorite desktop may do so. We have included the Gnome desktop in the VM as a selectable window manager; if you wish to use KDE you can install this via the yum package manager. See the next section on adding packages for more information. To switch your desktop to Gnome for the currently logged in user simply select the "Switch Desktop" menu item. Alternatively you may type switch-desktop in a terminal window. Either way you may now set the window manager to either Gnome or IceWM, or if you are really brave and like it lean and mean you could give TWM a try. If you do decide to run Gnome or KDE we recommend that you increase the amount of memory available to the VM from 256 to 384–512MB, you can do so by clicking "VMware Player" in the title bar ... Troubleshoot ... Change Memory Allocation. You will be presented with a dialog in order to adjust the memory allocation accordingly. It is safe to do this while the VM is currently running as any changes you make here will only take effect once you have "rebooted" the VM.

Adding Packages and Staying Current

The application that keeps CentOS up to date is called yum which is an acronym for "Yellowdog Updater Modified". The yum utility is a command line package management utility for RPM-compatible Linux operating systems such as CentOS and Fedora Core and is currently maintained as part of Duke University's Linux@DUKE project. Though yum is a command line utility there are other tools which provide a graphical user interface, we have included yumex on in the VM for your use. One of the very nice features of yum is its ability to automatically compute dependencies and figures out what things should occur to install packages, so when you say you want to install foo.rpm yum will figure out that foo requires bar to work and add bar.rpm automatically to the install transaction. Yum supports multiple repositories of packages; on the toolkit VM we have included the standard CentOS base as well as the excellent rpmforge package repository. To explore all of the installed and available packages you may select Software Updater from the menu or type `sudo yumex` from any normal terminal or shell. If you are in a root shell you do not need to supply the `sudo` command.

This might be a good time to install the rest of OpenOffice.org office application suite. We have already included Writer which is OpenOffice.org's word processing program we use in several examples later on, but to save disk space we did not incorporate the rest of this great suite of applications such as Calc (spreadsheet) and Impress (presentation). To get these components and others simply choose them in yumex or type the following in a normal shell: `sudo yum install OpenOffice*`. This will give you a complete list of the OpenOffice.org suite as well as any dependencies they have and give you the opportunity to install them.

Other System Setup Opportunities

CentOS provides both command line and GUI tools to add, change or remove various system level settings, here is a partial list of some that you might need to use:

- system-config-date – configure date and time settings, including using an internet based time server to automatically keep the system time up to date, a consideration when adding SOX related items anywhere in the system that are sensitive to time
- system-config-printer – configure printers
- system-config-network – configure network settings. The VM is initially setup to use DHCP and obtain an IP address automatically, however if you wish to set a static IP this would be the place to do it. Note, the applications included in the VM expect the hostname to be "itsox" so if you change this some things will most likely break without further configuration changes
- system-config-securitylevel – configure security settings. We have left the VM in a fairly permissive state however we do discuss the importance of firewalls and SE-Linux later in the book. You can change the defaults here; this may require a reboot of the VM.
- system-config-rootpassword – this allows you to change the password for the root user
- system-config-display – by default the VM is setup with a screen resolution of 1024×768. If you desire to change this here would be the place to do it.
- system-config-services – we have configured most applications and services that we need to be running to automatically start when you boot the ITSox2 Toolkit VM. There may be exceptions to this so if you need to modify what starts on boot this applet provides a mechanism to do so.

There are many more system configuration applets, to view a complete list of those installed in the VM type the following in a normal shell: `rpm -qa | grep system-config`. To see a list of all available applets whether they are installed or not type `sudo yum list system-config*`. Of course you can use yumex instead.

VM Spotlight – eGroupware

E-GroupWare

http://egroupware.org

The eGroupware suite is written in PHP and provides a framework in which applications can be written, providing common user management and access controls, a theme engine, and transparent database access. We have created portal site that was built from the SiteManager application as one way to demonstrate the power of eGroupware. The SiteManager application is a traditional content management system (CMS) similar to Xoops or Mambo and allows for websites to be built using modules and components, including most of the applications included in the eGroupware suite itself. The front page to the portal can be accessed anonymously however to work through most of the examples in the book you will need to login as one of the sample users listed in the next section. A toolbar icon represents each eGroupware application and we have provided a picture of this icon in each applicable section of the book for easy reference. Here is an overview of the eGroupware applications so you have some familiarity, and we will be using specific examples throughout the book as they apply to the Sox compliance requirements.

NOTE

All of the web-based applications discussed below and in subsequent chapters require an SSL encrypted connection from your web browser, please note the use of https://

eGroupware Applications

SiteManager

https://itsox2

The entire website for our example company has been constructed in the SiteManager application, which strives to be a user friendly and intuitive dynamic web authoring system. One of the features that make this extremely useful for our needs is that most of the eGroupware applications are also defined as useable modules and components; with fine grain access control this gives us the ability to create a very specific SOX oriented website for our company's internal use. SiteManager is able to generate dynamic content with distinct sections that various eGroupware users may edit depending on the permissions set in the application. This means that the generated website can have independent segments in which different people or groups are charged with design and maintenance. The site administrator generally sets the theme and creates headers, footers, and sidebars that will make up the site-wide look and feel, and individual sections can be set to be publicly viewable (anonymous access or private based on the defined ACL. From a themes perspective this CMS is compatible with the Mambo templates and there are probably hundreds if not thousands of free examples are available on the web. We will explore in detail the NuStuff portal in the next section but to give you an idea of some of the interesting modules that come with SiteManager in addition to virtually every built-in eGroupware application, here is a partial list:

- HTML – A content block is where you can author any valid HTML content
- File Contents – A content block is for displaying the contents of an existing file on the disk
- Applet – A content block to embed java or flash content
- Download – A content block that creates a download area for your website
- Amazon and Google – Special content blocks for searching third party websites

Combined with the applications available as components you should be able to create a sophisticated SOX oriented website for your own use, or modify the one that we have developed to suit your needs.

Home

https://itsox2/egroupware/home/index.php

The home application is the simplest of all eGroupware applications and provides a configurable "front page" view of the system once a user logs in. This user specific home page can be tailored to suit the individual's needs by changing your preferences, and although this application is not specifically used in the book it is provided with every default eGroupware instance. Some of the custom items that can make up your home page are:

- User Calendar
- New messages from the FelaMiMail email application
- Items from the InfoLog application
- Items from the News application
- Items from the AddressBook application

In order to change what appears on your home page and how they are displayed you will need to set the applicable preferences in each individual application that has a front-page hook. Administrators can set default and forced preferences for this item, users can change the order in which each section appears on their home page by using the arrow icons to the right of each block.

Preferences

http://itsox2/egroupware/preferences/index.php

Speaking of preferences, there is an application devoted to the definition and maintenance of site wide and per application settings. These will fall into one of the following categories:

- Forced Settings – All site and user settings can be "forced" to a particular value. If an item is set here then this is enforced site wide and only an Administrator will have the ability to change it, which obviously implies that administrative access is required to access this tab
- Default Settings – This is where you setup the default value for items and requires administrative rights to access. If an item is set here then this is initially enforced site wide however each individual user will have the ability to override the setting to suit his or her personal needs
- Your Settings – This is where you can set the value for items that have not been forced by an administrator. If an item is set here then this overrides the default setting.

TIP

eGroupware supports multiple languages so you can setup the preferences here that correspond to your particular location such as the language of your choice, country, time zone, date format, time format and how currency is displayed.

eGroupware also supports a themed environment however we only have one theme installed for the sake of disk space. As with the SiteManager application you can create your own theme to change the default look and feel of eGroupware if you so choose and once you have done this then you would set the theme parameters in the Preferences application as well.

Administration

http://itsox2/egroupware/admin/index.php

The Administration application is where all users and groups are managed as well as application defaults and ACL's. eGroupware applications can be globally managed here so that you can activate or deactivate applications at the site level, register new applications, and setup default access control lists. Users with administration rights to eGroupware such as Biff Johnson have the ability to allow groups of people access to various application data entered by the users. Figure 2.16 shows the ACL screen for the IT Services Staff group, specifically for the KnowledgeBase application:

Figure 2.16 KnowledgeBase ACL: EGW Administrators example

Knowledge Base - Preferences - ACL: [IT Services Staff] IT Services Staff Group			
a b c d e f g h i j k l m n o p q r s t u v w x y z **All**			
showing 1 - 28 of 28			
all fields ▾ [ ] Search			
Groups	**Read**	**Edit**	**Publish**
Default	☑	☐	☐
Executive Staff	☐	☐	☐
External Auditors	☐	☐	☐
External Consultants	☐	☐	☐
Facilities Staff	☐	☐	☐
Finance Staff	☐	☐	☐
Human Resources Staff	☐	☐	☐
IT Services Staff	☐	☑	☑
NoGroup	☐	☐	☐
users	**Read**	**Edit**	**Publish**
[anonymous] anonymous User	☐	☐	☐

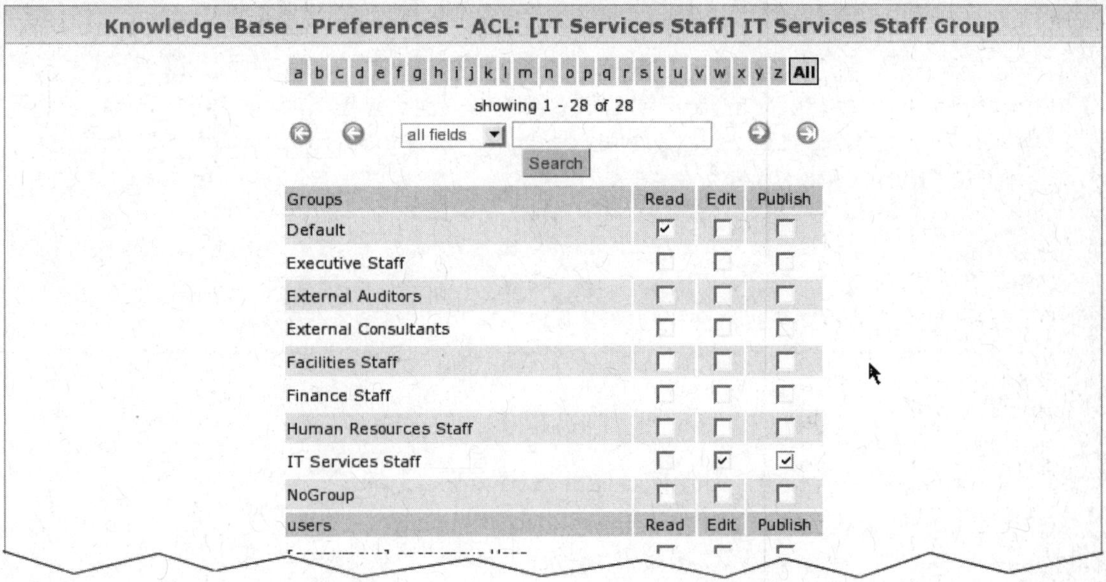

In the above example the permissions are set so that any article written by a member of the IT Services staff is readable by the "default" group (i.e. everyone) and all members of the IT Staff group may edit the article and publish as applicable. Rights for most applications are generally settable for the following types of access:

- read – grant read rights to non-private data
- edit – grant change rights to non-private data
- add – grant rights to add new data
- delete – grant change rights to non-private data

Optionally an eGroupware application may define other custom access control points as appropriate, the "publish" item in the previous example is one specific to the KnowledgeBase for instance. Individual users can also grant rights to their own data in some applications as well. Not every eGroupware application supports ACL's and not every type is implemented, however for those applications where ACL's are important to our Sox examples we will point this out in the appropriate sections of the book. In order to view and maintain ACL's on a per-application bases you start by selecting the user or group you wish to modify in the Admin module and then drill down to the ACL page for the application.

TIP

The use of ACLs is fairly pervasive in eGroupware and sometimes it is not immediately clear why a user cannot view or access a particular item as expected. When this is the case it usually boils down to having at right privileges defined for the item in question. We recommend keeping your ACL definition at the group level and avoid assigning user specific rights to reduce complexity. Of course there are exceptions to this rule as sometimes it is appropriate to grant individual access; one example might be granting write access to your calendar for your secretary that would normally manage this for you.

Also potentially important from an audit perspective detailed access logs are maintained by eGroupware and are viewable here as well as current sessions. The log shows login attempts that have failed; successful logins will show who logged in, when they logged out and how long they stayed on the site. Here is an example:

LoginID	IP	Login	Logout	Total
Brown	69.12.172.56	2007/07/10 - 09:56	2007/07/11 - 14:46	4:49:48
nfarley	69.12.172.56	2007/07/10 - 09:51	2007/07/10 - 09:56	0:05:37
kking	69.12.172.56	2007/07/10 - 09:46	2007/07/10 - 09:50	0:04:45
slundy	69.12.172.56	2007/07/10 - 09:44	2007/07/10 - 09:46	0:01:27
clahti	69.12.172.56	2007/07/10 - 09:08	2007/07/10 - 09:44	0:35:41
clahti	69.12.172.56	2007/07/10 - 08:13	**Bad login or password**	
nfarley	69.12.172.56	2007/07/10 - 07:16	2007/07/10 - 08:01	0:35:41

Within the Administration application there are a few sub-modules: PHP Information shows the currently installed PHP stack and configuration while System Information gives a synopsis of the currently running hardware such as memory, network, and file system statistics. Lastly you are also able to define a database backup schedule for the eGroupware application itself so you do not lose any of your valuable data, and this sub-module also gives you the ability to restore a previous backup set.

FelaMiMail Email Client

https://itsox2/egroupware/felamimail/index.php

FelaMiMail is the full-featured web email client that supports IMAP mailboxes. FelaMiMail is fully integrated into the eGroupware suite so that the AddressBook and system users are available to the client, additionally you can also access external applications such as an LDAP AddressBook. We have included this application so that you can view the IMAP mailboxes of system users provided by the Dovecot IMAP server running on the VM. Email has become an integral part of business communications and the importance of this will further illustrated as we discuss workflow notifications in later chapters. You can set new messages to display on your home page, and the mail client has the classic layout you would expect and provides a full set of features to allow you to use this as your primary email interface if you desire.

Calendar

https://itsox2/egroupware/calendar/index.php

The calendar application is one of those "meat and potatoes" kinds of functionality found in all groupware applications and provides individual and group shared calendaring. In your compliance efforts this is a useful tool to provide meeting schedules to all persons and groups that will be working on your Sox audit. The calendar supports the following views for both individual and group calendars:

- Day view
- Four days view
- Week view with weekend
- Week view without weekend
- Month view
- Planner by category
- Planner by user
- List view

Most of the views listed above are self explanatory however the planner bears further explanation. The planner shows a colored monthly calendar of a user or group of users, and can be filtered and sorted by a particular category. It is meant to give you an overview and help to manually schedule meetings and events.

TIP

Most eGroupware support the notion of categories. There are two separate kinds: global and local. Global categories are generally available to all applications where local categories are meant to supply an individual application or user. When defining categories you can assign an icon and/or color, and this information is used by the applications to provide visual feedback. Categories can be nested in a parent-child relationship, affording great flexibility in displaying information.

The user's chosen default calendar view will be display on his or her home page, the behavior and mode can be set in the preferences application. As you explore the calendar note the open to-do items from the InfoLog application, which are date-context sensitive.

AddressBook

https://itsox2/egroupware/addressbook/index.php

The AddressBook application contains information for all employees and external players in our Sox examples. As far as integration with other applications the links tab gives you the ability to reference InfoLog items to projects in the ProjectManager, Calendar entries, InfoLog items, Resource, TimeSheet entries, Tracker and Wiki items, therefore it becomes the central place to store information about the people involved in the compliance process. As with many eGroupware applications you can assign both global and local categories for contacts, and the application supports global and personal per-user contact lists.

The AddressBook also supports distribution lists and you can select these in the mail client to send mail to a list of people. Advanced searching and the ability to define custom fields round out the capabilities of this module.

InfoLog

https://itsox2/egroupware/infolog/index.php

The InfoLog application is an important application that covers items where tracking and record keeping is important. Links to the AddressBook can be established for items as well as external references and file attachments. There are several types of items you can track with InfoLog such as emails, to-do's, notes and phone calls. Each type also features the following functionality:

- Global and local categorization
- Links to AddressBook, Calendar, other InfoLog items,
- Integration and links to projects in the ProjectManager, Calendar entries, InfoLog items, Resource, TimeSheet entries, Tracker and Wiki items
- File attachments and URL references
- Delegation of items to other eGroupware users with status tracking.

This versatile application is well integrated with other modules where people are involved, most notably the ProjectManager application (see below), which uses the InfoLog entries as its task items. This application is can and should be used in an actual audit process to stay on top of things.

ProjectManager

https://itsox2/egroupware/projectmanager/index.php

The ProjectManager application is exactly what its name implies, a full fledged module supporting all phases of a project's element based management cycle including:

- Project startup activities such as order acceptance, project structuring and optional accounting and budgeting
- Project lifecycle management and control using milestones and reporting such as target vs. actual
- Progress tracking with subprojects, work packages, resource planning, process planning, dependent or constrained work items and costing
- Post project activities such as accounting and review reports

ProjectManager integrates seamlessly with many other eGroupware applications such as InfoLog, Tracker, TimeSheet, AddressBook and Calendar (a now familiar theme). This also supports user defined fields of information, global and local categories, Gantt charts, and owner/role based ACL's. We use this in following chapters to illustrate the tracking of projects that result from our "one page strategies". This application could be used to track virtually any project, including the SOX consultants and auditor's billable time and deliverables so that you can be sure you are getting what you are paying for.

We have added a very simple example project to introduce the features of ProjectManager. In chapters four and five we begin to use the ProjectManager application for our case study so we will not spend too much time with details here, this is just a quick preview of the major features. If you navigate to the ProjectManager page and select the "Example Simple Project" and poke around you will see many of the integration features with the other eGroupware applications. This project has the following attributes:

- Simple project with no budget or accounting
- Three project members, Christian Lahti, Rod Peterson and Biff Johnson
- Two milestones, Phase A and Phase B
- Six task items called Task 01 thru Task 06
- A planned start date of 2007/07/16 and planned end date of 2007/08/16
- The tasks are divided between the project members, and Task 02 has a constraint that Task 01 be completed first. Task 05 has been assigned to both Rod and Biff.

We added references to the Address book for the project members, other items that could be added in a real world example are subprojects, sub-tasks of each individual top-level task, items from the Resources application, TimeSheet entries to track actual time spent on tasks, and the list goes on. Go ahead and click on the Gantt chart to view the simple example so you have an idea of the power of this module. You can use other logins to view the assigned InfoLog tasks for each user as well. The Gantt chart for our mini project example is shown in Figure 2.17.

Figure 2.17 Example Simple Project Gantt Chart

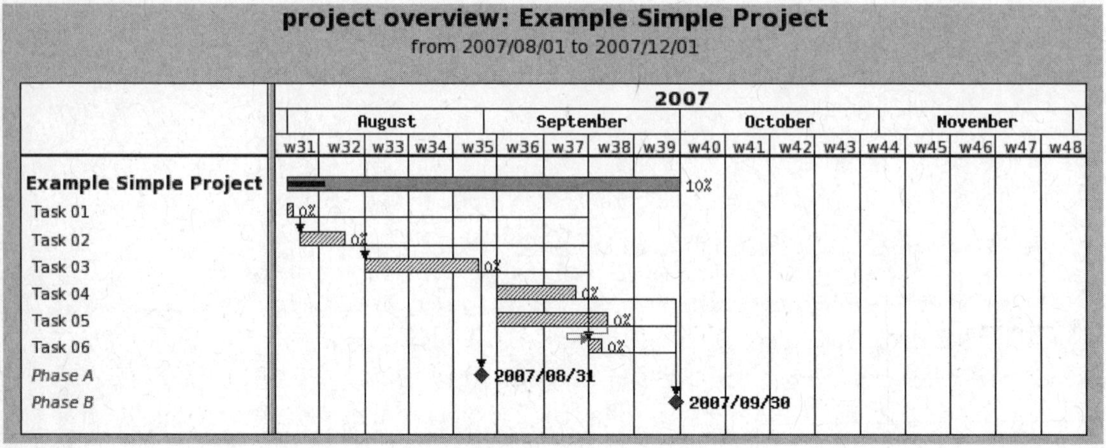

TIP

In order to view or change milestones click on the diamond shaped link in the Gantt chart.

Wiki

https://itsox2/egroupware/wiki/index.php

Wiki is a system of free-form collaboration tool in which anyone within the eGroupware portal (subject to optional access controls) can add and modify pages using a WYSWYG editor. This is another excellent example of another open source application born elsewhere as WikiTikiTavi (http://tavi.sourceforge.net) and integrated into eGroupware.

Wiki is based on an easy to learn markup language similar to html and in addition to the actual pages, content and links that are created, each page has a header and footer that provides information and allows actions to be taken. We are covering the general concepts of Wiki here since it is used in subsequent to support various aspects of SOX compliance activities.

General Wiki Concepts

WordsSmashedTogether create a page link. A "?" appears next to a new link that does not yet have a corresponding page, click it to create the new page. You do not have to use WordsSmashedTogether; you can use ((Hello World)) to create a new link. Here is a list of additional standard Wiki markups you can use as well:

- To create italic text, use `''hello world''` or `//hello world//`
- To create bold text, use `'''hello world'''` or `**hello world**`
- To create superscript text, use `^^hello world^^`
- To create subscript text, use `,,hello world,,`
- to create a horizontal line use ----
- to create a bullet list use

```
**bullet 1
**bullet 2
**bullet 3
```

Since Wiki comes with a WYSIWYG editor you really don't need to explicitly use the Wiki markup language, additionally you may also use regular HTML such as tables to further format your page. This module also supports the familiar ACL's on a per page basis and links to other applications, as well as a complete history view and change log so you can track who made edits to each individual page.

Lessons Learned

Structured Wikis

Earlier in the chapter many questions were raised regarding the role of collaboration software and its relationship to compliance. Let us look at one Wiki engine that is especially suited for SOX activities: TWiki. TWiki (http://twiki.org) is an open source Wiki for the enterprise which is specifically designed for businesses. It is a Wiki in its heart, where content can be shared collaboratively in an organic way. In addition,

Continued

TWiki has many features supporting the daily work, such as tracking projects, managing documents and meeting minutes, building glossaries and knowledge bases, and supporting workflows. TWiki is a structured Wiki that allows users to create their own Wiki applications. Someone with moderate technical skills can create simple applications, such as a call center status board, to-do lists, sign off sheets, as well as more complex applications, such as inventory systems, CRM applications, blogs, and discussion forums.

In a structured Wiki, structure can be added as needed. The "as needed" part is important. First and foremost, employees are enabled to document processes and to share their daily workflow in the free-form Wiki way. Secondly, employees can create structured Wiki application with forms, queries and reports to automate their daily workflow. This brings web 2.0 into the workplace. Users are in control of content and applications. Flexibility and freedom for employees brings a lot of productivity and transparency to the workplace, but it needs to be managed and nurtured properly. Data governance is important, especially when facing SOX. TWiki integrates with an enterprise directory (LDAP, AD, etc.), has fine grained access control, and version controls everything, including all meta data. You can verify who changed the access control settings of the signoff sheets.

Every successful Wiki deployment has at least one Wiki champion. This is a person who both understands the process of the work for a given project or business (the domain), and how to use a Wiki (best practices in collaboration). The Wiki champion is primarily concerned about the content; structure the content to make it easy for the users to navigate and find relevant content. In a structured Wiki, this champion also helps create Wiki applications. Structured Wikis are in the long tail of implementing business processes. The Wiki champions help automate business processes. When IT is responsible for the Wiki, they are mainly concerned about uptime and backup, not so much about the application layer. IT gladly hands off the responsibility for the content and Wiki applications to the lines of businesses. The open source TWiki is now backed by TWIKI.NET (http://www.twiki.net), a company offering subscriptions for a Certified TWiki distribution that includes support and automated software updates. This allows IT to deploy the enterprise Wiki in a managed way.

–Peter Thoeny

Bookmarks

https://itsox/egroupware/bookmarks/index.php

The bookmarks application tracks links to external websites similar to your browser favorites. This is an example of an application integrated with SiteManager, in which the links added to this application are displayed in each portal site as a SiteManager module. We have used the bookmarks application to provide convenient links grouped by chapter to all sites referred to in the text of this book. The Bookmark application supports ratings, public and private entries, and keywords.

Resources

https://itsox2/egroupware/resources/index.php

The Resources application is a management and booking tool integrated into both the Calendar and ProjectManager applications. Examples of resources would be meeting rooms, AV equipment, really anything you wish to track possession of over time. Each resource receives its own calendar and although we do not use this application in our subsequent SOX discussions in a production environment this might be useful for managing conferences with internal staff and external consultants.

TimeSheet

https://itsox2/egroupware/timesheet/index.php

TimeSheet is a time tracking application which is well integrated with ProjectManager. This module is designed to provide a mechanism for knowing how much time is spent on a particular task. This is very useful if you are paying consultants by the hour and wish to use an automated way of keeping tabs on their activities, which of course can be tied to InfoLog and ProjectManager items.

Tracker

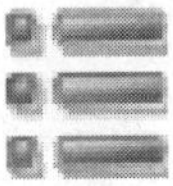

https://itsox2/egroupware/tracker/index.php

The Tracker application is designed for the reporting, tracking and resolution of problems. Traditional uses for this type of application are software defect tracking, RMA applications, case management and of course IT help desk. We will spotlight the Tracker in later chapters when we talk about a help desk application so we will not spend too much time on this right now. Of course this module has the usual integration with other eGroupware modules such as ProjectManager.

NewsAdmin

https://itsox2/egroupware/news_admin/index.php

The NewsAdmin application is another example of an eGroupware application integrated with SiteManager. In our example we use this application to make announcements on the portal site. In a production environment you might expand this to include announcements to your staff regarding SOX activities such as important meeting announcements, milestones, etc. With permissions and categories you can be sure that you are able to display information only to the people who should be viewing this, we defined a category called Main for our front page display items. Once a user is logged in then additional news items would appear based on their permission. Essentially this is a good tool to disseminate any kind of information and this also supports RSS feeds.

KnowledgeBase

https://itsox2/egroupware/phpbrain/index.php

The knowledge base is another example SiteManager integrated application. We use this in subsequent chapters to define implementation for applications, IT documentation and test procedures for SOX controls. The testing of your controls is the criteria on which your audit will be based and is a crucial deliverable. This module supports keywords, categories, file attachments, related article linking, and access controls so you ensure viewing of any sensitive information can be secured. Users can add questions and comments to the KnowledgeBase, relating these to existing articles to keep the information current and up to date or to clarify an aspect of the subject matter. A history log is kept for each item as well. KnowledgeBase is based on another Open Source project called PHP Brain and has been ported as an eGroupware application.

WorkFlow

https://itsox2/egroupware/workflow/index.php

The workflow application is one of the most important applications in the entire suite of tools, and we have done significant work in making this a useful tool for SOX compliance. WorkFlow essentially provides a mechanism for automating virtually any manual process. This application was originally based on Galaxia (http://tikiwiki.org/tiki-index.php?page=GalaxiaWorkflow) which was based on OpenFlow (http://www.openflow.it) a good example of the power of Open Source and people's ability to repurpose well written applications for their own needs without necessarily reinventing the wheel. Due to its importance to our SOX examples throughout the book we will cover the WorkFlow concepts thoroughly in the next chapter so we will not spend too much time highlighting the features of this module here.

Other Applications

- The following eGroupware applications were omitted from the VM in order to save valuable disk space. These were left out for various reasons, either they had no relevance to our discussion or there were issues with the stability of the code under the current 1.4 release. You can install any of these applications if you desire however your mileage may vary. Please see the eGroupware website for details on many of the applications not listed here.

Case Study: NuStuff Electronics, Setting the Stage

The Portal

https://itsox2

Building on the first chapter in which we outlined NuStuff Electronics' IT infrastructure, in this section we focus on the portal site purpose built to demonstrate our SOX examples throughout the book, based on the eGroupware open source groupware suite. Although there are many groupware applications available out there, we chose eGroupware for our examples because there were a variety of applications built-in and others we wrote and modified to make it a very useful toolkit for SOX compliance activities. To access the portal homepage launch the Firefox web browser from the menu, the homepage has already been set to the NuStuff portal as shown in Figure 2.18.

Figure 2.18 NuStuff Electronics sample company portal

The NuStuff SOX portal was completely designed in the SiteManager application, using several of the eGroupware applications to provide content. In our example we wanted to give the reader a central place to start from as they work through the book illustrations and concepts and also show the power of eGroupware as a platform we exploit for our specific uses. The portal is divided into three main sections as outlined below.

Main and Headers

- Home Page – This is the news application for eGroupware, these items are defined in the NewsAdmin application and appear on the front page, complete with the applicable ACL's for the items we wanted to be publicly available. If are logged into the portal site, additional news items will appear based on your group membership.
- EGW Portal – This menu selection takes you to the main eGroupware view, loading the Home application by default. This only appears if you have logged into the portal since it would not make sense to provide a link to eGroupware without being logged in.
- Site Index – This is a built-in feature of SiteManager for navigation.
- Login – This is the gateway to the eGroupware application suite, you must login to access this portion of the portal site.
- Administration – This block provides an interface to the SiteManager application to change the design of the portal. This was added to show that blocks can be present on a page but not appear unless the appropriate ACL is in place; in this case you would need to be a member of the eGroupware Administrators group in order to see this block.
- Google block – This is an example of a SiteManager module that provides search capabilities using an external provider. Note that since the ITSox2 Toolkit VM is not a public website Google cannot index this so if you select "this site" your searches will fail. There is another block for searching that you could add to the header that does perform searches on eGroupware content.
- Amazon block – This is another example SiteManager module which provides information from an external provider, in this case Amazon.com.

Launch Pad

- Fedora DS Console – This provides a link to the website provided by the locally running instance of Fedora Directory Server. We will be covering Fedora DS in detail in subsequent chapters, so we will not say too much about this here.

- Webmin – This transports you to the locally running instance of Webmin administration framework which is covered in detail later in the book, again we will not spend any time here just yet.
- Zabbix – This link will take you to the locally running instance of the Zabbix server. We will defer to later coverage of this application as well.
- KnowledgeTree DMS – This link will load the locally running instance of KnowledgeTree and as with all the other links in this section we will cover this in following material.

Reference

- Bookmarks and Links – This contains links to all references and sites made in the book and takes advantage of the Bookmarks application. This is divided into links by chapter and may include some bookmarks that do not appear in the book since this is constantly updated and may have new information subsequent to our chapter deadlines.
- Sample Configs – This contains any sample configurations and code listings in the book for convenient reference and download.
- KnowledgeBase – This is another example of an embedded eGroupware application, this displays all of the entries that are publicly accessible when browsing the portal anonymously. Of course if you login then additional KnowledgeBase articles appear based on your group membership.

NOTE

In order to logout of eGroupware and return to the portal site as an anonymous user select the icon. You will then be able to re-login as a different user, a procedure we will employ throughout the book as we discuss specific roles and wish to illustrate an example in the VM.

The Cast of Characters

https://itsox2/egroupware/addressbook/index.php

In order to demonstrate the concepts in the remainder of the book, we need to introduce you to the list of players that make up our sample company NuStuff Electronics. Here is a listing of the names and roles used in the remaining chapters:

Employee Listing

Name	Position
Mark Anderson	Finance Controller
Harry Black	Facilities Manager
Jane Brown	CEO
Ned Farley	IT System Administrator
Patricia Flannery	VP Human Resources
Nancy Green	IT Business Analyst
Biff Johnson	IT Director
Ken King	IT System Administrator
Marsha Lexington	HR Administrator
Fred Linton	CIO
Shawn Lundy	Accounts Payable
Megan Rand	IT Security Administrator
John Scott	IT Manager
Joe Smith	CFO
Brent Tooney	Accounts Receivable

SOX Auditor Listing

Name	Position
Molly Fairbanks	IT SOX Auditor
Charles Morrison	Finance SOX Auditor

IT SOX Consultant Listing

Name	Position
Christian Lahti	IT SOX Implementation Consultant
Rod Peterson	IT SOX Process Consultant

TIP

The password for all of the above employees and auditors is:

Letmein!1

The password for the SOX consultants is different because, well, it's our logins and we are not going to give you our password, right? It would be a violation of the Password Policy!

Group Listing

Group Name	Members
Default	All system users
Executive Staff	Jane Brown, Patricia Flannery, Fred Linton, Joe Smith
External Auditors	Molly Fairbanks, Charles Morrison
External Consultants	Christian Lahti, Rod Peterson
Facilities Staff	Harry Black
Finance Staff	Mark Anderson, Shawn Lundy, Joe Smith, Mary Wright
Human Resources Staff	Patricia Flannery, Marsha Lexington
IT Services Staff	Ned Farley, Nancy Green, Biff Johnson, Ken King, Megan Rand, John Scott
NoGroup	Anonymous SiteManager User

Summary

In this chapter we depart from our Sarbanes-Oxley discussion in order to install and explore the ITSox2 Toolkit VM. We covered is used on the ITSox2 Toolkit and how virtualization is a technology that allows multiple "computers" with different operating systems to run in isolation, side-by-side on the same physical machine with it's own hardware such as RAM, CPU, and networking components which are emulated as a generic set of components that the VM sees as normal devices and can use like any normal computer regardless of the actual underlying physical hardware components. Once the ITSox2 VM toolkit is installed we discuss the CentOS desktop and provide options for changing your window manager from the provided IceWM to Gnome, or KDE. We closed this section with a discussion of yum and yumex, which are the mechanisms used to keep a CentOS distribution up to date and for installing additional software.

Our VM Spotlight in this chapter was an overview of the eGroupware application suite. The eGroupware application provides common user management and access controls, a theme engine, and transparent database access, as well as many highly integrated applications or modules. In our case study section we introduce the NuStuff Electronics Sarbanes-Oxley portal designed with SiteManager. We outline the basic functionality of the Website and give a preview of how eGroupware's application suite can be leveraged to design a highly functional Website with minimal work. We also introduce the cast of characters, the fictional players in our Sarbanes-Oxley case study.

Solutions Fast Track

Virtualization vs. the Live-CD

- ☑ The Live-CD in the first edition, while highly functional had limitations for some people due to its nature, a read-only CD medium in which the file system is created in available memory.
- ☑ The ITSox2 Toolkit VM provides a virtual computing environment that affords a predictable and stable platform, performance and persistence of data. There are no hoops to jump through to save any changes you make to the VM since it is a full operating computer, running concurrently inside your host PC or laptop.
- ☑ VMware virtualization products were chosen as the underlying technology for the ITSox2 Toolkit VM since it is freely available and has broad host system support.

The CentOS Desktop

- ☑ The standard CentOS desktop was customized to use IceWM window manager due to its drastically reduced memory requirements as compared to Gnome or KDE.

- ☑ The yum and yumex applications are used to keep the system up to date and for installing additional software packages from both the standard CentOS repositories as well as any additional third party repository you wish to use.
- ☑ Gnome or KDE can be installed in the ITSox2 Toolkit VM however you will need to increase the available memory to the VM from 256 to 384–512MB of RAM. We recommend having 1GB of total system memory or more if you choose to do this.

VM Spotlight: eGroupware Application Suite

- ☑ Written in PHP, this is a framework which provides common user management and access controls, a theme engine, and transparent database access.
- ☑ The suite provides many highly integrated applications or modules that we leverage in our compliance efforts.
- ☑ Many of the eGroupware modules were born as best of breed applications elsewhere and integrated into the framework.

Case Study: NuStuff Electronics

- ☑ Building on the first chapter we introduce the NuStuff Electronics SOX portal, a website designed with eGroupware's SiteManager module.
- ☑ The portal provides a launch point for all of the other web-based applications featured in the book and is the entry point into the eGroupware suite.
- ☑ Many of the eGroupware applications are reused as SiteManager modules, providing content for our portal with minimal effort and a mechanism for keeping the portal content dynamic and fresh as new information is added into the eGroupware system.
- ☑ The NuStuff cast of characters provides an introduction to the fictional users who play a role in the SOX compliance process.

Frequently Asked Questions

Q: Is it necessary to install the ITSox2 Toolkit on my local system?

A: No, unlike the Live CD you are not required to install anything on your local system.

Q: Will the VM Player run on a MAC OS?

A: No, the incorporated VM Player runs on Linux and the following Windows platforms; Windows Vista, XP, Server 2003, and Server/Workstation 2000.

Q: Can I save my work as I go through the examples?

A: Absolutely. The ITSox2 VM Toolkit is a complete, running "computer" running in virtualization, complete with its own storage that will persist across reboots. It is important to keep in mind that due to size limitations the disk may only grow to a maximum of 8GB of data.

Q: You talk about using yum and yumex applications to keep the system up to date and for installing additional software packages, is there a way to automate this process?

A: Yum has a daemon (program) that you can set to run called `yum-updatesd` which will automatically download updates as they become available. To activate this, launch a shell and type `chkconfig yum-updatesd;service yum-updatesd start`. This will now run even if the system is rebooted.

Q: How can I get more information on Other System Setup Opportunities that weren't discussed?

A: The best place to start would be the CentOS and RedHat websites, http://centos.org and http://www.redhat.com respectively

Q: Is there anyway for me to modify your ITSox2 Toolkit?

A: Feel free to make any additions or changes you desire. The beauty of Open Source is that you have the source code to modify the application to fit your needs

Q: Can your ITSox2 VM Toolkit be freely distributed and used?

A: Yes, all of the components that make up the ITSox2 VM Toolkit are open source, all content that we have developed and/or provided are released under the Creative Commons license.

Chapter 3

SOX and Compliance Regulations

Solutions in this chapter:

- What is PCAOB
- PCAOB Audit Approach
- SOX Overview
- Sustainability Is the Key
- Enough Already
- VM Spotlight: Desktop Tools
- Case Study: Workflow Concepts

☑ Summary

☑ Solutions Fast Track

☑ Frequently Asked Questions

What is PCAOB

"It is not enough to do your best; you must know what to do, and then do your best."

–W. Edwards Deming, well known Quality Guru

The Public Company Accounting Oversight Board (PCAOB) is a private sector, non-profit corporation created by the Sarbanes-Oxley Act, to oversee the auditors of public companies. The purpose of the PCAOB is to 'protect the interests of investors and further the public interest in the preparation of informative, fair, and independent audit reports.' The PCAOB is set up as a private entity, but has several government-like regulatory functions, similar to the Securities and Exchange Commission (SEC) which is a Self Regulatory Organization (SRO) responsible for regulating the stock markets and other aspects of the financial markets in the United States. Section 101 of the Sarbanes-Oxley Act gives the PCAOB responsibility to:

- register public accounting firms;
- establish auditing, quality control, ethics, independence, and other standards relating to public company audits
- conduct inspections, investigations, and disciplinary proceedings of registered accounting firms
- enforce compliance with Sarbanes-Oxley

Along with these responsibilities to regulate audit-based services offered by audit firms, it is interesting to note that the PCAOB was also granted the authority to regulate non-audit services. Part of the PCAOB's power to set rules of the auditing industry include the power to regulate the non-audit services an audit firm provides to their clients. The PCAOB was given this authority in light of cases such as Enron and Worldcom, where auditors' independence was compromised due to the large fees audit firms were earning from these additional services to their clients. This is of particular note, because it negates a company's ability to use the same audit firm they use for non-SOX related audit. Furthermore, it regulates the interaction, level and type of communication conducted and limits the ability of the SOX audit firm to provide guidance.

PCAOB Audit Approach

The PCAOB endorses a "Top-Down" Risk Assessment (TDRA) approach for audit firm and auditor to use. This approach is a hierarchical framework that dictates the application of specific risk factors in determining the scope and evidence required in the assessment of internal control. During each phase of the audit assessment, qualitative or quantitative risk factors are used to define the scope and identify the required evidence. Since this is in all likelihood the approach your auditor will utilize, it is important that a company uses the same approach when they evaluate possible frameworks to assist in the evaluation process as they work through the particular components of their selected framework. Key TDRA considerations include:

1. identification of significant financial reporting elements
2. identification of material financial statement risks within accounts or disclosures
3. determination of which entity-level controls that address any risks
4. determination of which transaction-level controls that address risks in the absence of entity-level controls
5. determination of the nature, extent, and timing of evidence gathered to complete the assessment of in-scope controls

As part of the audit process, management is required to document how TDRA was interpreted and applied to derive the scope of controls tested. In addition, the sufficiency of evidence required (i.e., the timing, nature, and extent of control testing) is derived from the auditor's and managements TDRA.

SOX Overview

As a result of financial scandals at major Fortune 100 companies in 2001, Congress enacted the Sarbanes-Oxley Act of 2002, commonly referred to as SOX. This act not only affects how public companies report financials, but significantly impacts IT as well. Sarbanes-Oxley compliance requires more than documentation and/or establishment of financial controls; it also requires the assessment of a company's IT infrastructure, operations and personnel. Unfortunately, the requirements of the Sarbanes-Oxley Act of 2002 does not scale based on size or revenue of company. Small to medium-size companies and their IT organizations will face unique challenges, budgetary and personnel, in their effort to comply with the Sarbanes-Oxley Act of 2002.

The Transparency Test

The CFO Perspective

"It is not clear that the intent of SOX could not have been met with the requirements under Section 302. However, with the requirements included under Section 404, companies need a framework for implementation. COBIT provides a methodical approach to the IT function for Sarbanes-Oxley implementation and support. While using the framework provided, each company will need to customize the approach to their own size and complexity. A multi national, multi divisional organization is different from a single factory domestic company. The authors provide an example for this customization and rightfully point out that SOX will evolve over time, at least for the first few years."

–Steve Lanza

What Will SOX Accomplish?

There continues to be a lot of controversy and debate about this question. Although most people who are aware of the requirements to comply with SOX (Section 404) believe the intent was good, controversy remains over whether or not the existing 302 reporting requirements were sufficient. If you look at the wording and intent of Sections 302 and 404, you may see many similarities and subsequently understand why a controversy may exist as to whether (Section 404) SOX requirements and compliance was really necessary. The next two sections include an example of Sections 302 and 404 as they pertain to a company's Executive Management assertions.

Section 302

In accordance with Section 302, Executive Management of a public company:

1. are responsible for establishing and maintaining internal controls;
2. have designed such internal controls to ensure that material information relating to the issuer and its consolidated subsidiaries is made known to such officers by others within those entities, particularly during the period in which the periodic reports are being prepared;

Section 404

In accordance with Section 404, Executive Management of a public company:

1. are responsible for establishing and maintaining an adequate internal control structure and procedures for financial reporting;
2. must report the effectiveness of the internal control structure and procedures.

SOX Not Just a Dark Cloud

The initial response to Sarbanes-Oxley may be to consider it yet another drain on your already understaffed, overtaxed IT department. However, this does not necessarily have to be the case and would be a mistake. Whether SOX compliance is viewed as just another project, or a strategic opportunity for the IT department to reduce the project backlog, will be determined by how the CFO, CIO or IT Director positions SOX compliance with Executive Management.

TIP

When implementing new processes, procedures or applications for SOX compliance, the activities should add value to the business unit(s) or the overall business.

Be prepared for change. As auditors gain more knowledge about SOX, their interpretation will change and, subsequently, their requirements of your IT organization.

However, because a vast majority of companies will view SOX compliance as a Finance initiative and may not involve IT, or limit IT's involvement to the projects periphery, this may be easier said then done. Due to this "limited" perception of SOX compliance, the process of positioning with Executive Management to include IT within this initiative may in itself require significant effort. It will be well worth it. If properly executed, not only will the SOX compliance process give CFOs, CIOs and IT Directors an opportunity to address antiquated systems, personnel resource issues and documentation/process issues, it will also provide them the opportunity to forge stronger alliances with the business units. IT will be critical to the success of SOX compliance, and the support of the business units will be critical to the success of IT.

Lessons Learned

A New Tune

Historically speaking, prior to the enactment of the Sarbanes-Oxley Act, IT audits were fairly routine and straightforward. Both the auditor and you knew what was to be audited and why. So, as a matter of course the IT organization and the auditor would do the yearly audit dance of IT preparing and cleaning just prior to the audit, do the two-step for a while, resulting in the auditor producing a report that got little to no attention and in some cases not even a response. However, with the enactment of Sarbanes-Oxley there is a new tune and the dance has changed significantly. No longer will IT organizations be able to clean up their act prior to their audit. now they will need to show evidence of year-round adherence and execution of processes. If, as a result of an audit, observations are made that need to be addressed due to a risk control, IT will need to prepare a corrective action plan and execute to those accordingly.

Good News/Bad News

After nearly six years, Congress continues to express an interest in the Sarbanes-Oxley Act and its impact on public companies, in particularly small companies with a market capitalization of 75 million and less. As a result of this interest, there were two significant changes to compliance for the Public Company Accounting Reform and Investor Protection Act of 2002, better known as Section 404.

Good News

On May 24, 2007, the Board adopted Auditing Standard No. 5, if approved by the Securities and Exchange Commission, the new standard will supersede Auditing Standard No. 2, originally adopted by the Board in March 2004 and approved by the SEC in June 2004. This new standard will apply to audits of all companies required by SEC rules to obtain an audit of internal control. The new standard results from the Board's monitoring of auditors' implementation of Auditing Standard No. 2 through–among other things–inspections of internal control audits and public roundtable discussions held in April 2005 and May 2006. While the Board observed significant benefits produced by the audit, including higher quality financial reporting, it also noted that at times the related effort has appeared greater than necessary to conduct an effective audit. Based on these observations, and in light of the approaching date of December 15, 2007 for smaller companies to comply with the Act's internal control reporting requirements, the Board proposed a new standard on auditing internal control for public comment.

The PCAOB's adoption of the new standard (AS5) over the previous standard (AS2) was done to achieve four objectives. The objective of particular interest to small companies with a market capitalization of 75 million and less should be objective number three. The following is an excerpt from The Public Company Accounting Oversight Board today – May 24, 2007:

1. Focus the Internal Control Audit on the most important matters. The new standard focuses auditors on areas that present the greatest risk that a company's internal control will fail to prevent or detect a material misstatement in the financial statements. It does so by incorporating certain best practices designed to focus the audit on identifying material weaknesses in internal control before they result in material misstatements of financial statements. It also emphasizes the importance of auditing higher risk areas, such as the financial statement close process and controls designed to prevent fraud by management. At the same time, it provides auditors a range of alternatives for addressing lower risk areas, by more clearly demonstrating how to calibrate the nature, timing and extent of testing based on risk, as well as how to incorporate knowledge accumulated in previous years' audits into the auditors' assessment of risk. The auditors should use the work performed by companies' own personnel when appropriate.
2. Eliminate procedures that are unnecessary to achieve the intended benefits. The Board examined every area of the internal control audit to determine whether the previous standard encouraged auditors to perform procedures that are not necessary to achieve the intended benefits of the audit. As a result, the new standard does not include the previous standard's detailed requirements to evaluate management's own evaluation process and clarifies that an internal control audit does not require an opinion on the adequacy of management's process. As another example, the new standard refocuses the multi-location direction on risk rather than coverage by removing the requirement that auditors test a "large portion" of the company's operations or financial position.

3. Make the audit clearly scalable to fit the size and the complexity of any company. In coordination with the Board's ongoing project to develop guidance for auditors of smaller companies, the new standard explains how to tailor internal control audits to fit the size and complexity of the company being audited. The new standard does so by including notes throughout the standard on how to apply the principles in the standard to smaller, less complex companies, and by including a discussion of the relevant attributes of such companies as well, as less complex units of larger companies. The upcoming guidance for auditors of smaller companies will develop these themes even further.
4. Simplify the text of the standard. The Board's new standard is shorter and easier to read. This is in part because it uses simpler terms to describe procedures and definitions. It is also because the standard has been streamlined and reorganized to begin with the audit itself, to move definitions and other background information to appendices, and to avoid duplication by cross-referencing existing concepts and requirements that appear elsewhere in the Board's standards and relevant laws and SEC rules. For example, the new standard eliminates the previous standard's discussion of materiality, thus clarifying that the auditor's evaluation of materiality for purposes of an internal control audit is based on the same long-standing principles applicable to financial statement audits. Also, in order to better coordinate the final standard and the SEC's new rules and management guidance, the new standard conforms certain terms to the SEC's rules and guidance, such as the definition of "material weakness" and use of the term "entity-level controls" instead of "company-level controls."

Bad News

Recently, U.S. Congressman Scott Garrett proposed a bill titled The Small Business SOX Compliance Extension Act, with the intent of extending the filing deadline for small companies with a market capitalization of 75 million and less beyond December 15, 2007. U.S. Senators John Kerry and Olympia Snowe have lobbied for the extension as well. But at a meeting with the Congress, U.S. Treasury Secretary Henry Paulson reaffirmed the decision to not to do so.

Sustainability Is the Key

It is absolutely critical that SOX compliance be viewed as an ongoing process, rather than a point in time event. In cases where you will need to revise, develop and implement new procedures and controls, it will be vital to your continuing success that these are sustainable. You can rest assured that the auditors will return and will review evidence of the effectiveness of your ongoing controls—you must walk the walk. There are several things that you will need to keep in mind:

1. Can the frequency of review be maintained (weekly, monthly, bi-monthly quarterly, yearly, etc.)?
2. How much evidence of review will be maintained and how will it be stored?
3. How disruptive will the review process be to daily functions?
4. Can review evidence be systemically produced?
5. How much of the review process can be automated?

Finance and IT organizations of publicly traded companies should not be unfamiliar with audits or the necessity to have–even if non-documented–procedures required to manage key IT processes. In order to comply with SOX, the Finance and IT organizations of publicly traded companies will be required to not only formalize and document these processes, but to increase the number and granularity of their audit concerns. Finance and IT audits have traditionally been of a cursory nature and typically only covered the following areas:

1. Program change control
2. Segregation of duties in Finance and IT
3. Lack of user access controls and their periodic review
4. Weak password controls
5. Shared administration access rights

In order to comply with SOX, Finance and IT will find that not only have the areas of the audit process increased, but the nature and the complexity of the audit process have changed as well. No longer will an informal or even a loosely documented procedure suffice. Proof will now be the cornerstone in an organization's passing its SOX compliance. If an IT organization is to pass SOX compliance, it will need to show proof of formal documentation, management buy-off and sign-off and effectiveness of the implemented controls. Some examples of processes you will need to consider are:

1. Periodic review of effectiveness of controls
2. External security controls
3. External security change management controls
4. File & folder security
5. Control of access to sensitive financial data in non-production systems
6. Testing the backup & restore process
7. Physical access controls
8. Rapid response to employee and contractor terminations

9. Process for reporting, investigating and resolving security problems
10. Data retention policy

You might consider the following points to keep a bit of sanity in your compliance efforts. Try to keep compliance objectives as simple as possible so that no unnecessary work is generated as a result of overly complex controls. You should also seek to automate wherever possible. We cover many automation opportunities throughout the book but you should keep this in mind as an over-arching approach to your compliance activities. Doing so will save you many man-hours devoted to the process, especially over time. You should also get buy-in for new controls or documented processes from business owners and stakeholders as soon as possible. Inclusion and partnering with the business owners and stakeholders will better ensure adherence to the new controls or documented processes and minimize the possibility they might have a negative impact. Finally, your IT staff will initially need to reinforce new controls or processes with business units in order to facilitate the necessary behavioral change.

It is possible for an IT organization, even a small one, to attain SOX compliance. However, you will need to review your existing processes, document the ones that aren't documented and modify your processes to fit with your compliance frame work prior to execution.

Enough Already

In the years following the enactment of the Sarbanes-Oxley Act, there has not only been a revitalization of existing compliance acts and regulations, but a flurry of new ones as well. Unfortunately, given the nature of this observation there does not appear to be any relief in sight for IT organizations for the coming year. Subsequently if you are in an IT organization that has been fortunate enough not to have been affected by the Sarbanes-Oxley Act it is in all likelihood a mere matter of time before you will need to comply with either the Sarbanes-Oxley Act or another or another concocted by some regulatory body.

Other US Regulations/Acts In Brief

This section represents a brief summary of some of the Acts and Regulations that companies and their IT organization currently do or may have to contend with in the near future. Although some of these may not be strictly SOX-specific, some are related by their nature and bear at least a cursory glance.

- USA Patriot Act of 2001 – This act mainly eased restrictions and increased the ability of law enforcement agencies to search telephone and e-mail communications and medical, financial, and other records. The act also expanded the authority of law enforcement agencies to intercept wire, oral, and electronic communications relating to computer fraud and abuse offenses. Simply put, not only can law enforcement agencies intercept the stated information, they can also require that it be provided.

- eDiscovery of 2006 – Refers the act of discovery in a civil proceeding. Not only does eDiscovery encompass format data such as data contained Microsoft Office files, drawing, e-mails, Websites, it also includes raw data. eDiscovery has prompted companies to implement e-mail archiving, as well as data archiving systems so that they will be better able to respond to a subpoena for eDiscovery.
- Health Insurance Portability and Accountability Act (HIPAA) of 1996 – HIPPA is actually comprised of two titles. It is Title II that we are concerned with here as it also contains the provisions that addresses the security and privacy of health data, which is what an IT organization needs to focus on:

    1. Title I of HIPAA protects health insurance coverage for workers and their families when they change or lose their jobs.
    2. Title II of HIPAA, the Administrative Simplification requires the establishment of national standards for electronic health care transactions and national identifiers for providers, health insurance plans, and employers.

Compliance Around The Globe

If you are of the belief that compliance acts and regulatory requirements are unique to the United States you would be mistaken. Not only have international governments followed the US in regards to implementing Sarbanes-Oxley type acts they are also aligning themselves with other US compliance acts and regulatory requirements as it relates to privacy and reporting. In an effort not to belabor the SOX activities in other countries we will point out that similar SOX acts can be found in countries. Below are just a few examples:

- J-SOX of 2006 – Following in the wake of corporate scandals in the United States and such scandals as the Kanebo, Livedoor, and Murakami Fund episodes, Japan decided to implement their own version of SOX. The official name of Japan's SOX-like activities is Financial Instruments and Exchange Law, although it is commonly referred to unofficially as J-SOX. There is a fundamental difference between the American and Japanese version of SOX. The Japanese version focuses more on IT Controls and less on IT Governance than the American version.
- C-SOX of 2003 – C-SOX (or Bill 198) is Canada's version of SOX. However, unlike SOX Canada's version not only cover financial reporting and disclosure; it also includes areas of the government such as corporate disclosure, auto insurance and tax.
- European Data Protection Act of 1984 – This act was first implemented in 1984. In 1998, it was revised and its scope expanded. The European Data Protection Act is similar to HIPPA in the United States but it goes further than the protection of

medical records to encompass any data that could be used to identify a living person. The data can be anything from names, birthdays, anniversary dates, addresses, telephone numbers, Fax numbers, e-mail addresses, etc. However, unlike HIPPA, the European Data Protection Act only applies to information retained on a computer.

The question now becomes will companies' IT organizations be reactive in response to these compliance acts and regulatory requirements or will they choose to be proactive? If an IT organization is going to be proactive, they need to shift their thinking of these activities as discreet projects and start viewing compliance and regulatory acts as a single systemic IT process.

TIP

The primary focus of the SOX IT audit will be Information Security, Program Change and Data Backup & Recovery.

Compliance Acts and/or regulatory requirements should be view as a single IT process that utilize the same tools.

Where possible build on what you have, even if the process is not documented or formalized.

Processes that are currently outside the scope of SOX are Business Continuity Planning and operations that do not impact integrity/ access /reporting of financial data.

The relevance of diverging from a strictly US based Sarbanes-Oxley was to illustrate two points:

- Like the economy, compliance acts and regulatory requirements have become global.
- Compliance acts and regulatory requirements will not only be more pervasive in the United States, but internationally as well.

One of the goals of this book is to present a set of tools that will enable an IT organization to better understand their own unique compliance or regulatory requirements in such a systemic manner.

VM Spotlight: Desktop Tools

Sustainability will be the key to a company's ability to move beyond their initial efforts to comply with the Sarbanes-Oxley. In NuStuff Electronics effort to achieve SOX compliance, they determined they needed to identify tools that would assist them not only in the initial implementation, but also in sustaining activities as well. The tools listed below are the ones NuStuff Electronics chose to deploy on their desktops in their Sarbanes-Oxley compliance efforts. Here are the features of each tool and value of the particular tool that was selected.

OpenOffice

OpenOffice.org

http://openoffice.org

OpenOffice is an alternative to Microsoft Office that emulates both the major functionally, and to some degree, the look and feel of the Microsoft Office application suite. Not only can OpenOffice read and write most of the Microsoft Office file formats, it can read and write many more file format as well. OpenOffice is also able to run on more operating systems then Microsoft Office. There are versions available that run on Linux and Solaris, and ports are available or in progress for Mac OS X, OS/2 and other UNIX-like operating systems such as FreeBSD. As with Microsoft Office, OpenOffice is a compilation of applications that have been integrated in order to work closely together. Although OpenOffice is not a required application when it comes to Sarbanes-Oxley compliance, it does offer two unique advantages over Microsoft Office that you may find beneficial as part of the process:

- The ability to use and/or convert more file types then Microsoft Office.
- The ability to generate PDF files from with the application. This particular capability can be useful when generating and storing evidence in an electronic format.

The major OpenOffice components are:

Write

Write is the word processor in OpenOffice and is similar in look and feel to Microsoft Word. It provides a comparable set of functions and tools. Three of the significant advantages Writer has over Word are:

- Writer includes the ability to export Portable Document Format (PDF) files with no additional software
- Write provides a WYSIWYG editor for creating and editing Web pages
- Functions and number formats from Calc are available in Writer's tables

Calc

Calc is the spreadsheet application in OpenOffice. It is similar to Microsoft Excel, with equivalent features and functions. Calc provides some features you will not find in Excel. Calc automatically defines series for graphing, based on the layout of the user's data. Again, as with other OpenOffice applications, Calc is also capable of writing spreadsheets directly as a PDF file.

Impress

Impress is the presentation application in OpenOffice it is similar to Microsoft PowerPoint. Impress can export presentations to Adobe Flash (SWF) files allowing them to be played on any computer capable of running a Flash player. Impress does not have as wide a selection of ready made presentation designs when compared to PowerPoint. However, all of the expected functionality is present and you can either design your own templates or obtain many freely off of the Internet. The OpenOffice Website has many sample templates (see http://documentation.openoffice.org/Samples_Templates/User/template). Also, there is the OpenOffice Extras project, which is devoted to providing extra functionality for OpenOffice including templates (see http://ooextras.sourceforge.net for more information).

Base

Base is the presentation application in OpenOffice similar to Microsoft Access. As with Access, Base enables the creation and manipulation of databases, the building of forms and reports to provide easy access to and the manipulation of end-user data. Like Access, Base can be used as a front-end to a number of different database systems, including Access databases (JET), ODBC data sources and MySQL/PostgreSQL.

Draw

Draw is the vector graphics editor application in OpenOffice it is similar to Microsoft Publisher. Draw features versatile "connectors" between shapes, a range of line styles and facilitate building drawings such as flowcharts.

Firefox

http://www.mozilla.com

Mozilla Firefox was developed as an alternative to yet another Microsoft application, Internet Explorer. Firefox, like Internet Explorer, is a graphical Web browser. Unlike Internet Explorer, Firefox is a cross-platform browser, providing support for not only Microsoft Windows but also various versions other OS such as Mac OS-X, and Linux. Although unofficial and not supported the source code has been ported to additional operating systems, including FreeBSD, OS/2, Solaris, RISC OS, SkyOS, BeOS. The source code for Firefox is available under the terms of the Mozilla tri-license as free and open source software. In addition to the stated features, Firebox has demonstrated the ability to be a more secure browser then Internet Explorer and less inclined to be affected by viruses. Major standout features are:

- Tabbed Browsing – This gives you the ability to open Webpages in multiple tabs rather than windows and has proven to be quite fashionable as a feature. Internet Explorer has duplicated this functionality in IE7 due to the immense popularity introduced originally in Firefox.
- Plug-in Support – Firefox is extremely customizable through the availability of plugins. Some of the most popular are Adblock Plus, which blocks all advertisements on a site you don't want to see and Download Statusbar, which gives you fine grained control over files you download. One of our favorites, NoScript, gives you the ability to selectively block any script from executing in your browser.
- Spell Checker – When performing editing activities such as blog posts or Web-based email, Firefox features an integrated spell checker.
- Incremental Find – When searching using the integrated search feature, Firefox will suggest search terms for what you're looking for. Just start typing into the Search bar, and a drop down list of suggestions will appear. In terms of searching, Firefox gives you a choice of many search engines to choose from and comes preloaded with Google, Yahoo!, Amazon, eBay, Answers.com, and Creative Commons.
- Live Bookmarks and Bookmarks Toolbar – you can organize your favorite sites in a hierarchal tree and place links on a toolbar for easy access. A unique feature is the ability to view Web feeds such as news and blog headlines in the bookmarks toolbar or menu. You can quickly review the latest headlines from your favorite sites and click to go directly to the articles that interest you.
- Session Restore – If Firefox has to restart or closes when it comes back you can optionally choose to restore your previous session.
- RSS Web Feeds – Webfeeds are news items that you can subscribe to from various sources which update dynamically as new content appears on those Websites.
- Pop-up Blocker, Spyware and Phishing Protection - Firefox by default blocks annoying pop-ups and provides protection from Spyware and Phishing to help safeguard your financial information and protect you from identity theft. When you encounter a Web site that is a suspected forgery Firefox will warn you and offer to take you to a search page so you can find the real Web site you were looking for, or notify you of any download that is initiated giving you complete control over what is transferred to your system.

In short Firefox is a full-featured, shining example of community developed and supported software. Its strict adherence to standards-based HTML and extensibility makes this an excellent choice for daily use.

Evince

Evince http://www.gnome.org/projects/evince/

Simply a document viewer

Evince allows viewing of documents in multiple document formats. The file formats that are currently supported in Evince are pdf, postscript, djvu, tiff and dvi. The value of Evince is that it allows the GNOME Desktop to view various non-native document formats from single application. This capability is of particular importance if audit evidence is to be stored in PDF format. The key features of Evince are:

- search capabilities that display the number of results found and that highlights those results on the page
- thumbnails of pages provide a quick reference for where you'd like to go in a document
- in documents which support indexes Evince provides the option of viewing the document index for quick movement between sections
- with Evince you can select text in PDF files

Case Study: Workflow Concepts

One of the most important aspects of sustaining your compliance is the ability to consistently execute your processes over time. The ability to get the people who need to "do stuff" or "approve stuff" to actually do these things is essential to achieving and maintaining compliance year to year. When looking for workflow tools NuStuff Electronics discovered that there were two different types of tools available to help automate compliance tasks:

1. Entity based workflows that are driven by the states a document has to go through in order to reach a complete status.
2. Activity based workflows are driven by processes and workflow necessary to perform a function.

In looking at the functionality of workflow tools, NuStuff Electronics determined that they needed an activity-based workflow process tool for the processes they wished to automate. Because Galaxia is not only an activity-based workflow but an open source one as well, it is an ideal tool for implementing business processes utilized by NuStuff Electronics. Subsequently, they chose Galaxia as their workflow process tool to assist in their Sarbanes-Oxley compliance efforts. Before examining a simple Galaxia workflow process, we must first review the basic symbols associated with it. If you have had any experience with flowchart, you will find that some of the symbols of a workflow diagram may be different from other flow diagrams, the usage and the conception are the same.

Figure 3.1 Start Activity

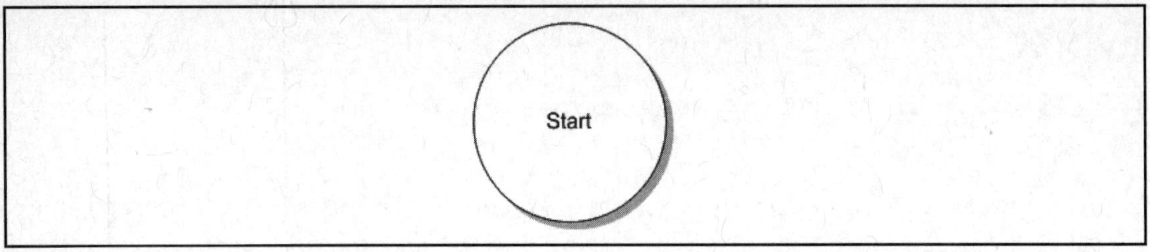

A circle denotes Start activities. All processes must have a minimum of one start activity. A Start activity is the only activity type that can be executed without having the presence of an instance; this is because instances are created when a start activity is executed.

Figure 3.2 End Activity

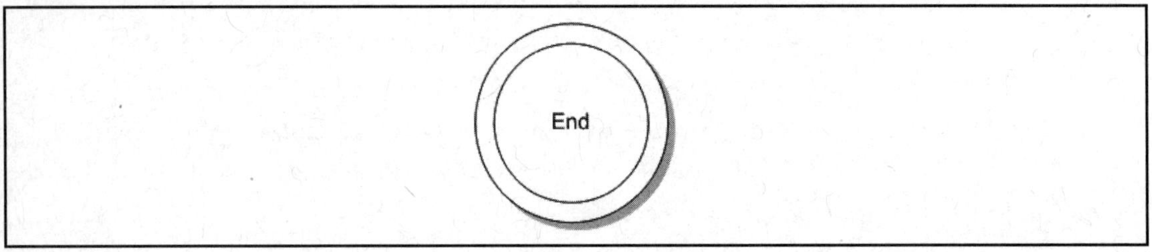

A double circle denotes End activities. An End activity, as one would expect, represents the end of a process. An End activity completes the process for a particular instance. All process must only have one end activity.

Figure 3.3 Normal Activity

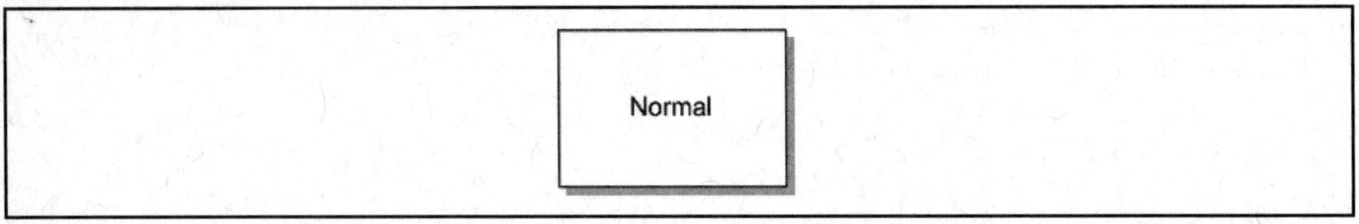

A rectangle denotes Normal activities. Normal activities don't have a particular meaning so they are used to represent tasks merely need to be done as a part of a process.

Figure 3.4 Switch Activity

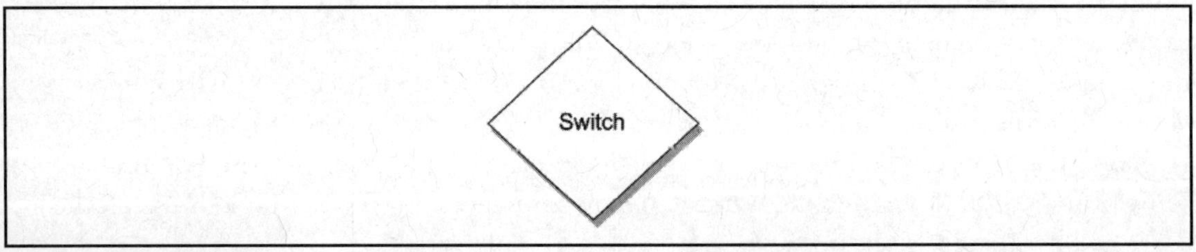

A diamond denotes Switch activities. A Switch activity represents a decision point in a process. When an instance reaches a Switch, activities are evaluated and depending on the conditions an instance can be routed to different activities.

Figure 3.5 Split Activity

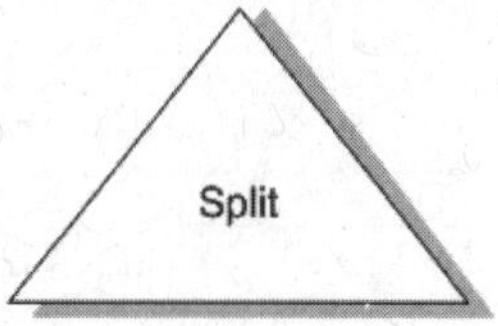

A triangle denotes Split activities. In Workflow, it is sometimes necessary for two or more activities to process independently in parallel. A Split activity can be used to Split instance and route it to many activities.

Figure 3.6 Join Activity

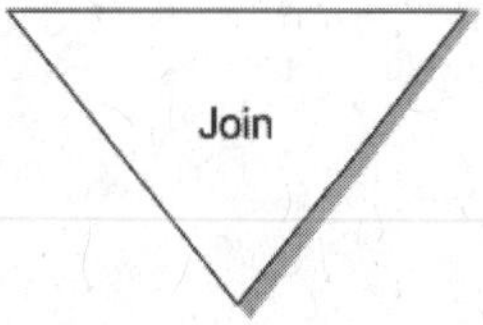

An inverted triangle denotes Join activities. A join activity is used to reconnect instances split by a Split activity. When an instance reaches a Join activity the engine confirms that the Split instance is present in at least one other activity.

Figure 3.7 Standalone Activity

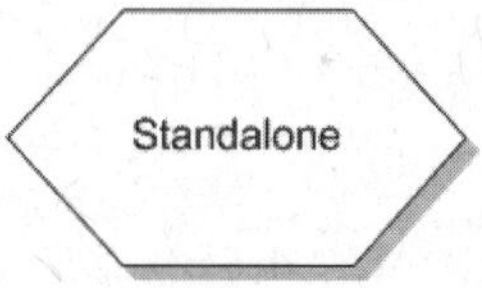

A hexagon denotes Standalone activities. A standalone activity is not related to instances so therefore they are not part of the normal flow of the process. These activities are well suited for activities that pertain to data management related processes such as, listings, adding items, removing items, etc. Any user with the appropriate permissions can execute a Standalone any time and activities.

By combining the individual components above we can create a trivial example of the "Foo" workflow process. Figure 3.8 illustrates this simple workflow in diagram terms:

Figure 3.8 Simple Galaxia Workflow Process

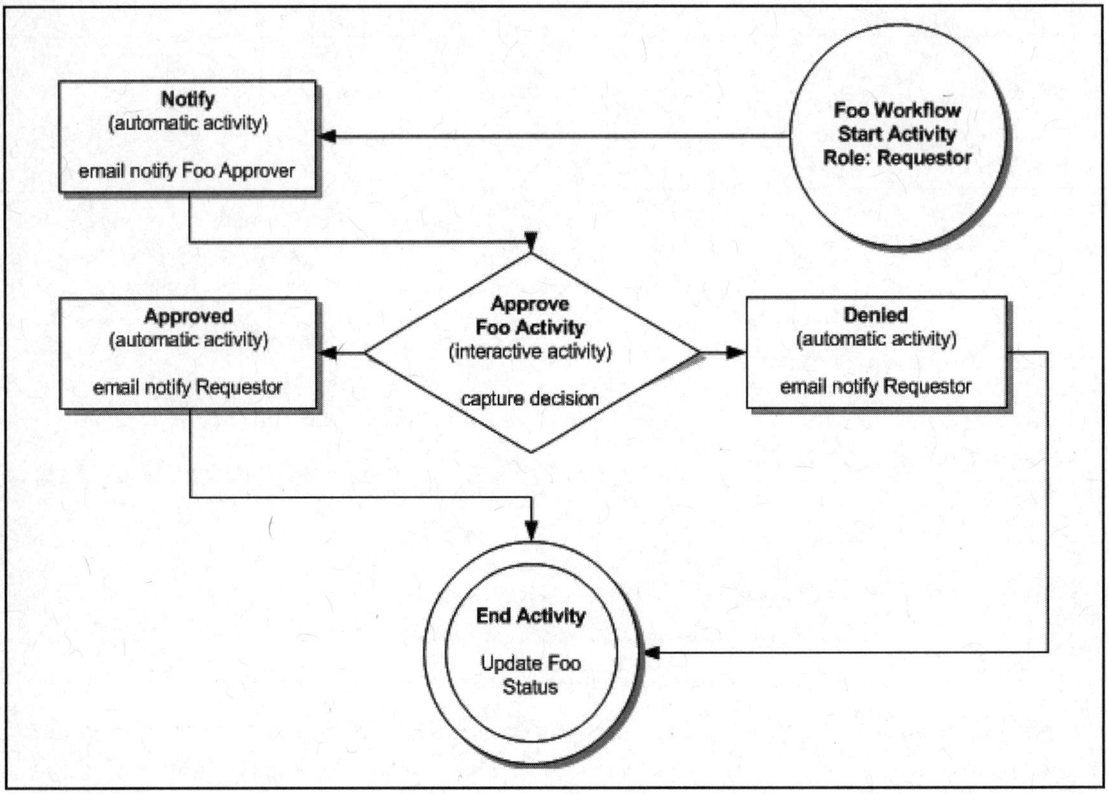

This workflow illustrates a simple process where a decision needs to be captured. There always needs to be a path from start to end. In the middle is where you define the activities and decisions for the process, and capture any pertinent information in the process. This serves as only an introduction to the workflow concepts, we will examine each in much greater detail as chapters progress and we delve into SOX specific workflows. If you would like to find additional information on Galaxia workflow symbols you can find it at http://workflow.tikiwiki.org/tiki-index.php.

Summary

In this chapter we discussed the purpose of the Public Company Accounting Oversight Board (PCAOB), how it came into existence and .the major responsibilities granted to the PCAOB by Section 101 of the Sarbanes-Oxley Act. For Sarbanes-Oxley audits the PCAOB endorses a "Top Down Risk Assessment (TDRA) as the approach for audit firm and auditor to use and how the approach is a hierarchical framework. Also in this chapter we discussed how Congress enacted the Sarbanes-Oxley Act of 2002 in an effort to prevent financial scandals such as the ones that occurred at Enron and MCI. Furthermore, how although Congress had the best of intentions when they enacted Sarbanes-Oxley Act of 2002; there are some fundamental issues with how the Act was drafted.

1. No IT specific wording for IT compliance
2. PCAOB adoption of AS5
3. No extension of filing deadline for small companies
4. Section 404 & 302 appear to overlap

Based on the aforementioned issues we established that small to medium size company's will face unique challenges in their effort to pursue compliance of Sarbanes-Oxley Act of 2002. Given the unique challenges that small to medium size company's have to contend with their ability to leverage Sarbanes-Oxley and position it correctly with Executive Management will be critical to their success. However, please keep in mind that most of the existing framework will probably be over-kill for most small to medium size companies and therefore may require some adjustment. Next, we looked at other domestic and international compliance acts and regulatory requirements that have or may come to affect the activities of companies IT organization. Although there is no one size fits all solution to fulfilling compliance acts and regulatory requirements, if IT organizations shifts their think to systemic compliance then they will be better able to manage overall compliance.

Solutions Fast Track

What is PCAOB

- ☑ The Public Company Accounting Oversight Board (PCAOB) is a private sector, non-profit corporation created by the Sarbanes-Oxley Act, to oversee the auditors of public companies.
- ☑ The purpose of the PCAOB is to 'protect the interests of investors and further the public interest in the preparation of informative, fair, and independent audit reports'.
- ☑ Section 101 of the Sarbanes-Oxley Act gives the PCAOB responsibility to:

a. registering public accounting firms;

b. establishing auditing, quality control, ethics, independence, and other standards relating to public company audits;

c. conducting inspections, investigations, and disciplinary proceedings of registered accounting firms; and

d. enforcing compliance with Sarbanes-Oxley.

☑ The PCAOB endorses a "Top Down Risk Assessment (TDRA) as the approach for audit firm and auditor to use.

☑ TDRA approach is a hierarchical framework dictates the application of specific risk factors in determining the scope and evidence required in the assessment of internal control.

☑ Since auditors will utilize it is important that a company uses the same approach when they evaluate possible frameworks and as they work through the particular components of their selected framework.

SOX Overview

☑ Sarbanes-Oxley Act of 2002 act not only affects how public companies report financials, but significantly impacts IT as well.

☑ Sarbanes-Oxley compliance requires more than documentation and/or establishment of financial controls, it also requires the assessment of a company's IT infrastructure, operations and personnel.

☑ Requirements of the Sarbanes-Oxley Act of 2002 do not scale based on size or revenue of company.

☑ Small to medium size companies (IT Dept) will face unique challenges, budgetary and personnel, in their effort to comply with Sarbanes-Oxley Act of 2002.

☑ A vast majority of companies will view SOX compliance as a Finance initiative and may not involve IT, or limit IT's involvement to the projects periphery.

☑ Limited perception of SOX compliance may make it difficult for CFOs, CIOs and IT Directors to position with Executive Management.

☑ The SOX compliance process will provide CFOs, CIOs and IT Directors the opportunity to forge stronger alliances with the business units.

Sustainability Is the Key

- ☑ It is absolutely critical that SOX compliance be viewed as an on-going process rather then a point in time event.
- ☑ Auditors will return periodically and will want to review evidence of the effectiveness of your on-going controls – you must walk the walk.
- ☑ Where you will need to revise, develop and implement new procedures keep in mind several things:
    a. Can the frequency of review be maintained (weekly, monthly, bi-monthly quarterly, yearly, etc.)?
    b. How much evidence of review will be maintained and how will it be stored?
    c. How disruptive will the review process be to daily functions?
    d. Can review evidence be systemically produced?
    e. How much of the review process can be automated?

Enough Already

- ☑ In the years following the enactment of the Sarbanes-Oxley Act there has not only been a revitalization of existing compliance acts and regulations, but a flurry of new ones as well.
- ☑ IT is merely a matter of time before every IT organization will need to comply with the Sarbanes-Oxley Act, or another act and/or regulatory body.

VM Spotlight: Desktop Applications

- ☑ In this section we include a look at some of the desktop applications that you might use or can assist in the developing of SOX related collateral.
- ☑ OpenOffice is an alternative to Microsoft Office that emulates both the major functionally and to some degree the look and feel of Microsoft Office application suite. OpenOffice runs on many operating systems and platforms, and gives the ability to read and save documents in several formats including PDF. Major components include:

  a. Write – the word processor
  b. Calc – the spreadsheet application
  c. Impress – the presentation application
  d. Base – the database application
  e. Draw – the vector graphics editor

☑ Firefox is a graphical Web browser like Internet Explorer, providing support for not only Microsoft Windows but also various versions other OSs such as Mac OS-X, and Linux. Some of Firefox's features include tabbed browsing, plug-in support, spell checker, incremental find, live bookmarks, bookmarks toolbar, persistent sessions, RSS Web feeds, pop-up blocker, spyware and anti-phishing protection

☑ Evince allows viewing of documents in multiple document formats such as PDF, postscript, djvu, tiff and dvi documents. Evince provides search capabilities, thumbnails of pages, and indexes and provides a perfect compliment to the PDF export functionality of OpenOffice

Case Study: Workflow Concepts

☑ Entity based workflows are driven by the states a document has to go through in order to reach a complete status

☑ Activity based workflows are driven by processes and workflow necessary to perform a function

☑ Start Activity – Every workflow starts with a Start Activity. This provides the entry point into the process and leads to the first activity

☑ End Activity – Every workflow ends with an End Activity. All workflows must have a path from start to end

☑ Normal Activity – This is a activity where you "do" something such as capture information

☑ Switch Activity – This is a decision point in the workflow, and can branch off in different directions based on the decisions made

☑ Split Activity – This is an activity where the processing can split into multiple paths for processing

☑ Join Activity – this is where a previous split joins back together, all splits must join at some point in the workflow

☑ Stand-Alone Activity – this is an activity where something is done or captured, but does not subsequently lead anywhere

Frequently Asked Questions

Q: Where can I find additional information about Sarbanes-Oxley?

A: There is a vast amount of information on the Internet about COBIT. We recommend the following sites as a good place to start:

- Information Guide to the Sarbanes-Oxley Act of 2002 http://www.sarbanes-oxley-101.com
- Public Company Accounting Oversight Board http://www.pcaob.com/

Q: Can SOX compliance be achieved without any automation?

A: Yes, it is possible but since auditors generally prefer evidence of a control to be system generated you might find it extremely difficult.

Q: Can I find additional information on Workflow on the Internet?

A: Yes, you can find additional information at http://workflow.tikiwiki.org/tiki-index.php.

Q: Does the PCAOB have regulatory authority beyond an audit firms SOX activity?

A: Yes, the can regulate the non-audit services an audit firm provide to their audit clients.

Q: Is there a definition as to what constitutes a small company?

A: Yes, companies with 75 million dollar market capitalization or below.

Q: Is it necessary that my company use the same tools as the ones in this chapter?

A: No, the tools that you use should be dependent upon the requirements of your company for SOX and non-SOX activities. However, it would be advantageous to seek out tools with similar capabilities. Here is a short list of more open source or free applications you can investigate:

- Opera Browser http://www.opera.com
- Apple Safari http://www.apple.com/safari
- Open Workflow Engine http://www.openwfe.org
- Dia diagramming software http://hans.breuer.org/dia/
- PDF Creator http://sector7g.wurzel6-webdesign.de/pdfcreator/index_en.htm

Chapter 4

What's In a Framework?

Solutions in this chapter:

- PCAOB Endorses COBIT
- Are The Developers of COBIT Controls Crazy? Is This Practical?
- The Top Contenders
- VM Spotlight: Project Plan
- Case Study: Frame Work Selection

☑ Summary

☑ Solutions Fast Track

☑ Frequently Asked Questions

PCAOB Endorses COBIT?

"It is the framework which changes with each new technology and not just the picture within the frame."

–Marshall McLuhan, Canadian Philosopher

Sarbanes-Oxley compliance will have a significant impact on the IT organization of most public companies. However, there is one enormous problem: there is no specific mention of IT in section 404 More importantly, there are no specifics given as to what controls have to be established within an IT organization in order to comply with Sarbanes-Oxley legislation.

If there is no specific mention in section 404 as to what IT needs to do in order to comply with Sarbanes-Oxley, then the logical question would be, "How can I comply with something without knowing what I need to do to comply?" Although there are various standards that a company can use for defining and documenting their internal controls–ITIL (IT Infrastructure Library), Six Sigma and COBIT, based on the endorsement of the PCAOB, the standard that the majority of auditors have adopted is COBIT. We will now define ITIL, Six Sigma, and COBIT.

ITIL is an international series of documents used to aid the implementation of a framework for IT Service Management. The intent of the framework is to define how Service Management is applied within specific organizations. Given that the framework consists of guidelines, it is agnostic of any application or platform and can therefore be applied in any organization.

Six Sigma, at many organizations, simply means a measure of quality that strives for near perfection. Six Sigma is a disciplined, data-driven approach and methodology for eliminating defects (driving toward six standard deviations between the mean and the nearest specification limit) in any process–from manufacturing to transactional and from product to service.

The COBIT acronym stands for Control Objectives for Information and related Technology. While the COBIT organization has been around since 1996, the guidelines and best practices they developed have almost become the de facto standard for auditors and SOX compliance. The adoption of the COBIT guidelines and practices as a de facto standard is likely due to the fact that the COBIT standards are not platform specific but rather platform independent. There are approximately 300 generic COBIT objectives, grouped under six COBIT Components. When reviewing and applying the COBIT guidelines and best practices to your environment, keep in mind that they will need to be tailored to your particular environment.

The Transparency Test

The Manager's Perspective

As Sr. Director of Engineering Services of a medium size software development company I manage several groups; Quality Assurance, Product Release, Customer Support and Information Technology. I believe I've setup the IT department so that it runs pretty well given tight budget and resource constraints that have remain in effect since the day we opened for business. Recently I was told that a SOX audit is a distinct possibility more likely sooner than later. I was asked if I knew what was involved in a SOX audit and my response was sure no problem. But in reality my experience with SOX has been mostly hearing in the news about how everyone dreads it. Now it appears that we get to share in the fun. My head has just developed an extra SOX induced throb.

Alright, that's enough self-pity. It's time to get up to speed on the details of SOX and find out what it means to get my IT department though the audit with as little overhead as possible. Years ago, I went through an ISO 9001 audit for a Japanese sponsored software development project I was involved in. The key to that success was like anything else, good preparation. So here I am just getting started. And what does everyone do these days when they want to learn about something? I googled "Sarbanes-Oxley Act" of course. Holy smokes! That's quite a list of documents, articles, blogs and forum entries all extolling in very verbose detail about some aspect of SOX. All very interesting to the wanna-be SOX auditors out there, but what I need are some straight facts about what I can do to prepare my IT department for the audit.

So, as you can see my experience parallels what the authors are saying in regards to the confusion and ambiguity surrounding Sarbanes-Oxley compliance. This being the case, and given my already heavy workload the author's objective of succinctly conveying Sarbanes-Oxley compliance requirements and reducing control objectives to conform with compliance rather then a framework I think will be key to my successful Sarbanes-Oxley compliance.

–Matt Evans

The Six COBIT Components

COBIT consists of six components:

Executive Summary: Explains the key concepts and principles

Framework: Foundation for approach and COBIT elements. Organizes the process model into the four domains:

1. Plan and organize
2. Acquire and implement
3. Deliver and support
4. Monitor and evaluate

Control Objective: Foundation for approach and COBIT elements. Organizes the process model into the four domains (discussed in a moment)

Control Practices: Identifies best practices and describes requirement for specific controls

Management Guidelines: Links business and IT objectives and provides tools to improve IT performance

Audit Guidlines: Provides guidance on how to evaluate controls, assess compliance and document risk with these characteristics:

1. Define "internal controls" over financial reporting.
2. Internally test and assess these controls.
3. Support external audits of controls.
4. Document compliance efforts.
5. Report any significant deficiencies or material weaknesses.

In conclusion, even though an IT organization is free to select any predefined standards they choose, or even one they develop, to assist them in obtaining Sarbanes-Oxley compliance, the mostly widely accepted standard is COBIT. Subsequently, you may find that selecting COBIT will be the path of least resistance to Sarbanes-Oxley compliance.

TIP

COBIT guidelines and best practices will need to be tailored to your environment.

Although the enormity of the COBIT guidelines and best practices may appear to be a daunting task, it can and should be distilled down to what is pertinent to your environment.

Entity Level Controls versus Control Objectives

Entity Level Controls consist of the policies, procedures, practices and organizational structures intended to assure the use of IT will enable the accomplishments of business objectives and that planned events will be prevented, or detected and corrected.

Control Objective is a statement of the desired result or purpose to be achieved by implementing control procedures for a particular IT activity. When developing and documenting your controls, there are several characteristics that you want to keep in mind so that your controls are as effective as possible:

Key Control Characteristics include the following:

- Employees are aware of their responsibilities for the control activities.
- The control is clearly understood.
- The control is effective in preventing, detecting or correcting risk.
- The operating effectiveness of the control activity is adequately evaluated on a regular basis.
- The standards and assertions required to execute the control are clearly understood.
- Deficiencies are identified and remedied in a timely manner.
- The performance of the control can be documented.
- The controls, policies and procedures are documented.

TIP

Although the goal is to implement a control that will be 100 percent effective, it is not a realistic goal. Therefore, the objective should be to implement the most effective control within your environment. Be prepared to explain to your auditors how your environment works and why a particular control is effective in your environment.

What Are the Four COBIT Domains?

We'll now briefly describe each of the four COBIT domains.

Planning and Organization

Planning is about developing strategic IT plans that support the business objectives. These plans should be forward-looking and in alignment with the company's planning intervals; that is, a two-, three, or five-year projection.

Acquisition and Implementation

Once the plans are developed and approved, there may be a need to acquire new applications or even acquire or develop a new staff skill-set to execute the plans. Upon completion of the

Acquisition phase, the plans now need to be implemented, which should include maintenance, testing, certifying, and identification of any changes needed to ensure continued availability of existing systems, as well as the new systems.

Delivery and Support

This phase ensures that not only do systems perform as expected upon implementation, but they continue to perform in accordance with expectations over time, usually managed via Service Level Agreements (SLAs). In this regard, systems can be related to infrastructure components or third-party services.

Monitoring

The monitoring phase uses SLAs or baselines established in the subsequent phases to allow an IT organization to not only gauge how they are performing against expectation, but also provides them with an opportunity to be proactive.

Are the Developers of COBIT Controls Crazy? Is this Practical?

A cursory review of the COBIT controls described in this section would convince any CEO, CFO or IT Director that not only will the implementation of COBIT controls be a daunting task, but that the developers of the controls must be crazy. Neither of the aforementioned assumptions necessarily has to be the case. Whether the task of implementing COBIT controls is daunting will depend on how much effort is put into filtering the COBIT controls. Keep in mind that although all the controls center on good, sound practices, even the largest and most well-run organization would not be able to implement all of them as defined by COBIT. It is a good idea, yes, but that doesn't make it practical. The keys to culling down COBIT controls center on a couple of questions:

1. Which controls are appropriate to my environment?
2. Of the appropriate controls which ones will maximize my efforts?

After you have successfully answered the previous questions you will be in a position to reduce the COBIT controls to a manageable and actionable list for your implementation. Prior to executing to your list of controls, it is advisable that you verify your assumptions with your auditors. Also, as part of your assessment process, you should identify all areas that were not appropriate for your environment and be prepared to justify and defend these exclusions to your auditors as part of the Gap and Remediation process.

Tables 4.1 through 4.4 outline a partial list of the COBIT Controls, which show some of the control objectives for each process cycle and what risk factor they relate to. For a complete listing see Appendix A.

Table 4.1 Planning and Organization

Control Objective		Risk
1	Management prepares strategic plans for IT that aligns business objectives with IT strategies. The planning approach includes mechanisms to solicit input from relevant internal and external stakeholders impacted by the IT strategic direction.	IT plans may not be present in the organization's long- and short-range plans. The organization's plans may not support IT.
2	Management obtains feedback from business process owners and users regarding the quality and usefulness of its IT plans for use in the ongoing risk assessment process.	IT plans may not be updated regularly.
3	An IT planning or steering committee exists to oversee the IT function and its activities. Committee membership includes representatives from senior management, user management, and the IT function.	IT plans may not be consistent with the organization's goals and may impair the achievement of business objectives.
4	The IT organization ensures that IT plans are communicated to business process owners and other relevant parties across the organization.	New business processes may conflict with current IT plans. Or new IT plans may conflict with current business processes.
5	IT management communicates its activities, challenges and risks on a regular basis with the CEO and CFO. This information is also shared with the board of directors.	IT activities may not be understood by management or business processes, so conflicts may not be known.
6	The IT organization monitors its progress against the strategic plan and reacts accordingly to meet established objectives.	Changes in the business or IT environment may impact IT plans without detection.
7	IT management has defined information capture, processing and reporting controls - including completeness, accuracy, validity and authorization - to support the quality and integrity of information used for financial and disclosure purposes.	IT architecture may not support the growth of the business or current business goals.
8	IT management has defined information classification standards in accordance with corporate security and privacy policies.	IT security levels may not be in compliance with regulatory or corporate policies regarding information protection.
9	IT management has defined, implemented and maintained security levels for each of the data classifications. These security levels	IT security levels may not be in compliance with regulatory or corporate policies regarding

Continued

Table 4.1 Continued

	Control Objective	Risk
	represent the appropriate (minimum) set of security and control measures of each of the classifications and are reevaluated periodically and modified accordingly.	information protection. IT security plans may not be updated regularly.
10	IT managers have adequate knowledge and experience to fulfill their responsibilities.	Information Systems data may not be reliable if systems are not functioning as intended or errors are not dealt with appropriately.

Table 4.2 Acquire and Implement

	Control Objective	Risk
1	The organization has a system development life cycle methodology that considers security, availability and processing integrity requirements of the organization.	Program development may not adhere to regulatory or corporate processes and procedures risking data integrity.
2	The system development life cycle methodology ensures that information systems are designed to include application controls that support complete, accurate, authorized and valid transaction processing.	Program implementations may not function as intended, risking the integrity of the calculations, data capture, data integrity, or the implementation of unauthorized processes.
3	The organization has an acquisition and planning process that aligns with its overall strategic direction.	New application selection may not support business and regulatory objectives.
4	The organization acquires software in accordance with its acquisition and planning process.	Business objectives may not be achieved or undetected processes may be installed into production systems.
5	Procedures exist to ensure that system software is installed and maintained in accordance with the organization's requirements.	Integrity of the implementation may not be achieved, and the program may not function as intended.
6	Procedures exist to ensure that system software changes are controlled in line with the organization's change management procedures.	Integrity of the implementation may not be achieved, the program may not function as intended, and unauthorized processes may be installed undetected.

Table 4.2 Continued

Control Objective		Risk
7	IT management ensures that the setup and implementation of system software do not jeopardize the security of the data and programs being stored on the system.	System program upgrades may change security settings and allow unauthorized access to protected information.
8	Procedures exist and are followed to ensure that infrastructure systems, including network devices and software, are installed and maintained in accordance with the acquisition and maintenance framework.	System program upgrades may interrupt production and network services corrupting data or other activities.
9	Procedures exist and are followed to ensure that infrastructure system changes are controlled in line with the organization's change management procedures.	System program upgrades may interrupt production and network services corrupting data or other activities.
10	The organization's system development life cycle methodology requires that user reference and support manuals (including documentation of controls) be prepared as part of every information system development or modification project.	Consistent application of application reporting and transaction processing may not occur jeopardizing the integrity of the data and financial statement reporting.

Table 4.3 Delivery and Support

Control Objective		Risk
1	Selection of vendors for outsourced services is performed in accordance with the organization's vendor management policy.	Financial data integrity may be compromised if the system is not functioning as intended.
2	A framework is defined to establish key performance indicators to manage service level agreements, both internally and externally.	Financial data integrity may be compromised if the system is not functioning as intended.
3	IT management ensures that, before selection, potential third parties are properly qualified through an assessment of their capability to deliver the required service and their financial viability.	Vendor viability may risk the delivery of programs and subsequent support of the application.

Continued

Table 4.3 Continued

	Control Objective	Risk
4	Third-party service contracts address the risks, security controls and procedures for information systems and networks in the contract between the parties.	Vendors have access to protected and sensitive data; confidentiality may be compromised.
5	Business continuity controls consider business risk related to third-party service providers in terms of continuity of service, and escrow contracts exist where appropriate.	If systems fail, data integrity may be compromised or business objectives may not be met.
6	Procedures exist and are followed to ensure that a formal contract is defined and agreed to for all third-party services before work is initiated, including definition of internal control requirements and acceptance of the organization's policies and procedures.	Confidentiality and achievement of business objectives may be breached.
7	A designated individual is responsible for regular monitoring and reporting on the achievement of the third-party service level performance criteria.	Vendor failures may go undetected.
8	A regular review of security, availability and processing integrity is performed for service level agreements and related contracts with third-party service providers.	Financial data integrity may be compromised if the system is not functioning as intended.
9	IT management monitors the performance and capacity levels of the systems.	Capacity to retain the source transaction information may be limited.
10	IT management has a process in place to respond to suboptimal performance and capacity measures in a timely manner.	Service level agreements may not be met.

Table 4.4 Monitor and Evaluate

	Control Objective	Risk
1	Performance indicators (e.g., benchmarks) from both internal and external sources are defined, and data are collected and reported regarding achievement of these benchmarks.	Breakdowns in performance may not be detected and corrected in a timely fashion.
2	IT management monitors its delivery of services to identify shortfalls and responds with actionable plans to improve.	Breakdowns in performance may not be detected and corrected in a timely fashion.
3	IT management monitors the effectiveness of internal controls in the normal course of operations through management and supervisory activities, comparisons and benchmarks.	Breakdowns in performance may not be detected and corrected in a timely fashion.
4	Serious deviations in the operation of internal control, including major security, availability and processing integrity events, are reported to senior management.	Breakdowns in performance may not be detected and corrected in a timely fashion.
5	Internal control assessments are performed periodically, using self-assessment or independent audit, to examine whether internal controls are operating satisfactorily.	Lack of independent assessment could cause structural control deficiencies to go undetected.
6	IT management obtains independent reviews prior to implementing significant IT systems that are directly linked to the organization's financial reporting environment.	System flaws could lead to errors that impact the financial reporting environment.
7	IT management obtains independent internal control reviews of third-party service providers (e.g., by obtaining and reviewing copies of SAS70, SysTrust or other independent audit reports).	Controls for outsourced IT assets and facilities may not be sufficient.

What's Controls Should I Use?

Let's explore how this intimidating list can be reduced based on the size and complexity of a particular company. For this process, we will look at our fictitious company, NuStuff Electronics. NuStuff Electronics is a successful semiconductor designer of baseband communication chips for original equipment manufacturers (OEM) of digital telephones. Operations span the globe with offices in, India, Japan, Singapore, the United Kingdom, and two offices in the United States. The majority of the design work is done in India, and research and development on new products is primarily done in the UK branch, with corporate headquarters in the US and the remaining offices performing sales and customer support. NuStuff out-sources its manufacturing needs to contract electronics fabrication firms and has approximately 800 employees worldwide. NuStuff has $60 million in assets and quarterly revenues averaging $20 million.

As important as what NUSTUFF Electronics IT infrastructure consists of and what they support, when it comes to determining what controls they will need to put in place, what they do not have and do not support is of equal importance. Here is a condensed version of NuStuff Electronics IT infrastructure:

Server Room (General, Sales, Support and Executive)

Based on their support server functions it is highly likely that NuStuff will need to address the same audit concerns as a larger company.

- SAN storage for network services and departmental file services
- Red Hat Advanced Server Linux and CentOS servers in a high-availability cluster for network services such as DNS, FTP, and HTTP
- Oracle Financials managed via outsource provider and a financial analyst on staff
- Fedora Directory Server LDAP and Samba for cross platform, single sign on authentication services
- Scalix for groupware/messaging services
- Astaro Firewall/VPN with dedicated interoffice IPSEC tunnels

Desktops (Sales, Support and Executive)

NuStuff Electronics' desktops for General & Administrative support applications and functionality inline any regular company.

For the most part, all non-engineering staff falls into this category from an IT desktop standpoint.

- Windows /XP desktops for general support staff, XP laptops for field sales
- Microsoft Office and Open Office for desktop applications

- Mozilla Firefox Web browser for Internet/intranet access
- Microsoft Outlook, Mozilla Thunderbird for e-mail clients
- Symantec Antivirus for virus and spyware prevention and detection
- Microsoft Visio for diagrams

Network Topology

Figure 4.1 diagrams NuStuff's corporate headquarters IT landscape. As can see as with a lot of IT organization today NuStuff Electronics has chosen as a cost saving strategy to outsource some of their IT function. This decision in itself is neither good nor bad but it will impact their Sarbanes-Oxley compliance activities.

Figure 4.1 NuStuff Electronics Network (Corporate Office)

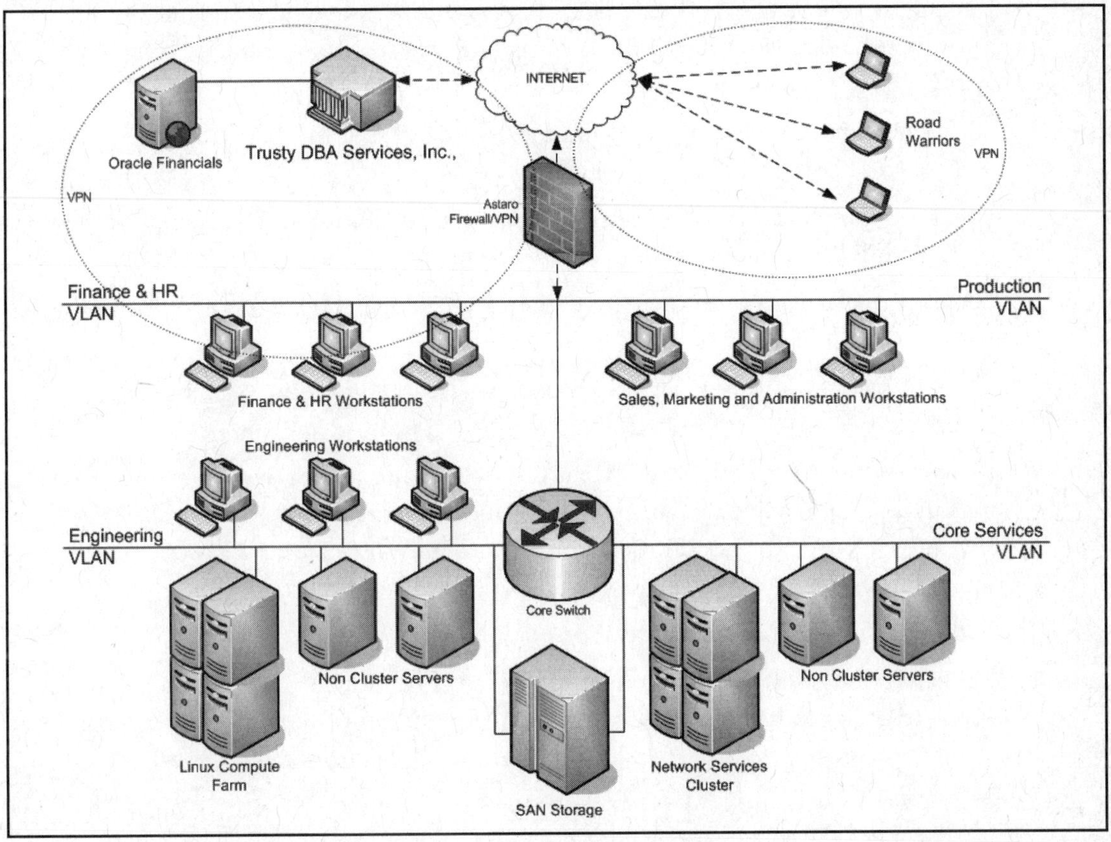

Table 4.5 Control Objectives Applicable for SOX Compliance at NuStuff Electronics

Domain	Removed Control Objective
Planning & Organization	N/A
Acquire & Implement	1,2& 10
Delivery & Support	5 & 10
Monitor & Evaluate	6

Planning and Organization

1. Management prepares strategic plans for IT that aligns business objectives with IT strategies.
2. Management obtains feedback from business process owners and users regarding the quality and usefulness of its IT plans for use in the ongoing risk assessment process.
3. An IT planning or steering committee exists to oversee the IT function and its activities.
4. The IT organization ensures that IT plans are communicated to business process owners and other relevant parties across the organization.
5. IT management communicates its activities, challenges, and risks on a regular basis with the CEO and CFO.
6. The IT organization monitors its progress against the strategic plan and reacts accordingly to meet established objectives.
7. IT management has defined information capture, processing and reporting controls, including completeness, accuracy, validity and authorization, to support the quality and integrity of information used for financial and disclosure purposes.
8. IT management has defined information classification standards in accordance with corporate security and privacy policies.
9. IT management has defined, implemented and maintained security levels for each of the data classifications. These security levels represent the appropriate (minimum) set of security and control measures of each of the classifications and are reevaluated.
10. IT managers have adequate knowledge and experience to fulfill their responsibilities.

Acquire and Implement

1. The organization has an acquisition and planning process that aligns with its overall strategic direction.
2. The organization acquires software in accordance with its acquisition and planning process.
3. Procedures exist to ensure that system software is installed and maintained in accordance with the organization's requirements.
4. Procedures exist to ensure that system software changes are controlled in line with the organization's change management procedures.
5. IT management ensures that the setup and implementation of system software do not jeopardize the security of the data and programs being stored on the system.
6. Procedures exist and are followed to ensure that infrastructure systems, including network devices and software, are installed and maintained in accordance with the acquisition and maintenance framework.
7. Procedures exist and are followed to ensure that infrastructure system changes are controlled in line with the organization's change management procedures.

Delivery & Support

1. Selection of vendors for outsourced services is performed in accordance with the organization's vendor management policy.
2. A framework is defined to establish key performance indicators to manage service level agreements, both internally and externally.
3. IT management ensures that, before selection, potential third parties are properly qualified through an assessment of their capability to deliver the required service and their financial viability.
4. Third-party service contracts address the risks, security controls and procedures for information systems and networks in the contract between the parties.
5. Procedures exist and are followed to ensure that a formal contract is defined and agreed to for all third-party services before work is initiated, including definition of internal control requirements and acceptance of the organization's policies and procedures.
6. A designated individual is responsible for regular monitoring and reporting on the achievement of the third-party service level performance criteria.
7. A regular review of security, availability and processing integrity is performed for service level agreements and related contracts with third-party service providers.
8. IT management monitors the performance and capacity levels of the systems.

Monitor & Evaluate

1. Performance indicators (e.g., benchmarks) from both internal and external sources are defined, and data are collected and reported regarding achievement of these benchmarks.
2. IT management monitors its delivery of services to identify shortfalls and responds with actionable plans to improve.
3. IT management monitors the effectiveness of internal controls in the normal course of operations through management and supervisory activities, comparisons and benchmarks.
4. Serious deviations in the operation of internal control, including major security, availability and processing integrity events, are reported to senior management.
5. Internal control assessments are performed periodically, using self-assessment or independent audit, to examine whether internal controls are operating satisfactorily.
6. IT management obtains independent internal control reviews of third-party service providers (e.g., by obtaining and reviewing copies of SAS70, SysTrust or other independent audit reports).

The Top Contenders

As we discussed previously in this chapter and chapter 3 there are many frameworks that IT organization have historically utilized like ITIL (IT Infrastructure Library), Six Sigma and COBIT. There are many more that we didn't mention such as ISO9000 and Deming. However, although for different reasons, it does appear that two frameworks are emerging as forerunners COBIT and ITIL. Since we have discussed COBIT previously in this chapter we will now touch upon ITIL.

ITILv2

The ITIL has been around for more than twenty years. It was conceived in the late 1980s by the British Central Computer and Telecommunication Agency (CCTA), now called the Office of Government Commerce (OGC). There is some debate whether the conception of the ITIL was forward-thinking or reactionary due to problems and inefficiencies within the CCTA. What isn't in debate, however, is that the ITIL framework is comprised of a collection of Best Known Methods (BKM) designed to enable the delivery of high-quality information technology services. In conjunction with the delivery of high-quality information technology services, the compilations of ITIL BKMs were selected to obtain high financial management. In the perfect situation, ITIL should enable organizations to more effectively manage IT costs, increase IT efficiencies and maximize existing IT resources. Because ITIL was conceived in Europe, there are significantly more implementations in Europe than there are in the United States. However, with all the regulatory and/or compliance acts in the United States, the tide may be shifting.

As with COBIT, please keep in mind that the ITIL framework is just that–a framework. Each company must evaluate their needs, their resources and whether they need to implement ITIL in its entirety or merely components of ITIL. It is likely that an ITIL purist and evangelist will have issues with the previous statement, arguing that a company will not be able to derive the full benefit of ITIL by not implementing it fully. To that we would say, so be it. The purpose of implementing any framework is not to implement it fully, but rather to implement a framework that addresses a business need. Simply put, as there are diverse differences in businesses and business models so too are there business requirements to implement a framework. The subcomponents of Service Delivery are:

- Availability Management - centers on serviceability, recoverability, and reliability. It gives criteria for managing and reporting application and infrastructure up-time and mean-time between failures.
- Capacity Management - centers on ensuring that the adequate and the appropriate resources are available based on upon service levels.
- Continuity Management - centers on an organization's ability to manage disaster recovery and outages.
- Financial Management - centers on an organizations ability to calculate infrastructure costs and practices that facilitate cost maximization.
- Service Level Management - centers around organization's ability to help users understand the trade-offs between support costs and service levels and driving and management of agreed upon Service Level Agreements (SLAs).

The subcomponents of Service Support are:

- Configuration Management - centers on an organization's ability to identify and control infrastructure assets or configuration items (CIs). ITIL CIs can be hardware, middleware, end-user applications,
- Change Management - centers on an organization's ability to manage change and describes how changes to configuration items should be done.
- Incident Management - centers on how an organization's identifies and deals with events that occur that are not part of the normal operational process. It also outlines how incidents should be classified and prioritized.
- Problem Management - centers on how an organization manages problems that occur in the environment and outlines how these types of events are handled and resolved via Requests For Change (RFC). Problem Management and Incident Management are closely connected, as problems, errors, defects, etc., are often the cause of incidents.

- Release Management - centers on how an organization manages the process of bringing software items into the operations environment. It outlines both the implementation and distribution of these activities.
- Service Desk - centers on how an organization manages incidents and problems through a centralized entity. Its focal points are items such as a service catalog and knowledge base for incident resolution.

Figure 4.2 diagrams the two main components of the ITIL model, i.e. Service Support and Service Delivery, and that the respective groups of their subcomponents come under the two main categories or Day-To-Day and Strategic.

Figure 4.2 ITIL Categorization

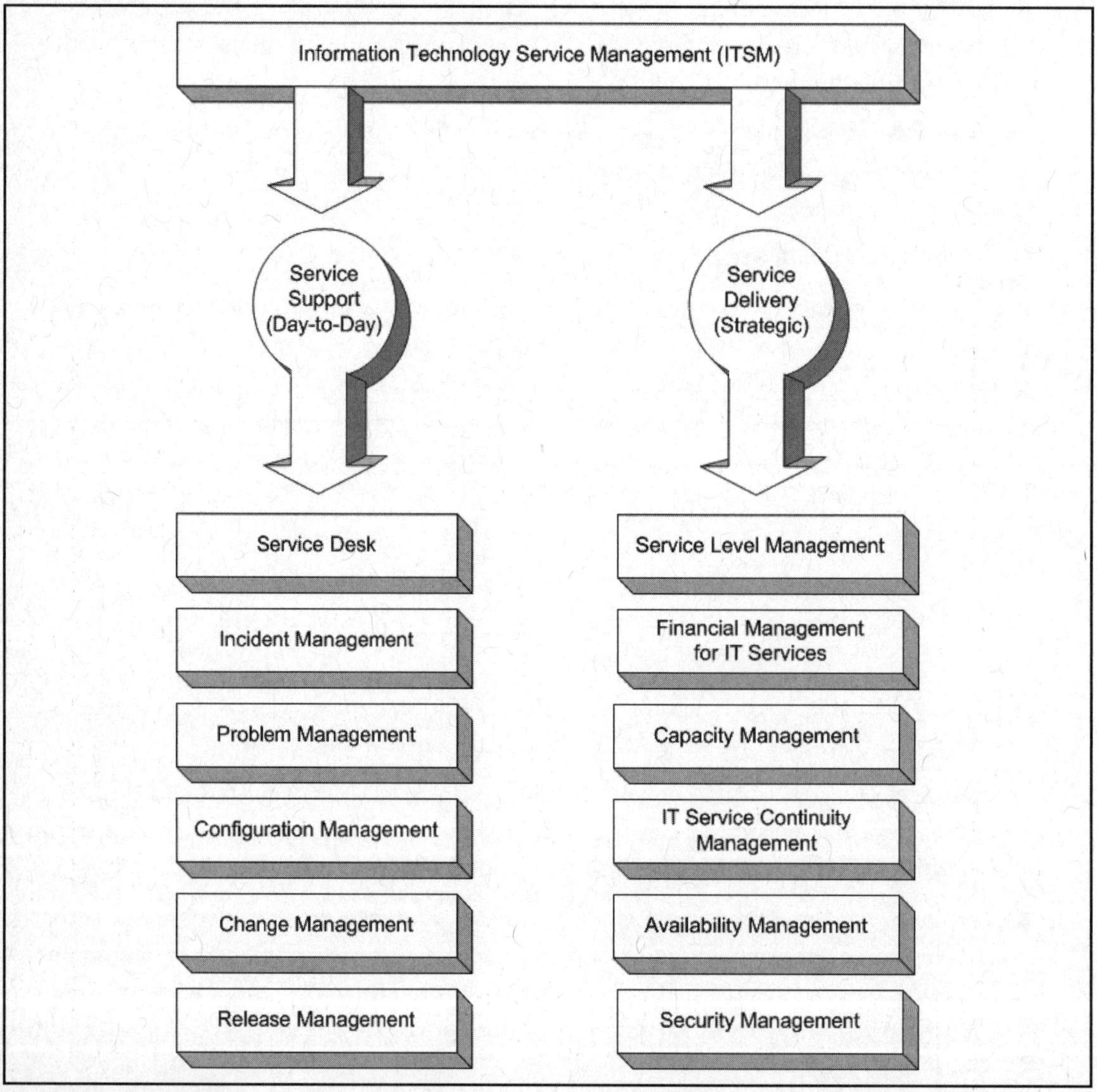

There Is No Panacea

As with COBIT, ITIL is humongous in scope and spans even volumes. As you might, guess the implementation of ITIL in its entirety might be a little daunting. To further exacerbate the problem, as with COBIT, ITIL gives only the "what" as it relates to what needs to be done but does not give you the "how." If this weren't bad enough, ITIL does not even give a prioritization for implementation so that at the very least one could correlate priory to value of component implemented. For the sake of clarity, all of the previous statements were made in relation to ITILv2, which was conceived more than twenty year ago.

In response to criticisms and shortcomings in ITILv2, ITIL released ITIL v3 in June 2007. One would have thought that one of the update's objectives would have been to remove complexity and simplify implementation – but au contraire. Not only did the new version not address the confusion with the implementation magnitude, additional elements were added into the framework– . Where the previous ITIL framework focused primarily on processes, v3 now revolves around services and has more emphasis placed on strategy and operations management. As a result, support and delivery processes are now divided over the lifecycles of service design, transition, and operations. As a result, there is concern among industry leaders that smaller environments will actually find the implementation of the ITILv3 methodology more challenging to implement then ITILv2.

Lessons Learned

Importance Of SAS70

If as in the case of NuStuff Electronics part of an IT Organizations strategy includes the outsourcing of services and some of these services are deemed material in their financial reporting process, these services will needed to be address in their compliance process. There are essentially two ways to address material outsourced services:

- Perform your own evaluation of the service organization's controls.
- If the service organization's is Type 2 SAS No. 70 certified have them provide it for the audit.

SAS 70 is an acronym for Statement on Auditing Standards No. 70. SAS 70 is an auditing statement issued by the Auditing Standards Board of the American Institute of Certified Public Accountants or the AICPA for short. Keep in mind when requesting a SAS 70 report from your service provider that there are two types of service auditor reports:

Continued

- Type I - report includes the auditor's opinion on the service organization's controls and the appropriateness of the controls to achieve the specific control objectives.
- Type II - report includes the information the Type I auditor's report and more importantly includes the auditor's opinion on whether the specific controls were operating effectively during the review period.

Keep in mind that you can fail Sarbanes-Oxley compliance if a service deemed material in the financial reporting process is outsourced and the service provider fails the internal assessment and/or can't provide a SAS 70. As in the case of NuStuff Electronics, moving forward, one of their requirements for potential outsource providers is Type 2 SAS No. 70 report.

VM Spotlight: Project Plan

As we discussed previously NuStuff Electronics' IT organization decided that eGroupware provided them with the best suite of applications and application integration. In this section, we walk you through the required steps in eGroupware to setup a simple project plan.

Select ProjectManager Icon from eGroupware Main Menu – see raised icon in Figure 4.3 graphic.

Figure 4.3 eGroupware Project Module

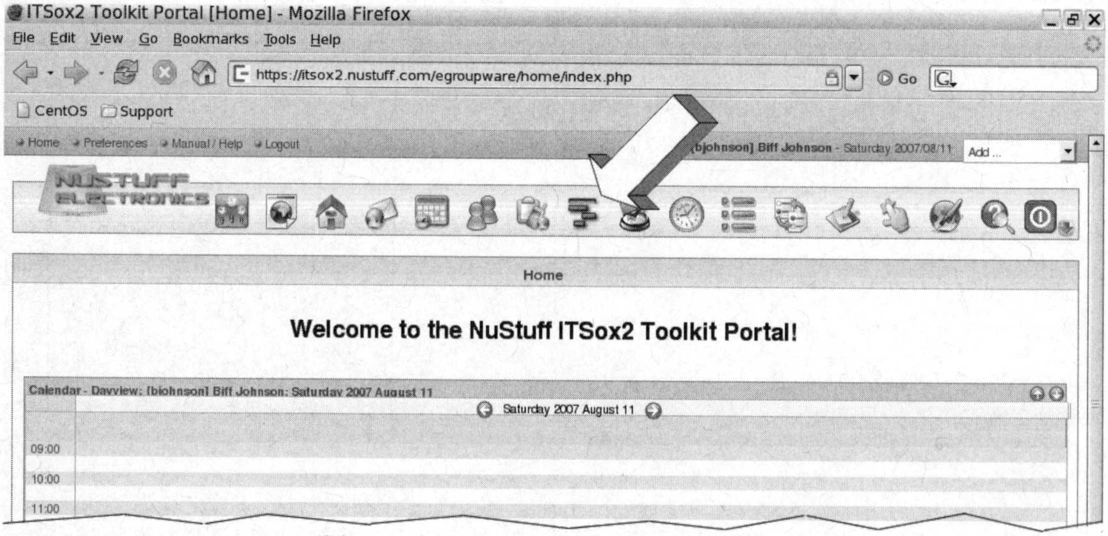

We are now going to add a new project for the SOX discovery phase. Select the "Add" icon from ProjectManager Main Menu – see raised icon in Figure 4.4.

Figure 4.4 ProjectManager Add Project

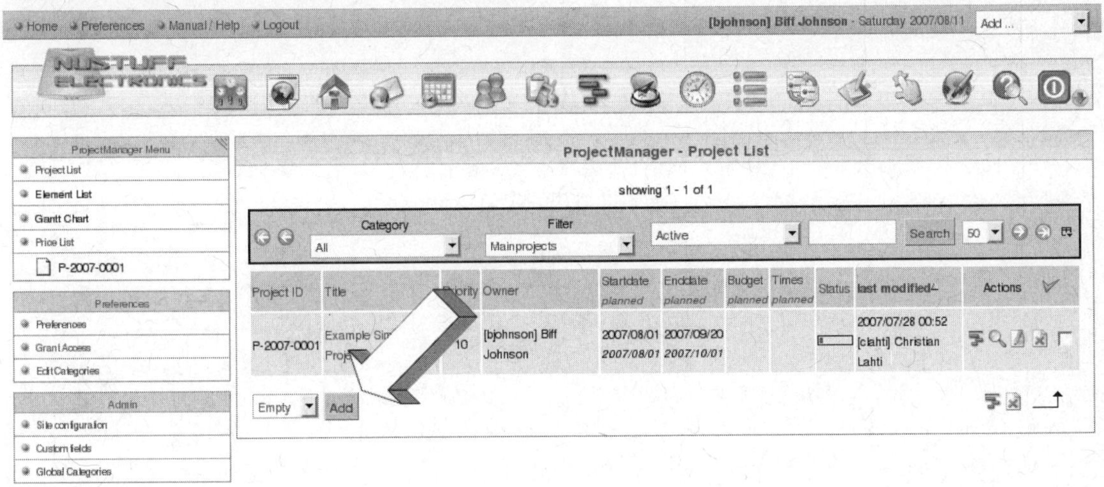

At this point in ProjectManager you will need to specify some of your project parameters. As with most project applications, ProjectManager requires you to specify items such as Title (project name), planned start and end dates, status, State, etc. For the purpose of this exercise, we will set Title (project name), planned start and end dates, status and State (see raised icon in Figure 4.5). Once you have specified your setting, select the "Save" icon.

Figure 4.5 Specify Project Parameters

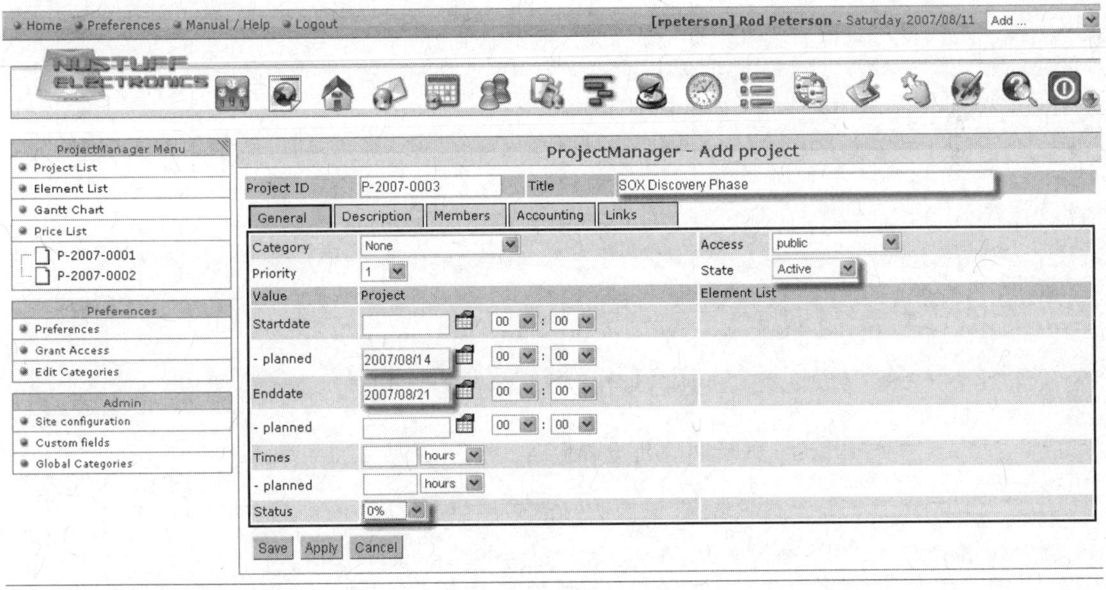

Once the project has been created and saved, you will want to begin adding various tasks. Project items are actually stored in various modules depending on what they are and associated with the project you just created. To add tasks to your project, click on your project title(see raised image in Figure 4.6).

Figure 4.6 Edit Project

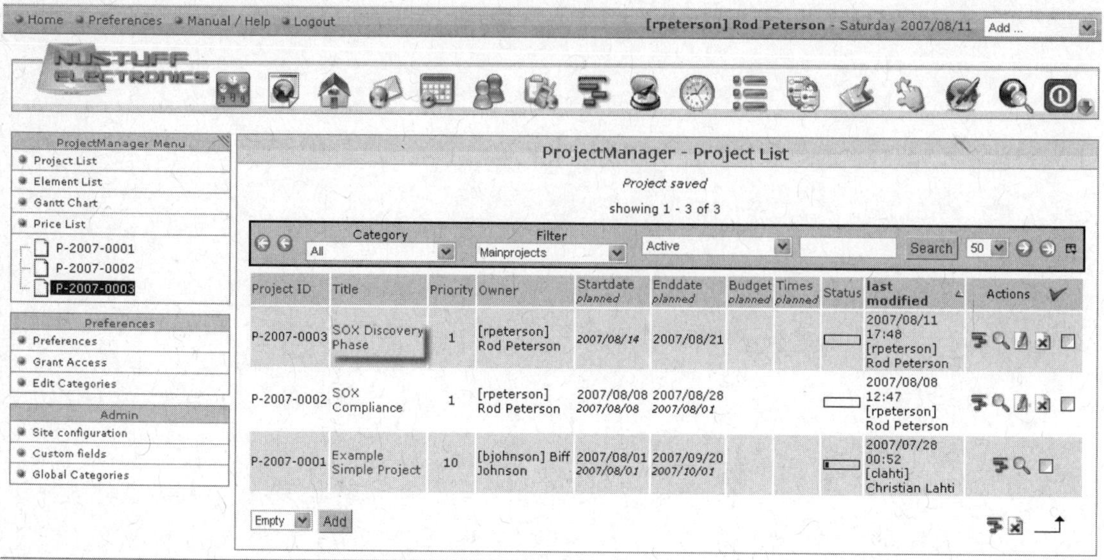

There are many options you can associate with your project, such as calendar and timesheet entries. You can also associate other projects as sub-projects. For the sake of this exercise, we are going to select InfoLog from the "Add new" box. InfoLog items are the "to-do" things you do and assign to other eGroupware users to complete (see raised image in Figure 4.7) and Select the "Add" button.

Figure 4.7 Add Project Tasks

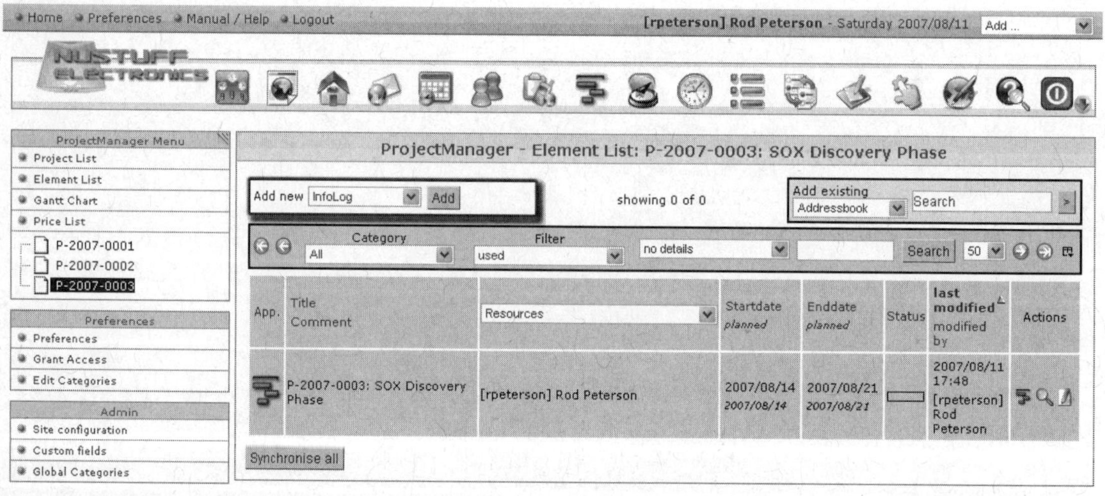

For the purpose of this exercise, we will set subject start and end dates, status, completed and also enter a description (see raised graphic in Figure 4.8). Once you have specified your setting, select the "Save" icon.

Figure 4.8 Define Task

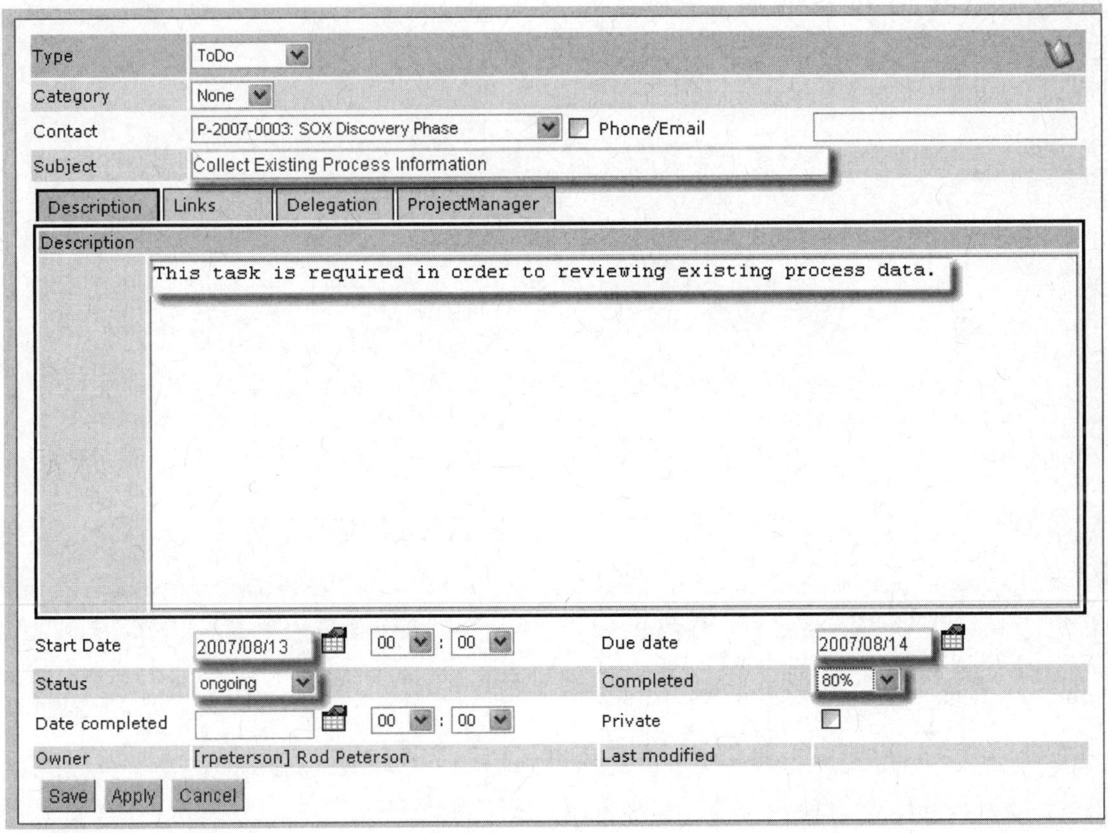

Congratulations! You have successfully created a basic project in ProjectManager (see raised graphic in Figure 4.9) Later in this book we will expand on the project plan and ProjectManager.

Figure 4.9 Project Task Added

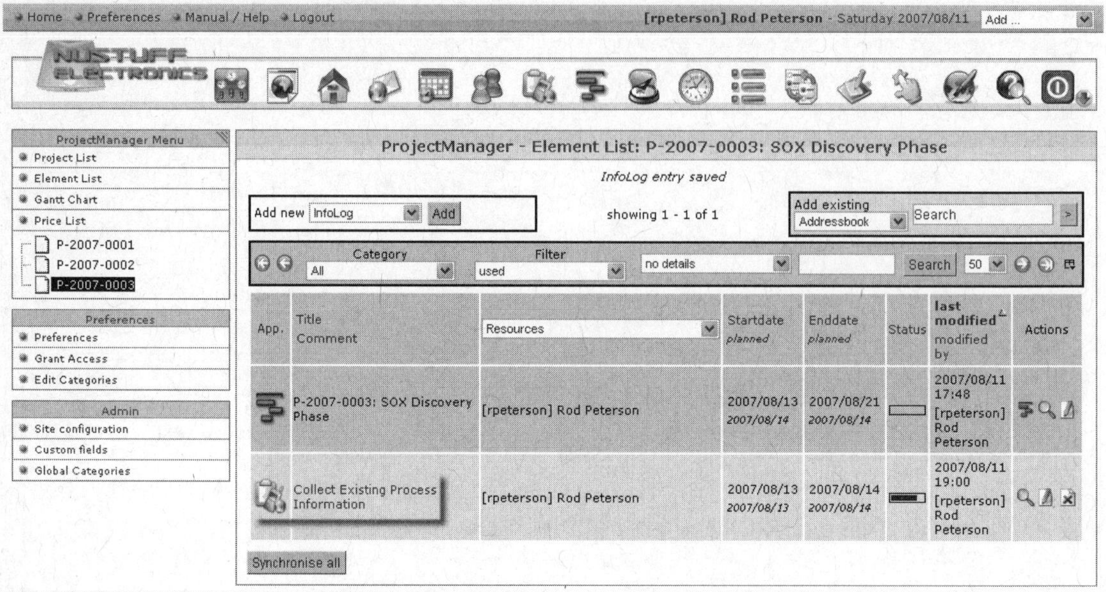

Case Study: Framework Selection

Now that we have provided you with summary information on the diverse frameworks being utilized by various companies and in different locations of the world you are probably asking yourself–which one should I use? If you are asking yourself this question, you are not alone. Companies like our fictitious NuStuff Electronics, as well as real companies, are struggling with this very question right now. Although we would like to be able to give a clear-cut definitive answer to this question, unfortunately there isn't one. As with the implementation of any framework, we must keep in mind some key thoughts

- There is no one size fits all.
- What's right for your organization?
- There are no hard and fast implementation rules.

If we keep the aforementioned thoughts in mind and return to our fictitious NuStuff Electronics, we can see that although NuStuff Electronics implemented COBIT in their environment, they could have just as easily implemented ITIL. However, when NuStuff Electronics looked at their business requirements, their existing environment and aligned these things to the COBIT and ITIL frameworks, they determined that for their needs ITIL had a deficiency and COBIT didn't. This is not to say that they couldn't have implemented ITIL and accomplish the same objective. NuStuff Electronics felt an implementation of ITIL would have required additional steps or augmentation.

Summary

The Sarbanes-Oxley Act of 2002 and COBIT have come to be synonymous with each other. Other standards exist, but COBIT has been most widely adopted by the PCAOB for audit firms. The COBIT guidelines are good standard operating procedures for IT organizations but they are not practical for company's to implement as written. Our fictitious company, NuStuff Electronics, demonstrated how–with planning and knowledge of their own operating environment–the COBIT guidelines could be culled down into something more manageable.

ITIL is another framework being used by IT organization's to achieve regulatory and/or act compliance. However, it too comes with certain problems. The updated ITILv3 serves as a model, showing that even when frameworks are modified they will not be able to fulfill everyone's needs. They will still require interpretation and customization. There is no panacea as far as frameworks are concerned.

Solutions Fast Track

PCAOB Endorses COBIT?

☑ There is no specific mention in section 404 as to what IT needs to do in order to comply with Sarbanes-Oxley.

☑ A company can use various predefined standards for defining and documenting their internal controls i.e.- ITIL (IT Infrastructure Library), Six Sigma, COBIT or even develop their own.

☑ The adoption of the COBIT guidelines and practices as a de-facto standard is likely due to the fact that the COBIT standards are not platform specific but rather platform independent.

☑ There are approximately 300 generic COBIT objectives, grouped under six COBIT Components.

☑ Entity Level Control - consist of the policies, procedures, practices and organizational structures.

☑ Control Objective - is a statement of the desired result or purpose to be achieved.

a. The control is effective in preventing, detecting or correcting risk.

b. The operating effectiveness of the control activity is adequately evaluated on a regular basis.

c. The standards and assertions required to execute the control are clearly understood.

d. Deficiencies are identified and remedied in a timely manner.

e. The performance of the control can be documented.

f. The controls, policies and procedures are documented.

☑ COBIT is comprised of four Domains

g. Planning and Organization

h. Acquisition and Implementation

i. Delivery and Support

j. Monitoring

Are The Developers of COBIT Controls Crazy? Is This Practical?

☑ COBIT controls may appear to any CEO, CFO or IT Director to be a daunting task, and that the developers of the controls must be crazy.

☑ Whether the task of implementing COBIT controls is daunting will depend upon how much effort is put into filtering the COBIT controls.

☑ The largest and best run organization would not be able to implement all of them as defined by COBIT.

☑ The keys to culling down COBIT controls center on a couple of questions:

k. Which controls are appropriate to my environment?

l. Of the appropriate controls which ones will maximize my efforts?

Top Contenders

☑ Two compliance frameworks are emerging as forerunners COBIT and ITIL.

☑ ITIL has been around for more than twenty years, it was conceived in the late 1980s by the British Central Computer and Telecommunication Agency (CCTA).

☑ ITIL framework is comprised of a collection of Best Known Methods (BKM) designed with the intent to enable the delivery of high quality information technology services.

☑ ITIL should enable organizations to more effectively manage IT costs, increase IT efficiencies and maximize existing IT resources.

- ☑ ITIL was conceived in Europe subsequently there are significantly more implementations in Europe than there are in the United States.
- ☑ ITIL has two main components ITIL are Service Delivery and Service Support.
- ☑ ITIL does not give implementation guidelines or prioritization of implementation.
- ☑ ITILv3 was introduced June 2007.
- ☑ ITILv3 may be more difficult for small companies to implement.

Frequently Asked Questions

Q: Where can I find additional information about COBIT?

A: There is a vast amount of information on the Internet about COBIT. I would recommend the following sites as a good place to start:

1. Information Systems Audit and Control Association http://www.isaca.org
2. Information Guide to the Sarbanes-Oxley Act of 2002 http://www.sarbanes-oxley-101.com
3. IT Governance Institute http://www.itgi.org/

Q: Is it really necessary that I use the COBIT guidelines?

A: No. You can follow any predefined standard or even use your own. However, bear in mind that if you choose to use a standard that your audit company is unfamiliar with you not only extend your compliance process but also jeopardize failing compliance.

Q: If I choose can I implement all of the COBIT guidelines?

A: Yes, if you have an unlimited budget and unlimited resources there is absolutely nothing stopping you from implementing all of the COBIT guidelines.

Q: Can SOX compliance be achieved without any automation?

A: Yes, it is possible but since generally prefer evidence of a control to be system generated you might find it extremely difficult.

Q: Do I have to implement all of ITIL?

A: No, as with COBIT how much of ITIL you elect to implement should be driven by your requirements.

Q: Which Framework is better - ITIL or COBIT?

A: It not questions of better, although different in some way both frameworks is based around quality concepts and promote IT controls.

Q: In what order should I implement ITIL?

A: As we discussed the ITIL gives neither implementation guidance nor prioritization, however, we would recommend that you implement the ITIL component that yields the most benefit to your organization in the shortest amount of time.

Q: Can I mix and match frameworks?

A: We would not recommend the random mixing and matching of frameworks but rather the augmentation of a framework with components of another framework if the implemented framework is found lacking.

Chapter 5

The Cost of Compliance

Solutions in this chapter:

- SOX and IT
- Compliance Issues
- What's In A Framework
- Assessing Your Infrastructure
- VM Spotlight: Fedora Directory Server
- Case Study: Costs

☑ Summary

☑ Solutions Fast Track

☑ Frequently Asked Questions

SOX and IT

"Corporate IT professionals lack a critical understanding of risk and compliance issues and pose a barrier to collaborating on compliance initiatives with audit and compliance professionals," based on a study by the Ponemon Institute."

–Robert Westervelt Excerpt from ComputerWeekly.com article published August 9, 2007

In case the intent of the above quote is not self evident, for our discussion it refers to the fact that IT's ability to understand compliance and compliance issues will be critical to the success of compliance. In chapter 3, we discussed that on May 24, 2007, the PCAOB adopted Auditing Standard No. 5, to replace Auditing Standard No. 2, and some of the benefits of this change. What we did not discuss however was why this is probably the most significant motivating factor for the new standard and one of the most significant changes as it relates to guidance for management. Over the last five years or so Sarbanes-Oxley has proven to be a financial drain on companies, not only small but large companies as well. For this reason the SEC working with the PCAOB developed the new audit standard AS5. In developing the new standard the SEC and the PCAOB determined that small companies should have leeway to use discretion in tailoring an evaluation that takes into account their individual circumstances. The following is an excerpt from Christopher Cox testimony before the Committee on Small Business, U.S. House of Representatives, on June 5, 2007.

"We expect the unduly high costs of implementing section 404 of the Act under the previous auditing standard will come down. They should come down because now, a company will be able to focus on the areas that present the greatest risk of material misstatements in the financials. This is what the law has always intended we be focused on. It's also what investors care about. It's what's important for achieving reliable financial reporting."

"Compliance costs should come down because the new SEC guidance that's been developed specifically for management will allow each small business to exercise significant judgment in designing an evaluation that is tailored to its individual circumstances. Unlike external auditors, management in a smaller company tends to work with its internal controls on a daily basis. They have a great deal of knowledge about how their firm operates. Our new guidance allows management to make use of that knowledge, which should lead to a much more efficient assessment process."

Section 404

Given the adoption of the new AS5 standard by the PCAOB and the wording of Section 404 it should make it easier for smaller companies to understand and come into compliance with Sarbanes-Oxley:

In accordance with Section 404 Executive Management of a public company:

- State the responsibility of management for establishing and maintaining an adequate internal control structure and procedures for financial reporting; and
- Contain an assessment, as of the end of the most recent fiscal year of the issuer, of the effectiveness of the internal control structure and procedures of the issuer for financial reporting.

Since it is the responsibility of the CEO and CFO of a company to provide the attestment of and sign-off on a company's SEC filing, an understanding by the CEO, CFO, CIO or IT Director of what compliance means and how best to comply is still crucial to the success or failure of the compliance efforts. In this chapter we will not only discuss the consequences of non-compliance and the benefits of compliance but we will also look at areas that have the potential to impede or facilitate your efforts to comply.

Why Comply?

If a company does not comply with the Sarbanes-Oxley Act, it will expose itself to the possibility of lawsuits and negative publicity. If a corporate officer, even if unintentionally, files an inaccurate certification, he or she is subject to a fine up to $1 million and 10 years in prison. If a corporate officer intentionally files an inaccurate certification, the fine can be as much as $5 million and possible 20 years in prison. When thinking about the severity of the consequences of noncompliance for corporation and corporate officers, we must remember that the intent, although arguably misguided, was to prevent occurrences such as those that happened at MCI and Enron—hence the stiff penalties for those at the top.

Therefore, the downside of not complying with the Sarbanes-Oxley Act can be pretty severe for a company's executive management. However, there is, perhaps not as tangible, an upside to complying. If your IT organization is typical, it is understaffed, has not done a technology assessment/refresh (applications/hardware) in quite some time, and activities like documenting and developing policies and procedures have been relegated to the backburner in deference to putting out current fires. By no means are we suggesting that the requirement to comply with Sarbanes-Oxley Act be used as a catchall or some sort of panacea to fix all the ills that exist in your IT organization. What we are suggesting is that because of the need to comply with the Sarbanes-Oxley Act, opportunities will present themselves to address both SOX deficiencies and other IT organization deficiencies. Moreover, with adequate research and planning, a CFO, CIO, or IT Director can capitalize on his or her compliance effort to address some of the aforementioned problems in the IT organization.

The Transparency Test

As Old As Time

Due to corporate wrongdoings of the Enron's of the world and other large corporations congress and other interested parties decided that we needed an accountability measure by which to determine culpability for corporate financial fraud. They named one of these measures Sarbanes-Oxley. In view of this measure a CIO/director of information technology would/should ask how does this impact me? Of course we all know the formal definition of the acronym CIO-Chief Information Officer. But for those of us in the business with the title CIO we often fondly refer to it as "Change Is Often" or "Career Is Over." If CIO/director of information technology is to be successful in I.T. they must drive change and change drives dirty laundry into the open whether it is a bad process or bad people. In all cases their will be those who wish to protect the "as is" and those who wish to protect the "as will be" in the end a CIO/director of information technology may be dismissed simply because he places to many people at risk while doing his job.

To quote Niccolo Machiavelli – "There is nothing more difficult to handle, more doubtful of success, and more dangerous to carry through, than initiating changes in an organization. The innovator makes enemies of all those who prospered under the old order, and only lukewarm support is forthcoming from those who would prosper under the new."

For the sake of this conversation lets take a moment to discern the reporting lines, if a CIO/director of information technology reports to the CFO his measure of success will most always be in dollars and cents-usually cut from the I.T. budget. If reporting lines are to the CEO then the other executives are immediately at odds with the CIO's progress and direction to be taken as it places them on there departments in a position of vulnerability and forces them out of there comfort zone, one could state that being in the I.T. management position is like being a "stepchild at a family picnic" neither side claims you. Now back to this Sarbanes-Oxley thing, simply stated Sarbanes-Oxley will inevitably be a measure by which the success of a CIO/director of information technology is determined. However, by the mere nature of Sarbanes-Oxley, whether as a result of the inherent change or the discomfort of the perceived encroachment of IT on other departmental territory a CIO/director of information technology will be highly visible and vulnerable at the same time during the process of Sarbanes-Oxley compliance. Although there is no silver bullet, the successful CIO/director of information technology will have to be able to manage change, not only process changes but changes in perceptions. This phenomenon alone forces the CIO to go outside his own circle of management to build support for his objectives – which in this case is Sarbanes-Oxley compliance.

–Bill Haag

Compliance Issues

There are many issues that could potentially jeopardize your ability to apply with Sarbanes-Oxley, some of which we will examine later in this section, the one that even with new audit standard AS5 will be the inclination to over implement. The inclination is definitely an understandable one since one of the biggest misconceptions of Sarbanes-Oxley compliance is that a company needs to evaluate and have controls for every process in their IT organization, which is not the case. As a matter of fact a company needn't evaluate or have controls for some of the financial processes or processes that interface with their financial systems as long as the processes or interfaces are not material to the financial reporting process. Therefore, if there is one caution that we would render it would be to take your time in the assessment process of your financial process so that you are better able to hone in on what processes are material in your financial reporting process.

NOTE

With adequate research and planning, Sarbanes-Oxley compliance can be used to correct additional IT organization deficiencies beyond SOX. Failure to comply may negatively affect both a company's executive management and the company.

The Human Factor

At this point, we will assume that executive management's understanding of the need to comply with SOX is sufficient motivation to drive the necessary changes in their company. Hence, we will not provide reasons for the change, but rather guidance for change and identify what might be an unthought-of obstacle. In general, many people are adverse to change, especially change within the organization or company for which they work. Some primary reasons for resistance to change include:

- Fear of the unknown
- The belief that things are fine as they are
- Do not understand what's driving the change
- Believe the change is just another exercise that can be ignored
- WIFM (What's In It For Me) comes into play

Again assuming executive management has come onboard the change process can be focused on lower level of employees at the company. To do so successfully, communication is a key factor. Communication can and should be deployed in various ways and repeated consistently and often, via memos issued by executive management, employee meetings, informal meetings, and, if feasible, one-on-one conversations with individual employees. Sun Tzu once said, *"If you know the enemy and know yourself, you need not fear the result of a hundred battles. If you know yourself but not the enemy, for every victory gained you will also suffer a defeat. If you know neither the enemy nor yourself, you will succumb in every battle."*

This might seem to be a strange quote to find in a book addressing Sarbanes-Oxley compliance, and in particular, in a section addressing change. If you were to ask any CFO, CIO, or IT director responsible for implementing Sarbanes-Oxley who "enemy" was in reference to, he or she would probably say "the end user." These individuals would be in for at best a surprise, or at worst, a gotcha that could hamper their ability to obtain compliance. The biggest threat in this regard will more than likely come from within the IT organization itself. As a rule, within any company or organization exist what we refer to as *personal processes*. These processes are usually formed on an individual basis with an underpinning of *quid pro quo* (Latin for "something for something," used to generally describe the mutual agreement between two parties in which each party provides goods or service in return for goods or service). If you are not aware that these processes exist and that they can be extremely detrimental to your compliance process, regardless of how good your controls in place are, your controls will almost certainly fail. To combat personal processes, here are some simple guidelines to keep in mind:

- Over-communicate to the IT staff what you are doing, what their roles are, and the consequences of failure
- Get buy-in on the plan and what needs to be done
- Wherever possible, delegate decisions to the IT staff
- Bear in mind that change is a process, and usually a slow one
- Retain responsibility for reaching the ultimate goal
- Lead by example
- If necessary, demonstrate any consequences with staff and user community

TIP

IT staff may not openly embrace change. Communicate, *ad nauseam*, the need for SOX compliance and the consequences of noncompliance. Understand and manage the personal processes and don't make assumptions when it comes to change.

Walk the Talk

Table 5.1 is an example of the type of policy NuStuff Electronics might use in its activities to comply with the Sarbanes-Oxley Act, and although NuStuff Electronics is fictional, the procedure and the areas it defines are not. At first glance, the policy appears to be a standard procedure that would have been used by any IT organization prior to Sarbanes-Oxley, so what the big deal? If the Sarbanes-Oxley Act was never drafted, your IT organization would be audited based on traditional audit practices and guidelines. The auditors would give this policy a cursory review, at best, more to ensure that you have a documented procedure than anything else. Now, here is the big deal, aside from being an actual policy used to obtain Sarbanes-Oxley compliance, it contains two very important areas that were added to make the procedure acceptable for Sarbanes-Oxley compliance—5.0 Review and 6.0 Enforcement. Although it is important that you adhere to what you state in your policy, be particularly cautious about what you stipulate in Review; if you can't or won't adhere to it, don't state it. Whether during initial testing after the remediation phase or during a subsequent compliance audit, the auditor will want to see evidence of the effectiveness of the control.

As stated previously in this book, COBIT is merely a detailed set of "Best Known Methods" for IT. Therefore, there is no need to discuss this particular policy's password parameters (password length, password age, etc.); those would need to be tailored to fit your environment. However, we would like to make note of some of the particular areas in the policy to which you must pay special attention:

- **Scope:** If your company has other locations, whether domestic or international, it is critical that you define what location your policy affects and which it does not.
- **Review:** There are two concerns that you must define well regarding the "Review":
- **The frequency of review:** If the interval between reviews is too long, the auditor will perceive the control as ineffective.
- **Connect the review of evidence to the CFO:** Although the IT manager may perform a review monthly, the CFO will perform a review of the effectiveness of the control and evidence on a quarterly basis.
- **Enforcement:** You may not have to define consequences in all policies, but if you do, you want to ensure that Human Resources review them before formalizing and publishing your modified or new policy. Generally speaking, it is a good idea to have information readily available to all employees, using whatever mechanism works best in your environment. Moreover, the employment packet for a new hire should contain a document (that the new employee will sign) stating that he or she has read and understands the corporate policies.

NOTE

It is also a good idea to periodically re-distribute corporate policies and have existing employees review and sign a document stating they have read and understand the policies.

Table 5.1 An Outline of a SOX Compliance Policy

NuStuff Electronics Company	
Information Technology	
Section	
Policy Title	Password Control Policy
Approved By	Joe Manager
Policy Number	123
Issue Date	04/15/2005
Supercedes	

1.0 Overview

Passwords are an important aspect of computer security, and are the front line of protection for employees' accounts. A poorly chosen password may result in the compromise of NuStuff's entire corporate network. As such, all NuStuff employees (including contractors and vendors with access to NuStuff systems) are responsible for taking the appropriate steps, as outlined here, to select and secure their passwords.

2.0 Purpose

The purpose of this policy is to establish a standard for the creation of strong passwords, the protection of those passwords, and the frequency of change.

3.0 Scope

The scope of this policy includes all employees who have or are responsible for an account (or any form of access that supports or requires a password) on any system that resides at any NuStuff facility, has access to the NuStuff network, or stores any nonpublic NuStuff information.

Table 5.1 Continued

4.0 Policy

4.1 General

All system-level passwords (Win2000/NT) admin, application administration accounts, etc.) must be changed on at least a quarterly basis.

All employee-level and system-level passwords must conform to the guidelines described here.

4.2 Guidelines

4.2.1 General Password Electronics Guidelines

Passwords are used for various purposes at NuStuff, some of which include user level accounts, Web accounts, e-mail accounts, screensaver protection, voicemail password, and local router logins. Since very few systems have support for one-time tokens (dynamic passwords that are only used once), everyone should be aware of how to select strong passwords.

Poor, weak passwords have the following characteristics:

- The password contains less than eight characters.
- The password is a word found in a dictionary (English or foreign).
- Names of family, pets, friends, coworkers, fantasy characters, etc.
- Computer terms and names, commands, sites, companies, hardware, software.
- The words *sanjose*, "*sanfran*," or any derivation.
- Birthdays and other personal information such as addresses and phone numbers.
- Word or number patterns like aaabbb, qwerty, zyxwvuts, 123321, etc.
- Any of the above spelled backward.
- Any of the above preceded or followed by a digit (e.g., secret1, 1secret).

Strong passwords have the following characteristics:

- Contain both upper- and lowercase characters (e.g., a–z, A–Z).
- Have digits, punctuation characters, and letters (e.g., 0–9! @#$%^&*()_+|~-=\'{}[]:";'<>?,./).
- Are at least eight alphanumeric characters long.

Continued

Table 5.1 Continued

- Not a word in any language, slang, dialect, jargon, etc.
- Are not based on personal information, names of family, etc.
- Passwords should never be written down or stored online. Try to create passwords that can be easily remembered. One way to do this is to create a password based on a song title, affirmation, or other phrase. For example, the phrase might be "This May Be One Way To Remember," and the password could be "TmB1w2R!", "Tmb1W>r~", or some other variation.

4.2.2 Password Protection Standards

Do not use the same password for NuStuff accounts as for other non-NuStuff access (e.g., personal ISP account, option trading, benefits, etc.). Where possible, do not use the same password for various NuStuff access needs.

List of "don'ts":

- Don't reveal a password over the phone to ANYONE.
- Don't reveal a password in an e-mail message.
- Don't reveal a password to the boss.
- Don't talk about a password in front of others.
- Don't hint at the format of a password (e.g., "my family name").
- Don't reveal a password on questionnaires or security forms.
- Don't share a password with family members.
- Don't reveal a password to coworkers while on vacation.
- Do not use the "Remember Password" feature of applications (e.g., Eudora, Outlook, Netscape Messenger).
- Do not write passwords down and store them anywhere in your office. Do not store passwords in a file on ANY computer system (including Palm Pilots or similar devices) without encryption.

4.3 System Policy

■ Maximum Password Age	90 Days
■ Minimum Password Length	8 Characters
■ Minimum Password Age	5 Days
■ Password Uniqueness	5 Passwords
■ Account Lockout	10 Bad Login Attempts
■ Lockout Duration	Forever (until admin unlocks)

Table 5.1 Continued

Change passwords at least once every six months (except system-level passwords, which must be changed quarterly). The recommended change interval is every four months.

If you suspect an account or password has been compromised, report the incident to IT.

5.0 Review

IT Director to annually review current Windows Password policy to ensure current policies are still enforced.

6.0 Enforcement

Any employee found to have violated this policy might be subject to disciplinary action, up to and including termination of employment.

Who Are You and What Do You Need

For the sake of this section we need to establish a common understanding of two terms:

- Authentication – A control, usually system or application, based on username and password by which identify of an individual is verified
- Authorization/Access – A process by which, once authentication is verified, an individual is given access to system elements based on identity.

As you can see authentication is very different from authorization/access, this distinction is very important as we cover this section. Therefore, lets try and simplify the definition even further by saying that authentication is usually control by a system or application and authorization/access is usually a process whereby the criteria of who, what and why is evaluated and a decision is made to grant or deny authorization/access. Undeniably there are many aspects to compiling with Sarbanes-Oxley and in all likelihood just about all of the other compliance acts or regulatory requirements we have discussed and some we have not. But with out a doubt authentication and authorization/access will be two of the most important ones in your ability to achieve compliance. However, for the sake of this section we will only be focusing on authentication. This being said we would like to offer some suggestion and guidelines as to what you consider when looking at your current authentication system or when evaluating a new one.

- Practicable – authentication must be practicable and sustainable not only from an administrative perspective but from a users perspective as well. After all if it is too cumbersome to your user community they will look for ways to circumvent it, whether it is via a manful process, automate process or through a user revolt. Ideally, it should support single user sign-on, it should be robust but simple for the users to use, it should require minimal maintenance overhead and it should be scalable.
- Secure – You may be saying security, duh, but along with the system having to be able to offer a certain amount of assuredness that end-user are not able to perform authentication without providing the right credentials. It is also important that the system be capable of providing encryption for things such as ids and passwords that are continually sent across the wire. Although encryption on it own cannot stop sniffing and attack by using sniffed id and passwords or forged credentials. One of the last things your authentication system should do is prohibit the ability to the support staff from easily access authentication data or circumventing the system. This last item will be of particular interest during an audit and to your auditor.
- Platform Independent – An ideal authentication system should have the ability to support a variety of OSs.
- Protocol – As with OSs, an authentication system should support a variety of protocols.
- Reporting – This may seem to be a no brainier but it is one that can't be emphasized enough since this capability will be of paramount importance when you are working with your auditor and they are requesting evidence of compliance. At a minimum your system should be able to report on unauthorized attempts, and compromises.

It is our hope that you now have a better understanding of why your authentication system is and/or will be so important in your compliance efforts.

What's In A Framework?

If we go back to the quality practice of PDCA (Plan, Do, Check and Act) developed in the 1950s, we might start to see similarities between COBIT, ITIL and the Quality discipline. At their core, COBIT and ITIL are no more than a quality discipline, much like PDCA, Deming, or Juran, of which the major objective is to have a closed looped process that drives continuous improvement. If we were to apply the quality approach to a control objective, it would look something like Figure 5.1. The concept of Figure 5.1 could be applied to a manual process; however, it tends to work best when automation is used. The define policy/control objective would drive the requirements, functionality, and configuration of a application or tool, which would in turn drive the functionality and configuration of a application or tool for monitoring/reporting, whereas the cycle continues to repeat.

Figure 5.1 Quality Control Process Cycle

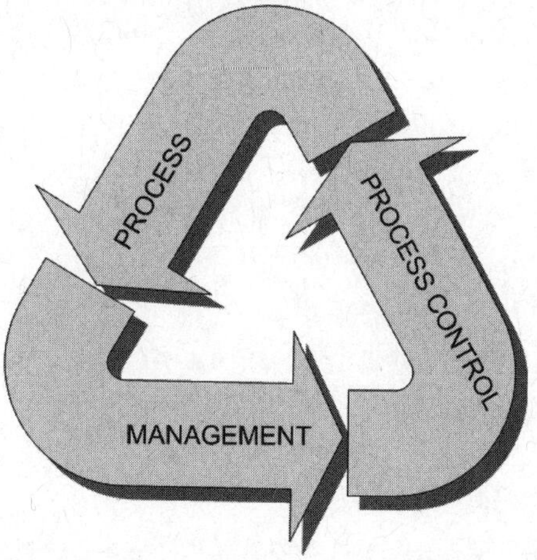

As we have stated, at the heart of most frameworks is fundamentally a quality process but to leave you with the impressions that this means that there are no differences between frameworks would be a disservice. Two prime examples of frameworks that although quality is at the core the intended focus at the inception was different. Where the ITIL framework centers on the efficiency, and effectiveness of IT processes and of course the quality of IT services from a customers perspective. The COBIT framework centers on compliance and audit ability of IT processes, not necessarily with regards to Sarbanes-Oxley compliance but rather in general. Simply put COBIT is about "Saying what you do", "Doing what you say" and "Proving it". Even though COBIT and ITIL intents were different they do indeed align at a level that by which Sarbanes-Oxley can be achieved with ITIL with as stated previously with a few modifications. In the subsequent chapters of the book we will attempt to present the alignment of COBIT and ITIL in a logical manager to convey what elements should be consider for Sarbanes-Oxley compliance.

TIP

Think of COBIT as a quality process where Definition/Implementation, Execution and Managing/Monitoring of Control Objective should follow a closed loop process. Simply put, your policies should not contain any activities your organization can't sustain.

Assessing Your Infrastructure

One of the positive aspects is that you may very likely have the hardware in-house and experience minimal acquisition needs to leverage open source for your SOX needs. Another is you may very likely have some in-house expertise in order to more easily incorporate the ideas of the remaining chapters in the book. One of the disadvantages is that your need for documentation of the various systems may be greater than a purely OTC environment because early in the process the auditors we encountered had much more knowledge of proprietary systems. Therefore, it seemed like a long road to consensus that an environment with open source could pass the certification process. As more information about the auditing process becomes standardized and auditing firms gain more core competency in this arena, it is expected that this will become less of a factor in the future. Some of the major infrastructure-related areas that may come onto the radar screen of SOX compliance include:

- System security access controls
- System availability reliability
- Data backup and retention

Open Source to Support Proprietary Systems

Although you may not have the in-house expertise, don't despair; there is plenty of opportunity to utilize OSS in non-OSS environments, and we will show specific technologies and methods that you can employ to accomplish this goal. We will be taking a look specifically at:

- Workflow and approval mechanisms
- Document and revision control
- Policy and procedure management
- Reporting, monitoring, and escalation

Regardless of whether you already have open source software deployed or wish to augment your environment, you need to have the ability to certify your methods and configurations in a demonstrable way. The way to accomplish this is with internal controls. These controls universally will fall into two categories of prevention controls and detection controls; however, all controls share the following characteristics and constraints:

- Testable a well-defined test is in place to validate the control
- Repeatable the same test yields the same result every time
- Sustainable the ability to maintain your controls and certification processes over time

Lessons Learned

InfoSec and SOX, A Marriage of Convenience

I have to secure my client's intellectual property against threats both external (viruses, hackers, worms, etc.) and internal (disgruntled employees, accidental disclosure, and malfeasance). To do so successfully I need to ensure that my role, as well as the roles of others, is performed in adherence to InfoSec principles. Two of these are segregation of duties and audit trails. If no single person can control a process from end to end then no single person can manipulate that process to their own ends. Similarly, if a user knows that all activity is tracked by the system, they are less likely to commit any act in which they would incriminate themselves. Therefore half of the work of protecting the system is done in these two simple implementations.

The trouble is the cost. There is a cost associated with having two people do what one person is capable of doing on their own. There is a cost of taking time to review audit trails to scan for suspicious activity. There is the cost of storing all the logs of user activity. The challenge before every InfoSec manager is to convince upper management that the risk outweighs the cost of prevention.

SOX has been very helpful with this challenge. Prior to 2002 these principles of data protection were just that; principles. SOX turned good advice into law. By requiring these measures in finance and all that interacts with finance (essentially everything in any modern company whose goal is profit) we see a more robust governance of intellectual property throughout the company. Preparing for a SOX audit is tiresome and difficult, especially if that preparation requires a major re-organization of the company. When planning the changes, seek out your InfoSec people. We've been espousing this stuff for years.

–John T. Scott

VM Spotlight: Fedora Directory Server

http://directory.fedoraproject.org/

A directory server provides a centralized directory service for your intranet, network, and/or extranet with information that systems and client applications can "lookup" information and is somewhat analogous to a phone book. A directory server can typically integrate with existing systems and applications and acts as a centralized repository for the consolidation of employee, customer, supplier, partner, or any other information you wish to make available. You can extend Directory Server to manage user profiles and preferences as well as user authentication. Why is a directory server important in a book geared toward a discussion of SOX? We use LDAP to provide a single point of user/group authentication for systems and applications and it is these features that we focus on for our SOX examples.

Fedora Directory Server (FDS) is an enterprise-class Open Source LDAP server for Linux. Formerly Netscape Directory Server, RedHat purchased this product in 2004 and after rewriting a few proprietary components re-released this under a somewhat modified version of GPL v2. Due to its Netscape roots the core code handles many of the largest LDAP deployments in the world and has been hardened by real-world use. FDS is a mature, fully V3 compliant LDAP server and its major features include:

- 4-Way Multi-Master Replication to provide fault tolerance and high write performance.
- Scalability: thousands of operations per second, tens of thousands of concurrent users, tens of millions of entries, hundreds of gigabytes of data.
- Extensive documentation, including Installation and Deployment guides, How top's and user submitted examples via the project Wiki.
- External Directory Synchronization – FDS can interoperate with other information stores, one example is user and group synchronization with Microsoft's Active Directory Server.
- Secure Authentication and Transport – FDS has a robust security model and supports LDAP over SSLv3 and TLSv1. Secure passwords and integration with other and authentication mechanisms such as Kerberos are supported via a standard SASL interface (we discuss Kerberos and SASL in a bit more detail in chapter eight).
- Support for LDAPv3 – This is the current LDAP standard, mainly covered by RFCs (http://www.ietf.org/rfc) numbers 2251 through 2256. There is currently an IETF working group which is responsible for maintaining these and they are updating and revising the existing RFCs.
- On-line, zero downtime LDAP-based update of schema, configuration, and management information – FDS can be fully administered and configured via standard LDAP operations. This provides maximum flexibility and uptime.
- In-tree Access Control Information (ACIs) – FDS has a robust fine-grained security model in which access can be controlled for every aspect and element of the directory (more on this below).
- Graphical console for all facets of user, group, and server management (also covered in detail later in the chapter).

- Plug-in Support – FDS can be extended to do whatever you need via a well defined plug-in interface, in fact many of the features available in the standard FDS distribution are implemented as plug-ins.

Before continuing with our look at FDS in particular it is best we spend a bit of time discussing LDAP in general for those of you who may not be familiar. This section is not meant to be an exhaustive look at LDAP, or even FDS for that matter since there are many books and documentation online to serve this purpose. We do want to cover the basic concepts so the reader has a basic understanding of the concepts and terminology so we don't start throwing around unfamiliar acronyms. If you are already aware of LDAP concepts you can safely skip this section.

LDAP Overview

Lightweight Directory Access Protocol (LDAP) provides a common language that client applications and servers use to communicate with one another. LDAP is the "lightweight" version of the Directory Access Protocol (DAP) used by the ISO X.500 standard. DAP, originally developed at the University of Michigan, gives any application access to the directory via an extensible and robust information framework, but at an expensive administrative cost. DAP uses a communications layer (OSI stack) that is not the Internet standard TCP/IP protocol and has complicated directory-naming conventions. LDAP on the other hand is a derivative of this very expansive protocol and preserves the best features of DAP while reducing administrative overhead by using an open directory access protocol running over TCP/IP and simplified encoding methods. It retains the X.500 standard data model and can support millions of entries. LDAP is primarily used as an information store whose main characteristics are:

- Optimized for very fast read operations – LDAP is designed to service mostly read operations as the intent is to provide an information store with relatively static data. Although updates can be performed from the client side, by far the most operations are "read" in nature and LDAP is designed with this in mind.
- Hierarchical storage – LDAP differs from other information stores (notably relational databases) in the fact that the information is stored in a parent/child Directory Information Tree (DIT). Databases traditionally store their information in "tables" where elements (fields) are related each other by use of primary and foreign keys. See Figure 5.2 for an example.
- Standardized LDAP protocol – much like ODBC is a standard client access definition for databases LDAP provides a clear and well defined interface for access. Any cross platform LDAP client using this standard should be able to talk to any LDAP server in the same way and expect the same results.

- Logically Centralized and Physically Distributed – LDAP provides mechanisms to create the appearance of a wholly centralized information store to the client, however pieces of the tree can physically reside on multiple servers and locations and LDAP provides "referrals" to these locations if a client requests information that is not physically on the server that it originally made the request to.

NOTE

There are many examples of directories that exist that you might not think of as specifically being directories. Here are a few instances, some of are standard LDAP, some are "sort of", and some change the LDAP protocol by use of proprietary extensions, but for the most part all of these are LDAP servers in one way or another and support standard protocol operations:
Windows NT Directory Services (NTDS) for Windows NT
Active Directory for Windows 2000/2003
Novell Directory Services (NDS) for Novell NetWare
Open Directory: Apple's Mac OS X Server
Apache Directory Server: Apache Software Foundation
Oracle Internet Directory: (OID) is Oracle's directory service

Figure 5.2 LDAP Hierarchical Structure

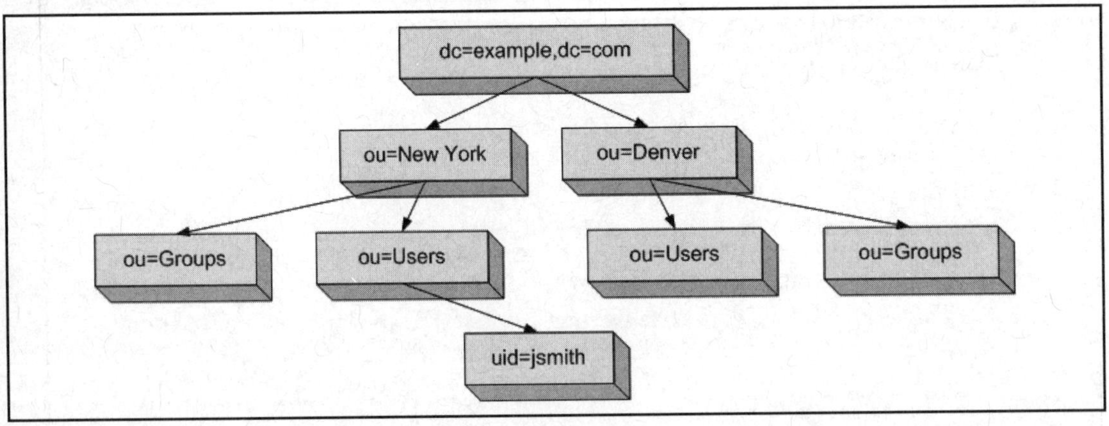

A directory is a tree of directory entries. The structure of this information store stars with the Schema, which is a collection of ObjectClasses that are supported by the directory server or instance. ObjectClasses describe the collection of attributes that may or must comprise an entry.

Finally, Attributes (such as uid and mail) contain various types of data and are subject to Syntax Rules and Attribute Matching Rules (e.g. case-insensitive matching). An attribute has a name (an attribute type or attribute description) and may have one or more values depending on the rules defined for the attribute. OIDs (also used in SNMP) are assigned numeric representations of ObjectClasses and Attributes (see Figure 5.3).

Each entry has a unique identifier called its Distinguished Name (DN). This consists of its Relative Distinguished Name (RDN) constructed from some attribute(s) in the entry, followed by the parent entry's DN. Be aware that a DN may change over the lifetime of the entry, for instance when entries are moved within a tree. To reliably and unambiguously identify entries, a GUID Globally Unique Identifier may be provided in the set of the entry's attributes. The DN is the name of the entry; it is neither an attribute nor actually part of the entry itself. For example "uid=jsmith could be the RDN, and "ou=users,ou=New York,dc=example,dc=com" is the DN of parent. A server holds a subtree starting from a specific entry, e.g. "dc=example,dc=com" and its children. An analogy for DN and RDN could be a fully qualified path to a file `/usr/share/doc/foo.txt`, where the DN is the full path and the RDN is the relative filename.

TIP

When a client makes a standard LDAP request the server rarely defines any ordering or sorting of values if more than one entry is returned in the result.

Figure 5.3 LDAP Directory Objects

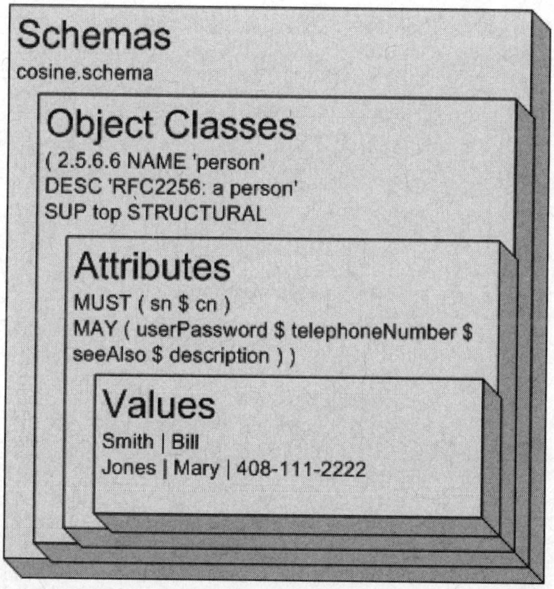

In brief here is a description of the elements that make up and objectclass:

- NAME – name of entry
- DESC – description, usually includes the RFC this entry pertains to
- SUP – (superior) derived from this other attribute type
- STRUCTURAL – corresponds to a real world object. Entries must belong to a structural object class, so most object classes are structural
- OPERATIONAL – Defines classes and attributes that the server does not return in a client query unless specifically asked for by name
- AUXILIARY – attributes an entry may have. This does not represent a real world object, but represents additional attributes that can be associated with a structural object class as supplement. May belong to zero or more structural object classes
- ABSTRACT – defined only as a superclass or template for other (structural) object classes. An entry may not belong to an abstract object class
- MUST – required attributes
- MAY – optional attributes

ObjectClasses also support inheritance (See Figure 5.4). In our example the person objectclass defines the required attributes sn and cn (MUST) and a few optional attributes such as description (MAY). The organzationalPerson objectclass further extends the person class to include many other optional attributes that were not originally part of the parent class. In this way entries can be comprised of "stacked" objectclasses.

Figure 5.4 LDAP ObjectClass Inheritance

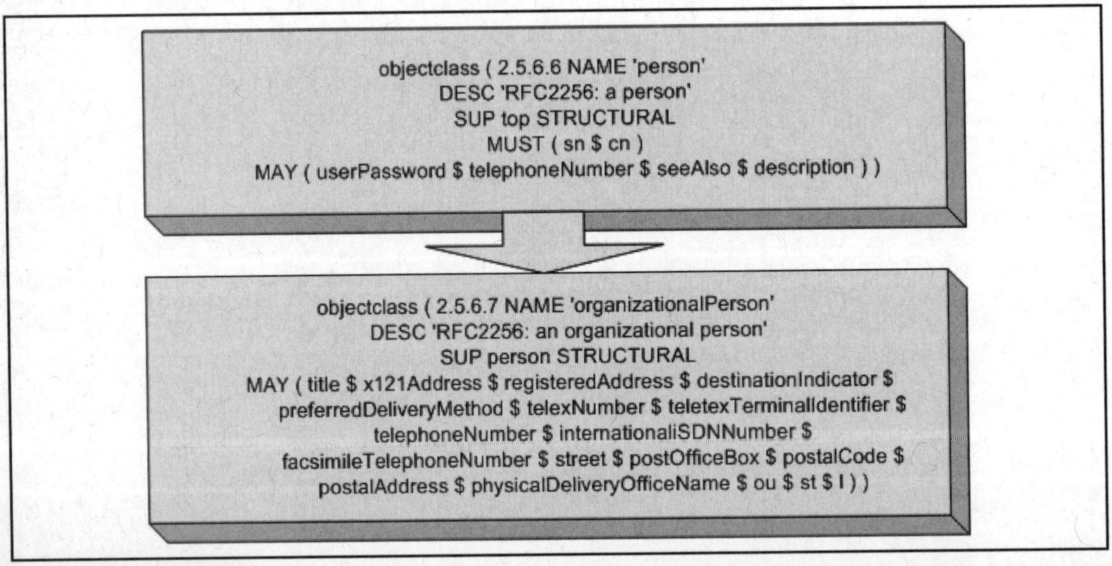
objectclass (2.5.6.6 NAME 'person'
DESC 'RFC2256: a person'
SUP top STRUCTURAL
MUST (sn $ cn)
MAY (userPassword $ telephoneNumber $ seeAlso $ description))
objectclass (2.5.6.7 NAME 'organizationalPerson'
DESC 'RFC2256: an organizational person'
SUP person STRUCTURAL
MAY (title $ x121Address $ registeredAddress $ destinationIndicator $
preferredDeliveryMethod $ telexNumber $ teletexTerminalIdentifier $
telephoneNumber $ internationaliSDNNumber $
facsimileTelephoneNumber $ street $ postOfficeBox $ postalCode $
postalAddress $ physicalDeliveryOfficeName $ ou $ st $ l))

Another objectclass concept is multi-parent attributes (Figure 5.5)

Figure 5.5 LDAP Multi Parent Attributes

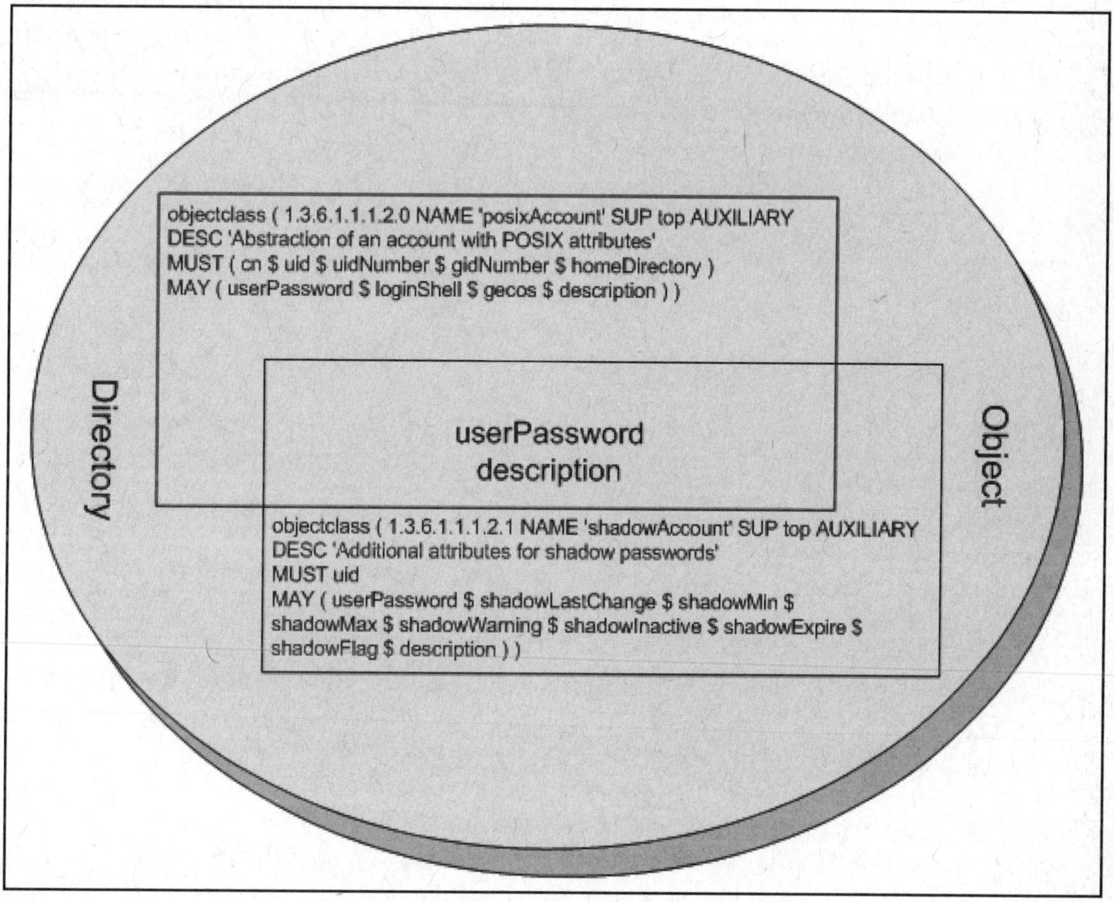

TIP

One thing most vendors agree on is the use of LDIF (Lightweight Directory Interchange Format) which is an ASCII file format used to exchange data and enable the synchronization of that data between vendor implementations. LDIF files are used to import and export information in the DIT and are comprised of lines using simple "attribute:value" notation. LDIF can be used to add, modify, or delete data in any compliant LDAP information store. Even Microsoft provides a tool for this purpose called ldifde (http://support.microsoft.com/kb/237677)

Fedora Directory Server in Detail

FDS will run on any machine (such as our virtual machine) with a minimum of 256 MB. You should however plan from 512 MB to 1 GB of RAM for best performance on production systems depending on the size of your network and expected load. You will need approximately 300 MB of disk space for a minimal installation. For production systems you should plan at least 2 GB to support the product binaries, databases, and log files (log files require 1 GB by default); 4GB and greater may be required for very large directories.

FDS is geared toward the RedHat and Fedora releases of Linux. This is not to say that this will not run on other distributions like Ubuntu or Novell SUSE; however the FDS project supplies precompiled binaries in RPM format geared toward the two primary distributions.

TIP

Ubuntu and Debian distributions can install FDS RPMs using the "alien" command. There is a how-to for these distributions here: http://www.directory.fedora.redhat.com/wiki/Howto:DebianUbuntu

Since there is extensive documentation online for FDS we will not spend too much time here discussing installation and configuration, the most complete set of documents is for RedHat Directory Server, which is the directory server officially supported by RedHat for its enterprise line of products. For the most part the documentation is interchangeable and applicable to FDS; see http://www.redhat.com/docs/manuals/dir-server/ for a complete list. We have deployed FDS in the ITSox2 VM Toolkit as the authentication "backend" for the CentOS operating system and all other application that require authentication, thus providing a single password to remember for all of the above. The following sections cover the rest of our exploration of FDS.

The Fedora Directory Server Console

Virtually all administrative tasks can be performed via the Java console application provided with FDS. In order to bring this up on the VM simply select this from the menu or in a root shell type:

```
cd /opt/fedora-ds
./start-console
```

After logging in with the administrator credentials (admin/letmein!1) you should see the console as depicted in Figure 5.6.

Figure 5.6 Fedora Directory Server Console

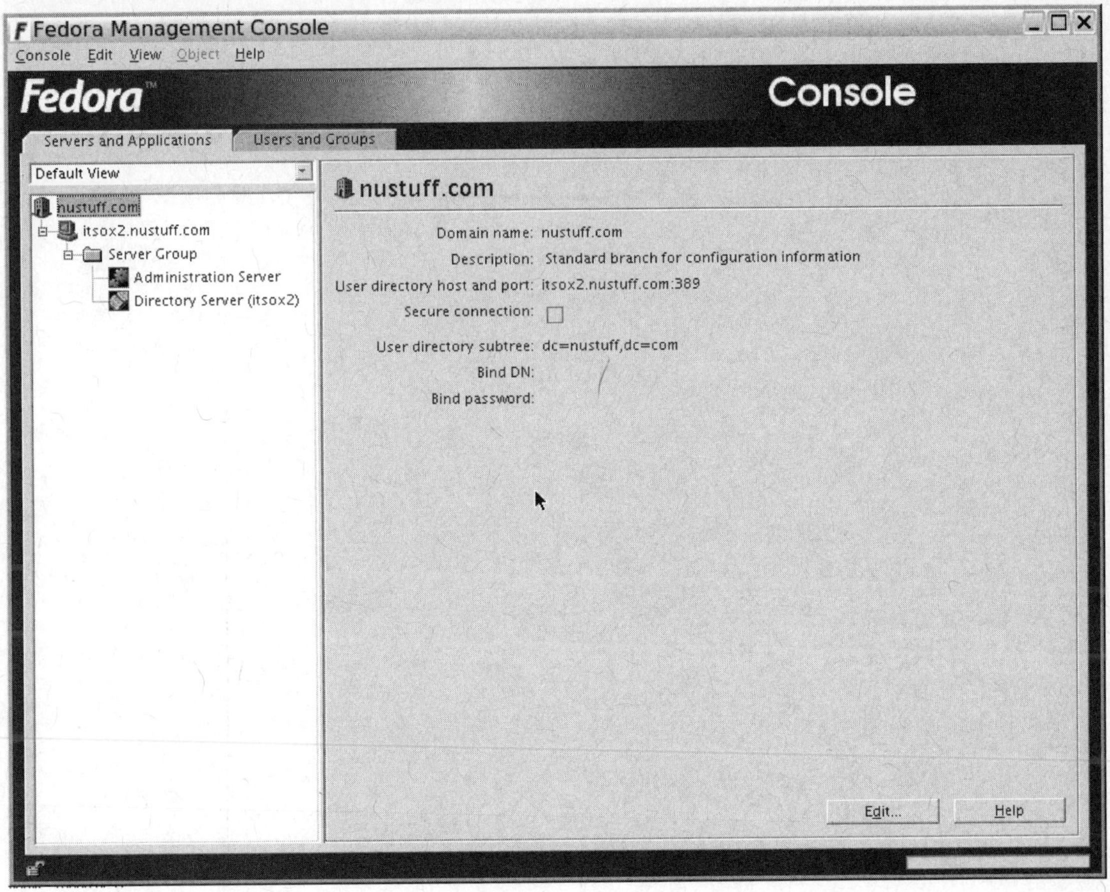

From here you can access both the Administration Server and the Directory Server instance for nustuff.com. The Administration Server gives you the ability to manage the local LDAP instance such as stopping and starting the services, configure logging and manage server certificates. The more interesting pieces for our discussion lie in the Directory Server instance, where you can manage all aspect of the information store, even if pieces of this were to reside on other physical servers. Here is a closer look.

Managing Fedora Directory Server

In a bit of redundant functionality the first tab of the instance view allows you to manage the running instance that you connected to. It is essentially a clone of the Administration Server functionality and provides access to all server based management tasks (Figure 5.7).

Figure 5.7 Fedora Directory Server Tasks Tab

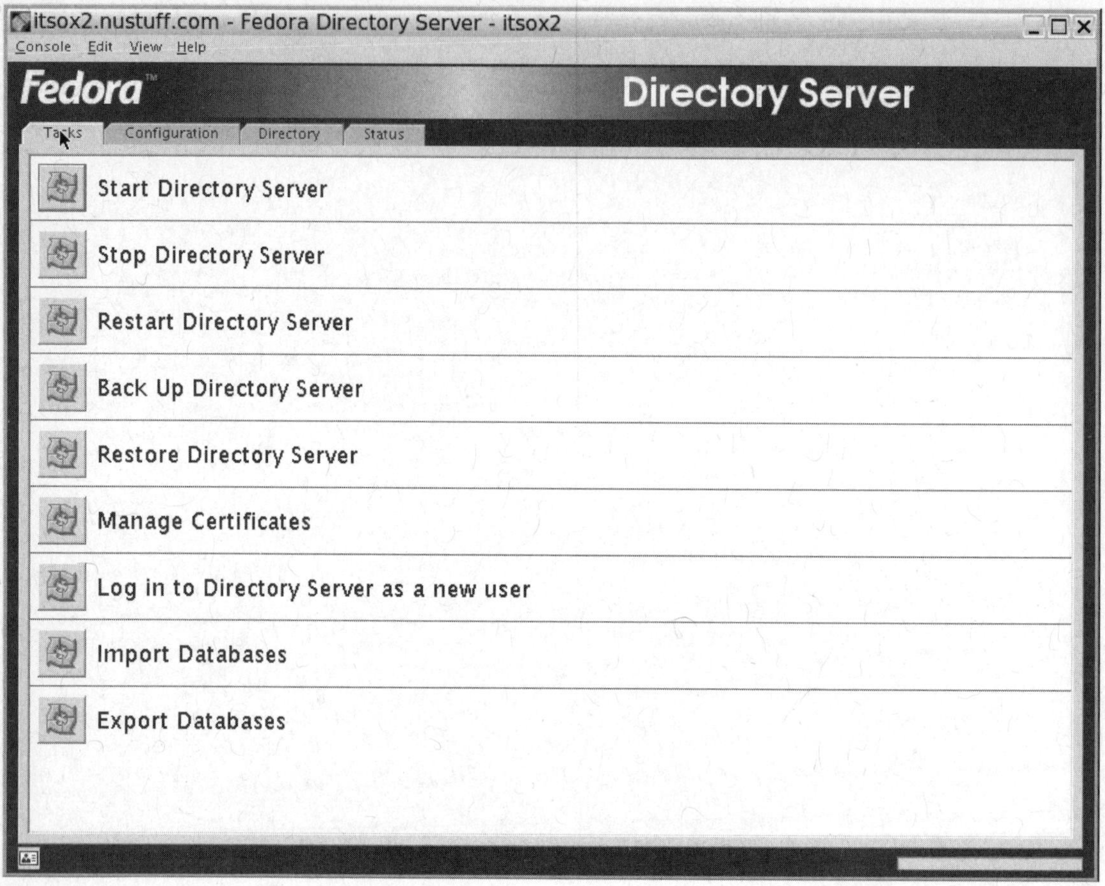

You can start, stop or restart directory services as well as backup and restore the information store. You can also manage server certificates so that you can secure client connections with either SSL or TLS encryption. One more interesting note is the import and export features listed here use the LDIF format for maximum interoperability with other LDAP vendors. In fact almost every aspect of FDS can be managed or manipulated using LDIF files!

Configuring Fedora Directory Server

The configuration tab is where you define and manage your schema, replication targets, logging options and plug-ins (Figure 5.8).

Figure 5.8 Fedora Directory Server Configuration Tab

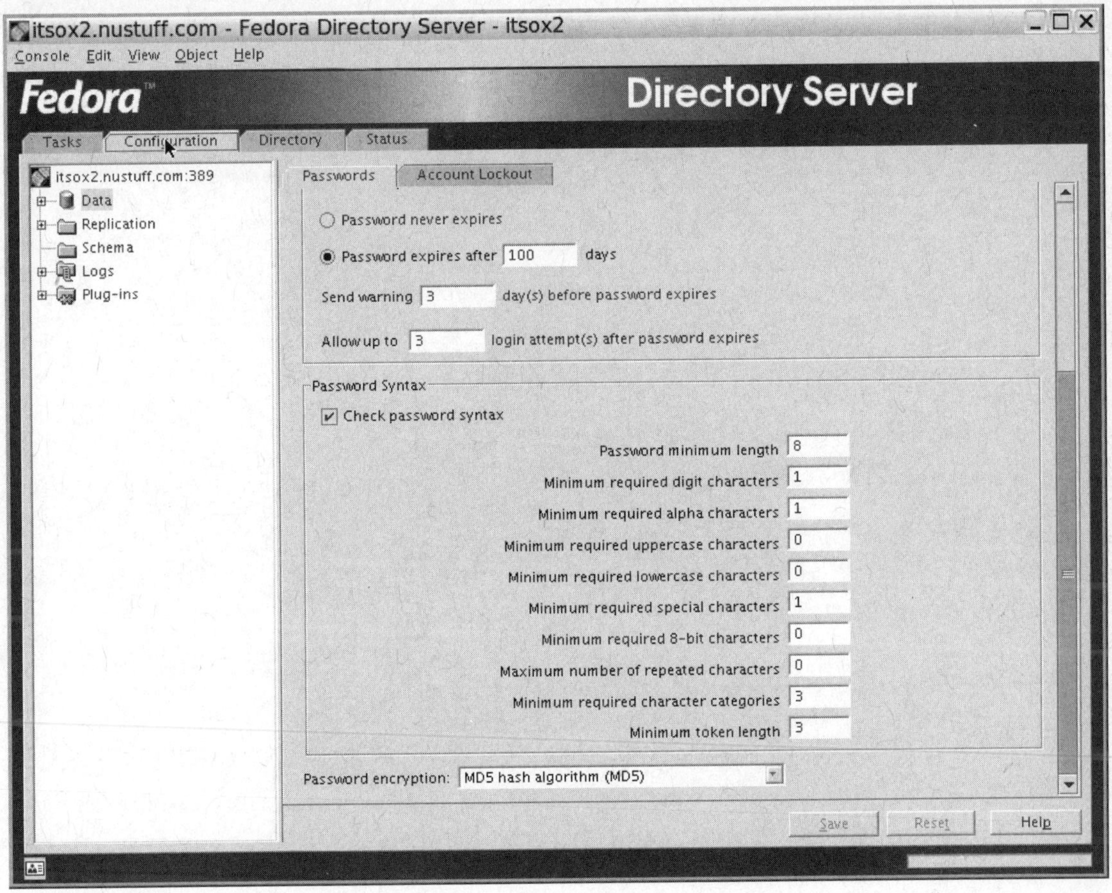

Speaking of passwords, what good is having a corporate policy defined if you cannot enforce it? FDS provides complete policy options for virtually every conceivable diabolical policy requirements you care to impose via the configuration tab, starting with the Global Password Policy. Password policies in FDS are very flexible and can be defined at the global, subtree, or even individual entry level. Figures 5.8 through 5.11 shows all of the options that can be set for your policy, in this case the Global policy is displayed when you click on the "data" node. As you can see it is easy for us to implement (and enforce) the policy we defined earlier in the chapter.

Figure 5.9 Fedora Directory Server Password Policy Detail

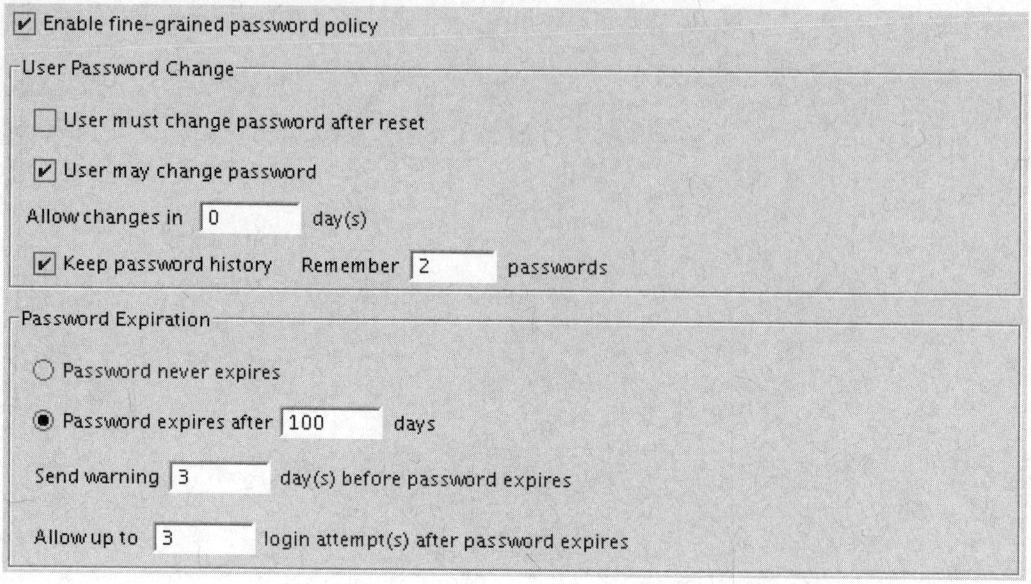

This section provides the following options

- Enable fine-grained password policy – turns on or off policy checking in general
- User must change password after reset – if this is selected and administrator changes an entry password, the next authentication attempt will require the password to be changed
- User may change password – if this is *not* selected the user will not be able to change his or her password after an administrator set it
- Password Expiration – turns options turns on or off password expiration after X number of days
- Send Warning – a properly configured client can be warned of an impending expiration if this is set
- Allow up to X login attempts – will allow a properly configured client to log in for a specified number of times before locking out

Figure 5.10 Fedora Directory Server Password Syntax Checking

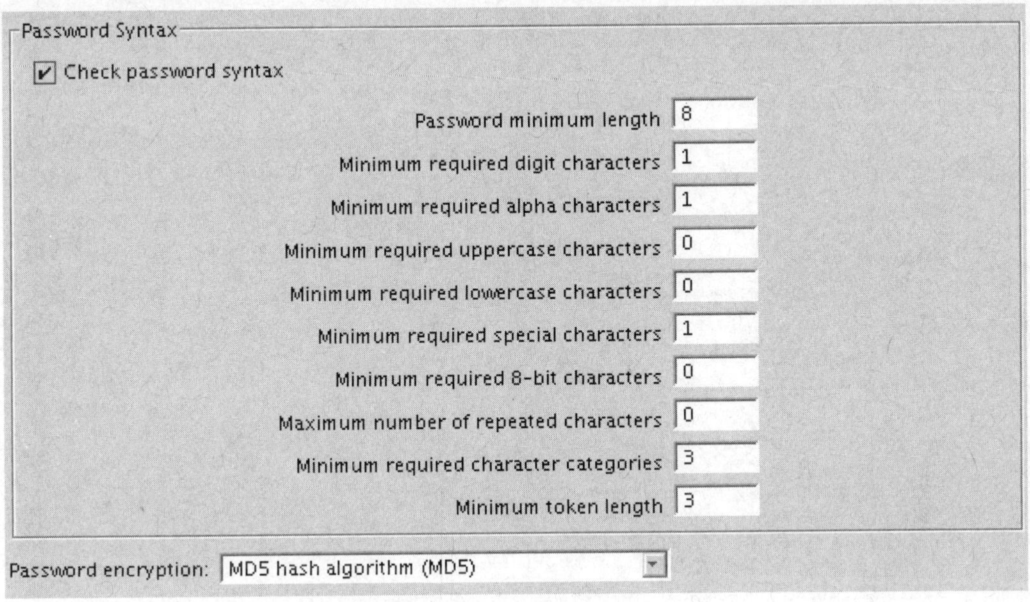

This section provides password quality checks:

- Check password syntax – turns on or off quality checks in general
- Password minimum length – defaults to 8
- Minimum required digit characters – 0–9
- Minimum required alpha characters – a–z, A–Z
- Minimum required uppercase characters – A–Z
- Minimum required lowercase characters – a–z
- Minimum required special characters – !@#$, etc.
- Minimum required 8-bit characters – characters beyond A–Z and 1–9 in the ASCII set
- Maximum number of repeated characters – times the same char can be immediately repeated such as aaabbb
- Minimum required character categories – categories are lower, upper, digit, special, and 8-bit
- Minimum token length – The smallest attribute value length that will be used for "trivial" words checking (i.e. a user named "bob" cannot use "bob" in his password)
- Password Encryption – the algorithm used by the server to ultimately store the password.

Figure 5.11 Fedora Directory Server Account Lockout

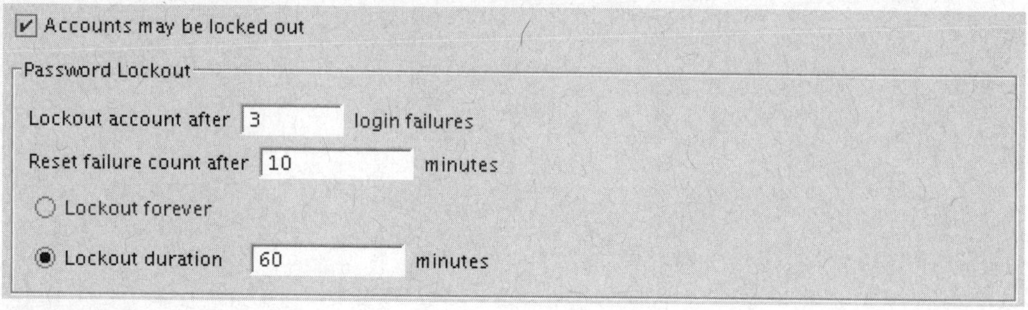

This section defines the behavior for locking accounts:

- Accounts may be locked out – turns on or off locking in general
- Lockout after X login failures – repeated failures that cause a lock condition
- Reset failure count after X minutes – resets the failed count after specified duration
- Lockout forever – requires an administrator to unlock the account
- Lockout duration X minutes – automatically unlocks and account after specified duration

WARNING

Although clients who wish to authenticate against FDS can send the encrypted password to the server for comparison, in order for password policies to work properly a password *change* request must be passed to the server in clear text. This is because the server cannot decode a previously encrypted password to see if it violates any of the policy settings. Due to this fact it is absolutely crucial that you employ the use of SSL or TLS to provide on-the-wire security so that anyone using a network traffic "sniffer" cannot see the password as clear text. We have implemented TLS on the ITSox2 Toolkit VM to demonstrate this concept and the FDS website provides a comprehensive SSL/TLS how-to (http://directory.fedoraproject.org/wiki/Howto:SSL)

Viewing and Updating the Directory

The next tab provides the mechanisms for managing the actual directory entries. There are usually three main contexts in FDS, although you can define more (Figure 5.12). In our VM instance these are:

- o=NetscapeRoot – this is where you typically define administrative accounts and groups together with their preferences
- dc=nustuff,dc=com – this is the actual "root" of the DIT for managing our users and groups. You can define multiple roots for the server however clients typically connect to a single root at any one time thus roots are isolated from one another
- cn=config – this is where you turn on or off various server configuration settings and preferences. Most of the GUI functions of the console such as setting the Global password policy actually set configuration items stored here

Figure 5.12 Fedora Directory Server Directory View

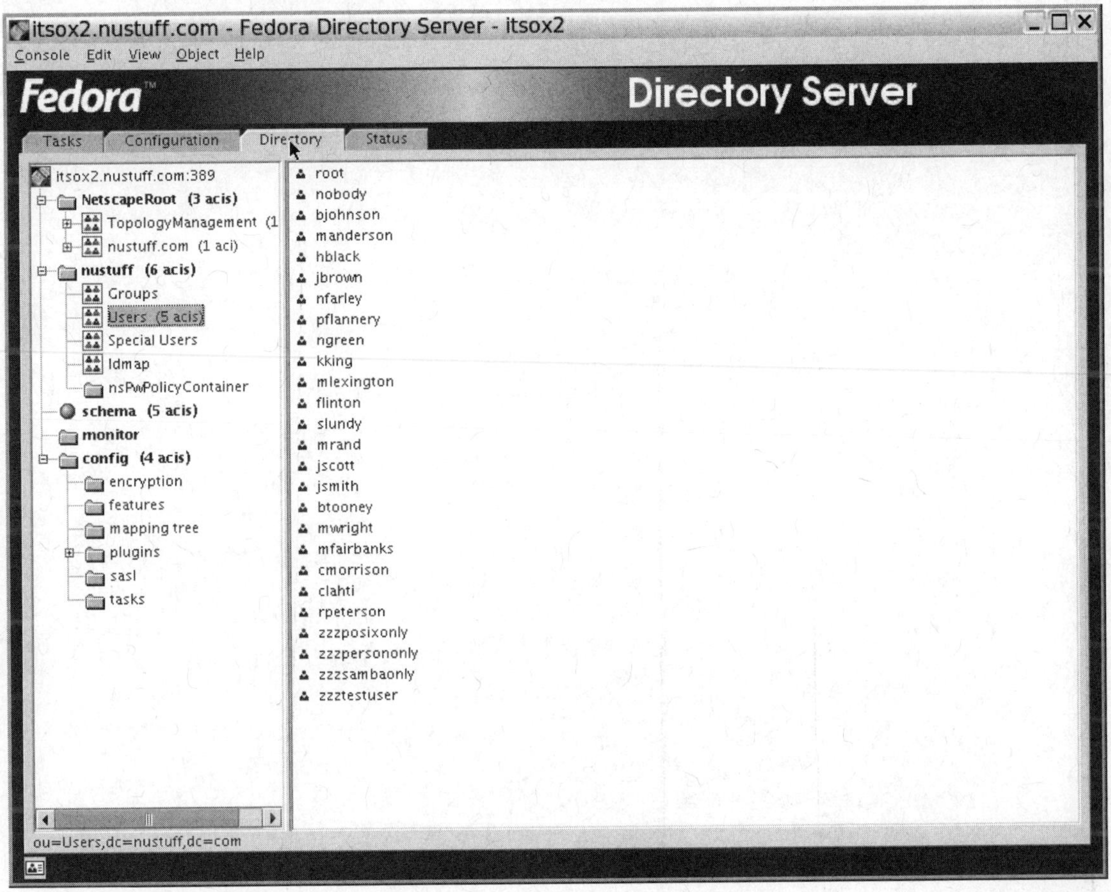

One important aspect of managing access to your directory is through the use of Access Control Information (ACI) objects. The console provides a mechanism for securing every aspect of this via the ACI editor depicted in Figures 5.13 and 5.14.

Figure 5.13 Fedora Directory Server ACI Visual Editor

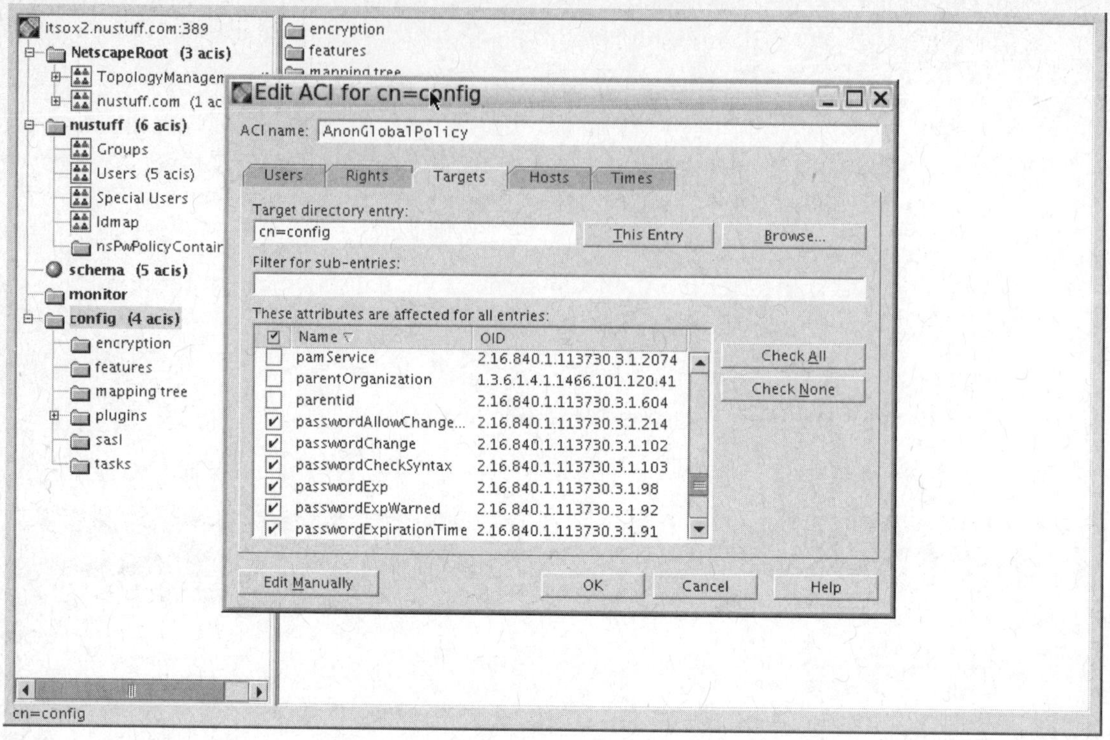

Figure 5.14 Fedora Directory Server ACI Manual Editor

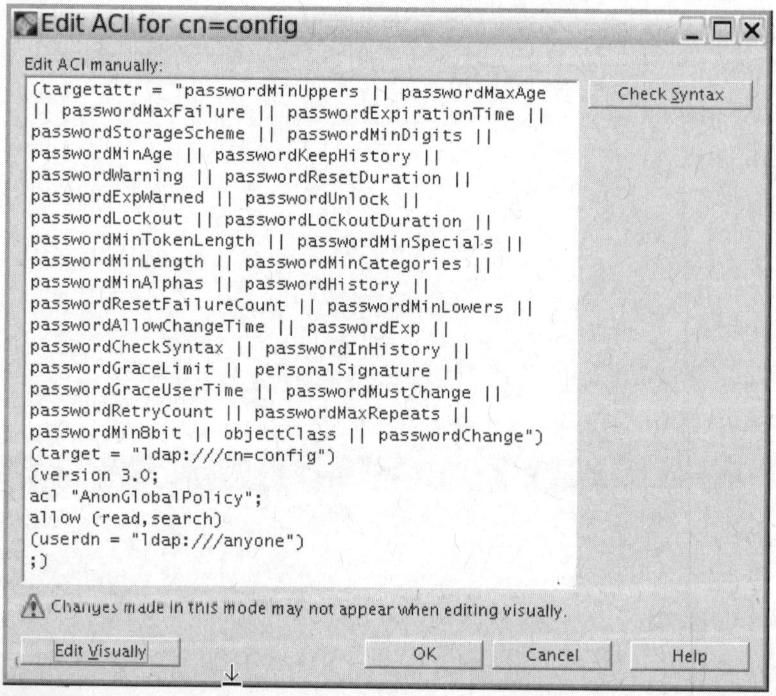

Using the editor you can manage user and host access to the DIT, even specifying which times these operations can be performed. Every attribute is available for securing access rights including:

- Read – a client can see the values of the targeted attribute(s)
- Compare – a client can perform a server-side comparison of the targeted attribute(s)
- Search – a client can determine if the targeted attribute(s) even exist
- Selfwrite – a client can write to the targeted attribute(s) if they are the "owner"
- Write – a client can modify the targeted attribute(s)
- Delete – a client can remove the targeted attribute(s)
- Add – a client can add the targeted attribute(s)
- Proxy – a client can authenticate as another user for this entry

We used the editor to add the ACI to grant read permissions to the Global password policy attributes under cn=config. This was necessary so that users can view the current policy in force for their account, and provides an excellent example of when it is appropriate to apply a custom ACI other than the defaults provided by a typical FDS installation.

Managing Users and Groups

Although the console does provide facilities for managing users and groups, we found the functionality somewhat lacking in certain critical areas (see sidebar). Due to this we did what any other proponent of Open Source would do, we rolled our own suite of management tools called the FDS Toolkit and released this as Open Source under the GPLv2. The project resides at SourceForge.Net, the premier website for hosting Open Source projects (http://fdstools.sourceforge.net).

When we originally restructured our IT infrastructure in 2000, our primary goal was use as much Open Source technology as possible and practical. At the time we chose to use OpenLDAP as our main authentication store since it was very mature and robust and there was much by way of community support surrounding the project. We also chose to replace our Microsoft NT domain entirely with Samba v2x so that we could eliminate for the most part all Windows servers in the environment but still provide domain authentication services to Windows clients. Along the way we also discovered many client tools in the Open Source arena that worked well with OpenLDAP in terms of access and management, including:

- OpenLDAP client – the suite of tools that are part of the OpenLDAP distribution itself provides excellent command line tools for searching and updating the DIT (http://www.openldap.org/)

- smbldap-tools – a very good LDAP toolkit to manage Posix and Samba attributes, written by IDEALX (http://sourceforge.net/projects/smbldap-tools/)
- phpLdapAdmin – a web based tool for managing LDAP directories, entries and attributes written in PHP. (http://phpldapadmin.sourceforge.net/)
- LDAP Browser – A java based tool for viewing and modifying the DIT (http://www.mcs.anl.gov/~gawor/ldap)

There are many other tools out there but these became our part of our standard arsenal of tools. Then along came SOX. We had reasonably good policies and controls in place and had written a web application for our administrators to maintain our particular DIT; however SOX added the need for us to demonstrate certain things such as password policy *enforcement.* Although OpenLDAP provides mechanisms for doing password quality checks this is somewhat difficult to implement. In spite of this we were able to get the controls in place we needed to pass our SOX audit. Then in 2006 we took a good look at Fedora Directory Server once this became Open Source. We really liked the features available in this very mature directory server, including multi-master replication for our overseas offices and the comprehensive management console. We then decided to replace our OpenLDAP implementation; however this led us to discover:

- Most client tools are geared toward either LDAP generically or OpenLDAP specifically. We found subtle and not so subtle problems with each of the above mentioned tools, mainly due to the Password Policies and ACI's we defined in our FDS implementation
- The FDS console is an excellent tool for managing the server itself, but there were no facilities to manage either Posix shadow accounts (the mechanism most Unix/ Linux client use for password expiration settings), or Samba V3 attributes (necessary for us to move our Samba domain to use FDS)

Since this was the case we looked at modifying some existing client tools such as the smbldap-tools which had served us very well with our OpenLDAP implementation. After a thorough review of the code, however we felt that it just didn't have many of the features we wanted such as logging and password policy management, and it would be a significant effort to mould this into the tool we wanted. Thus FDS Toolkit was born. (http://fdstools.sourceforge.net) The point of this discussion is that when evaluating the suitability of any application, whether it is Open Source or proprietary, functionality and the ability for it to be used by mere mortals should ultimately factor into any decision to deploy that product. The beauty of Open Source is that if you find something close but not quite, it is entirely within your power to change that by adding the functionality yourself.

The toolkit contains all the necessary functionality to manage users and groups in FDS, including both a command line tool and a web based GUI (we cover the GUI in more detail in chapter 7 when we spotlight Webmin). The command line tools reside in `/usr/sbin` and provide the following for user and group management:

- fdsgroupadd – provides the ability to add or remove FDS groups, optionally including Posix, Samba and AIX objectclasses
- fdsgrouplist – lists groups with flexible searching options
- fdsgroupmod – provides the ability to modify existing group attributes and behavior with regards to Posix, Samba and AIX objectclasses
- fdsgroupshow – provides group details including user membership
- fdsgroupstatus – provides the ability to view or change the status of groups, including complete removal from FDS
- fdsgroupuser – provides the ability to manage existing user memberships for groups
- fdspasswd – provides a mechanism for changing password attributes for to Posix, Samba and AIX objectclasses
- fdsuseradd – provides the ability to add or remove FDS users, optionally including Posix, Samba and AIX objectclasses
- fdsusergroup – provides the ability to manage existing group memberships for users
- fdsuserlist – lists users with flexible searching options
- fdsusershow – provides user details including group membership
- fdsuserstatus – provides the ability to view or change the status of users such as disabling and unlocking, including complete removal from FDS

NOTE

You must be root to use all of the above command line tools, with the exception of fdspasswd which resides in `/bin` and can be used by any system user to change his or her FDS password.

In Summary Fedora Directory Server provides an important piece of the ITSox2 Toolkit, providing a secure cross system and cross application authentication store with robust features that are directly applicable to IT SOX compliance. In fact since many applications both Open Source and proprietary can use LDAP as a means of authenticating users, FDS would make an excellent choice for the main centralized store for any IT environment in general, including windows via either Active Directory synchronization or Samba.

Case Study: Costs

As it was necessary for us to define NuStuff Electronics IT infrastructure environment in chapter 2 it is equally important that we now define the people and budgetary resource or lack thereof that as a small company NuStuff Electronics has at their disposal. As we discussed in chapter 2 NuStuff Electronics has office its corporate office in San Jose and remote offices in Boston, London, India, Japan and Singapore. The employee resources in each respective office are as follows:

- San Jose – two employees (IT Manager & Sys Admin)
- Boston – no IT employees (office supported remotely from Corp)
- London – no IT employees (office supported remotely from Corp)
- India – one employee (Sys Admin)
- Japan – no IT employees (office supported remotely from Corp)
- Singapore – one employee (Sys Admin)

Just as NuStuff Electronics, as it is with most small companies, has a shortage of IT staff it also has a limited budget. As a matter of fact the other than staff, regular hardware and software maintenance NuStuff Electronics does not have budgetary funds for any projects, especially one as cumbersome and complex as Sarbanes-Oxley compliance. The following is an excerpt from an article by Thomas Claburn in InformationWeek, published August 2, 2007:

"Since the Sarbanes-Oxley Act became law in 2002, the average cost of compliance for companies with less than $1 billion in annual revenue has risen by over $1.7 million to reach $2.8 million in 2006, a 171% increase since 2003 but about the same as 2005."

Although the costs associated with compliance with Sarbanes-Oxley seem to have stabilized in 2005, 2.8 million dollars is still a significant amount of money for a small company to expend. If you recall NuStuff Electronics has annual revenue of 60 million dollars and therefore 2.8 million represents approx 5% of NuStuff Electronics annual revenues. Whew, that is a lot of money for any company large or small. Anyway, we hope by now that if you didn't have an appreciation of the predicament as a small company you face that you at least have an awareness of the cost associated with Sarbanes-Oxley costs. It is these costs that drove NuStuff Electronics to seek out creative ways to achieve Sarbanes-Oxley compliance while simultaneously requiring minimal to no budgetary expenditures. With this goal in mind and already having utilized open source application in other areas of their IT infrastructure NuStuff turn to open source yet again to solve their business problem. For the reasons stated previously and others NuStuff decided to implement Fedora Directory Server in order to minimize compliance costs.

Old Habits Are Hard To Break

During the course of this book we will touch inevitably touch on various aspects of this chapter, as well as "Prevent Controls and "Detect Controls". However, if there is not a common understanding and goal within the IT Organization responsible for executing your existing and new processes and procedures, your audit will almost assuredly be destined to fail.

Although NuStuff Electronics IT Organization already had various procedures as with most IT Organizations they weren't strictly followed and needed to be modified. So as part of the process of preparing for their Sarbanes-Oxley audit NuStuff Electronics IT Organization updated and re-implemented and tested all of their procedures. Thinking they had done all that needed to be done NuStuff Electronics was fairly confident that they would not have any problems related to IT execution. They couldn't have been more wrong. As part of their audit one of the auditors required access to NuStuff's network in order to test a control so, he asked the Admin for the required access. Without a hesitation the Admin granted the auditor access to the network without following any of the procedures that had been communicated to the auditors. Fortunately for NuStuff this incident happen during their self-assessment phase and was correct prior to their real audit. The main point to remember is that you can never test too much to see if the processes that have been defined are the ones people are actually following.

The above lesson to be learned is as a result of NuStuff's 1st year compliance process and although the compliance processes have become easier with time the concept of compliance continues to be a struggle. Subsequently, there is this constant battle between old habits and new behavior; therefore it is critical that one is ever vigilante in reminding IT Staff and management of the "what" and "how" of Sarbanes-Oxley compliance.

Summary

In this chapter, we discussed the need for SOX compliance and the possible consequences of noncompliance—lawsuits, negative publicity for the company, and fines for executive management. Although it's important to understand the consequences, it is paramount that you realize that with planning, knowledge, and a great deal of work, it is possible to comply with the Sarbanes-Oxley Act, even in a small company.

We also discussed how the inclination to over implement was something that IT organization needed to be aware of and to avoid this inclination. To further our discussion we talked about what we described as the "The Human Factor" and "Walk The Talk." While perhaps not as interesting to IT people some of the other material we covered, it is just as vital that you evaluate and manage these items as they pertain to your environment. In "The Human Factor" section, we discussed reasons why employees, including IT staff—Quid Pro Quo, may not embrace the changes that SOX compliance may bring, and steps to try to effectively negate any resistance to change. Given the importance of change management, we re-emphasize these steps here:

- Over-communicate to the IT staff what you are doing, what their roles are, and the consequences of failure.
- Get buy-in on the plan and what needs to be done.
- Wherever possible, delegate decisions to the IT staff.
- Bear in mind that change is a process, and usually a slow one.
- Retain responsibility for reaching the ultimate goal.
- Lead by example.
- If necessary, demonstrate any consequences with staff and user community.

Finally, we discussed "Walk the Talk," and the importance of keep your policies/procedures and or controls as simple as possible. Also you have to be to sustain whatever you implement, so if you can't sustain it or won't, do not define it. Also included in this section was a review of an example of a procedure that can be used as part of SOX compliance and discussed some of the key attributes an authentication system should possess. Then we proceeded to discuss how most frameworks that are use today have roots back to some of the quality frameworks that were developed in the 1950s. From frameworks we moved into assessing your IT infrastructure and key points to consider when looking at IT personnel and potential open source applications.

Our VM Spotlight in this chapter was a comprehensive overview of the Fedora Directory Server and LDAP. As part of our discussion we talked about how Fedora Directory Server (FDS) is an enterprise-class Open Source LDAP server for Linux and some of the key capabilities of Fedora Directory Server we discussed:

- 4-Way Multi-Master Replication to provide fault tolerance and high write performance
- External Directory Synchronization – FDS can interoperate with other information stores, one example is user and group synchronization with Microsoft's Active Directory Server
- Secure Authentication and Transport – FDS has a robust security model and supports LDAP over SSLv3 and TLSv1. Secure passwords and integration with other and authentication mechanisms such as Kerberos are supported via a standard SASL interface (we discuss Kerberos and SASL in a bit more detail in chapter eight).
- Support for LDAPv3 – This is the current LDAP standard, mainly covered by RFCs (http://www.ietf.org/rfc) numbers 2251 through 2256.
- On-line, zero downtime LDAP-based update of schema, configuration, and management information.

Solutions Fast Track

SOX and IT

☑ On May 24, 2007, the PCAOB adopted Auditing Standard No. 5, to replace Auditing Standard No. 2.

☑ Over the last five years or so Sarbanes-Oxley has proven to be a financial drain on companies, not only small but large companies as well.

☑ It is the responsibility of the CEO and CFO of a company to provide the attestment and sign off on the company's SEC filing.

☑ CFO, CIO, and IT Directors must understand what compliance means.

☑ There are significant consequences of noncompliance.

Compliance Issues

☑ Even with new audit standard AS5 there will still be an inclination to over implement Sarbanes-Oxley controls.

☑ Companies needn't evaluate or have controls for some of the financial processes or processes that interface with their financial systems as long as the processes or interfaces are not material to the financial reporting process.

☑ Caution should be taken in the assessment process of your financial process so that you are better able to hone in on what processes are material in your financial reporting process.

Assessing Your Infrastructure

- ☑ Open source is platform agnostic, most prominent projects run on most platforms including Windows.
- ☑ Having a proprietary infrastructure does not prevent the use of open source for Sarbanes-Oxley compliance, many projects can be used to augment your environment and assist in making the compliance process less painful.
- ☑ Having open source already deployed can survive an audit; all IT environments have the same compliance requirements regardless of platform.

What's In A Framework

- ☑ At there core, COBIT and ITIL are no more than a quality discipline, much like PDCA, Deming, or Juran, of which the major objective is to have a closed looped process that drives continuous improvement.
- ☑ The define policy/control objective would drive the requirements, functionality, and configuration of a application or tool, which would in turn drive the functionality and configuration of a application or tool for monitoring/reporting, whereas the cycle continues to repeat.

VM Spotlight: Fedora Directory Server

- ☑ Fedora Directory Server (FDS) is an enterprise-class Open Source LDAP server for Linux.
- ☑ Formerly Netscape Directory Server, RedHat purchased this product in 2004 and after rewriting a few proprietary components re-released this under a somewhat modified version of GPL v2.
- ☑ Due to its Netscape roots Fedora's core code handles many of the largest LDAP deployments in the world and has been hardened by real-world use.
- ☑ 4-Way Multi-Master Replication that provides fault tolerance and high write performance.
- ☑ Scalability: capable of thousands of operations per second, tens of thousands of concurrent users, tens of millions of entries, hundreds of gigabytes of data.
- ☑ Extensive documentation, including Installation and Deployment guides.
- ☑ External Directory Synchronization – FDS can interoperate with other information stores, for example, user and group synchronization with Microsoft's Active Directory Server.

- ☑ Secure Authentication and Transport – FDS has a robust security model and supports LDAP over SSLv3 and TLSv1.
- ☑ Support for LDAPv3 – This is the current LDAP standard, mainly covered by RFCs (http://www.ietf.org/rfc) numbers 2251 through 2256.
- ☑ On-line, zero downtime LDAP-based update of schema, configuration, and management information – FDS can be fully administered and configured via standard LDAP operations.
- ☑ In-tree Access Control Information (ACIs) – FDS has a robust fine-grained security model in which access can be controlled for every aspect and element of the directory.
- ☑ Graphical console for all facets of user, group, and server management.
- ☑ Lightweight Directory Access Protocol (LDAP) provides a common language that client applications and servers use to communicate with one another.
- ☑ LDAP is the "lightweight" version of the Directory Access Protocol (DAP) used by the ISO X.500 standard.
- ☑ DAP, originally developed at the University of Michigan, gives any application access to the directory via an extensible and robust information framework, but at an expensive administrative cost.

Case Study: Costs

- ☑ The average cost of compliance for companies with less than $1 billion in annual revenue has risen by over $1.7 million to reach $2.8 million in 2006.
- ☑ Small companies still face unique financial challenges when complying with Sarbanes-Oxley.
- ☑ NuStuff Electronics utilized open source to achieve Sarbanes-Oxley compliance while simultaneously incurring minimal to no budgetary expenditures.

Frequently Asked Questions

Q: Are there any other repercussions if a company fails to comply with the Sarbanes-Oxley Act?

A: Yes. In addition to possible litigation and negative publicity, a company could possibly be de-listed if it fails to comply with the Sarbanes-Oxley Act.

Q: Can I really use some of my exiting policies?

A: Yes. You may have to make some slight modifications, but as long as they are documented and support a control, they can be used.

Q: As a result of the adoption of AS5 will compliance cost really come down?

A: AS5 gives companied the ability to reduce compliance costs but whether a company actual realizes cost reduction will depend on their implementation.

Q: Have penalties been applied to any company that did not comply?

A: Yes. For specifics you can visit the PCAOB site: http://www.pcaob.com/Enforcement/Disciplinary_Proceedings/index.aspx

Q: Other then cost can over implementing really be detrimental?

A: Yes, other then cost over implementing will be a barrier to sustaining your defined activities.

Q: I do not have to implement controls for all of my IT processes?

A: Yes, only those processes that are material to the financial reporting.

Q: Will people really be that a barrier to compliance?

A: Yes, but with careful management you can negate this issue by:

- Over-communicating to the IT staff what you are doing, what their roles are, and the consequences of failure.
- Getting buy-in on the plan and what needs to be done.
- Wherever possible, delegating decisions to the IT staff.
- Bear in mind that change is a process, and usually a slow one.
- Retaining responsibility for reaching the ultimate goal.
- Leading by example.
- If necessary, demonstrating any consequences with staff and use community.

Chapter 6

What's First?

Solutions in this chapter:

- The Work Starts Here
- What Work?
- Planning and Organization
- Working The List
- Policy Definition and Management
- VM Spotlight: KnowledgeTree Document Management
- Case Study: NuStuff Electronics

☑ Summary

☑ Solutions Fast Track

☑ Frequently Asked Questions

The Work Starts Here

"Most business men generally are so busy coping with immediate and piecemeal matters that there is a lamentable tendency to let the "long run" or future take care of itself".

–Gustav Metzman

We often are so busy "putting out fires," so to speak, that we find it difficult to do the planning that would prevent those fires from occurring in the first place. As a prominent educator has expressed it, Americans generally "spend so much time on things that are urgent that we have none left to spend on those that are important." When we ran across this quote in our efforts to ensure that we gave credit where credit was due we immediately knew that it was the right quote to use in Chapter 6. The words "business men" easily lend themselves to be substituted by "IT Professionals" and in doing so sums up the existence of most IT Professional and articulates why COBIT identified Planning and Organization as one of their Domains and conversely ITIL identification of similar components in Service Delivery. Because as part of daily life within IT organizations, fires generally tend to get the highest priority and unfortunately the areas where IT can really add value to a company, such as planning and documentation, are usually relegated to the back-burner.

In this chapter we will look at the numerous control objectives in the COBIT Planning & Organization domain and ITIL Service Delivery counterparts and based on our experience offer suggestions on how a small to medium size company might be able to reduce them to a manageable process. Prior to delving into the various control objectives of the COBIT and ITIL components, we would like to reiterate that although COBIT is the defacto standard which the majority of the audit firms have adopted and the practices defined within COBIT are generally good practices to have Most large companies would find the implementation and sustaining activities daunting if not impossible, this applies to ITIL as well. Therefore, as part of this chapter we will focus more on illustrating how with the appropriate processes and documentation a small to medium size company can effectively comply with relevant COBIT controls or ITIL guidelines while not over burdening their IT organization with controls to the point where documentation and paperwork become their main focus rather than providing value to the company.

Previously, we discussed the opportunity for a CFO, CIO, or IT Director to capitalize on their Sarbanes-Oxley Act compliance effort to position their IT Organization as a strategic advantage in their company; well it all starts here. If you were to ask the majority of the CFOs, CIOs, or IT Directors how IT was perceived at their company most of them would say Executive Management views IT as a necessary evil, non-value overhead, or even worse, they just fix computers don't they? But if you were to ask them, if it were an ideal world how would you like your IT Organization to be perceived, the answer would be vastly different. The majority of CFOs, CIOs, or IT Directors would say that they believe their IT Organization can be a utilized as a strategic advantage to the company, one capable of not only improving employee productivity but also capable of contributing to the bottom line of the company.

Now, we are surely not suggesting that the Sarbanes-Oxley Act is some sort of magic wand that will transform a company's opinion of its IT Organization regardless of its competence or effectiveness, but what we are suggesting is that if aforementioned issues are not barriers then COBIT or ITIL and SOX compliance could provide the bridge from a reactive day-to-day IT Organization to strategic IT organization. Given that policies are a part of every framework the majority of the works resides in this chapter, whether the framework is COBIT or ITIL because this where you will need to examine what policies, processes, and practices your company currently has (documented or not) and which one will be needed. Later in this chapter we will provide you with some examples or process and policies to assist you in your process.

What Work?

If there is one thing that we think needs to be over emphasized is the concept that COBIT, ITIL and SOX are distinctly different entities. Again, COBIT and ITIL are a set of platform agnostic guidelines developed to allow an IT Organization to implement BKMs (Best Known Practices) while Sarbanes-Oxley Section 404's focus is on internal controls over financial reporting, period. Therefore in conjunction with developing an overall IT strategy you will need to look at your various IT function to determine two things:

1. Which IT activities are relevant to your financial reporting process and
2. Of these function, which ones are significant t the financial report process.

This activity is not only critical for work within the COBIT Planning and Organization Domain and the ITIL Service Delivery, but it will also be paramount to your successful execution of the subsequent Domains. In addition to enabling you to define your work activities, it will also enable you to define the scope of your SOX audit. Defining the scope of your audit will not only enable you to keep your activities on track but more importantly it will enable you to keep your SOX audit and auditors on track, which will probably be the most difficult task.

The major tasks that you will need to accomplish are:

- Development of IT plans process
- Development of IT plans
- Determination of your IT activities on financial reporting systems
- Determination of significances of your IT activities on financial reporting systems

As every IT environment is different the goal of this chapter and subsequent chapters concerning the COBIT Domains and the ITIL Service Delivery is to illustrate that the COBIT and ITIL guidelines can and should be customized to achieve SOX compliance. Keep in mind, for the sake of illustration purposes, that the application of these guidelines for this chapter and subsequent chapters were applied to our fictitious company NuStuff Electronics. Consequently, the stated guidelines should be used as examples to guide you through the process of your planning and scoping activities as you determine the specifics of your environment.

Planning and Organization

In Chapter four we established a high level definition for the COBIT "Planning & Organization Domain" which was "Planning is about developing strategic IT plans that support the business objectives. These plans should be forward looking and in alignment with the company's planning intervals; that is, a two, three, or five-year projection". Also in chapter four we provided an overview of ITIL and broke ITIL down into two separate subcomponents. Now as part of Chapter 6 we will look at the specifics of each of the control objectives of the COBIT "Planning & Organization Domain", the corresponding ITIL categories and attempt to summarize and distill the various control objectives to lend themselves more to the structure of a small to medium size company.

Given the diversity of functions an IT Organization may support based on their particular company's requirements, it would be presumptuous of us to assume that we could create a one-size fit all template for you to use. So, that is not our intent but rather our goal is to give an example, based on our experience, as to how the COBIT and ITIL guidelines can be pared down for SOX compliance to better accommodate the resources and structure of a small to medium size company and enable you to apply the same concepts to your particular needs. To accomplish this we will use our fictitious NuStuff Electronics as the company that needs to come into compliance with the Sarbanes-Oxley Act. If a particular control objective of the COBIT "Planning & Organization Domain" or corresponding ITIL categories were not applicable to NuStuff Electronics and/or generally would not apply to a small to medium size company it has not been listed as part of the guidelines below. For a complete list of the COBIT guidelines, please see Appendix A for ITIL reference www.ITIL.com. Although some of the items identified in this chapter and subsequent chapters concerning the COBIT Domains or the corresponding ITIL components will not strictly adhere to the guidelines previously given, as stated as one of the main objective of this book, they will assist you in repositioning your IT Organization.

Prior to proceeding to we would establish that the following information as it relates to the correlation of COBIT and ITIL has been derived based on our Sarbanes-Oxley compliance experience, subsequently there may be more areas that can be correlated. We would also like to reiterate that although NuStuff Electronics elected to deploy COBIT and we are structuring Chapters 6, 7, 8 and 9 with the COBIT Domains, we are not endorsing COBIT as a framework but merely utilizing the COBIT Domains to provide structure for our correlations. Further, in our correlation process we observed cased where we did not see a correlation to COBIT from ITIL, these areas will be clearly identified as one that you will need to augment ITIL to achieve Sarbanes-Oxley compliance.

We begin with Section 1 which deals with the overarching strategy and planning process. Although none of the items in this section can be eliminated as part of the effort to simplify the

Define a Strategic IT Plan control objective, it may not be as complicated and time consuming to satisfy as may be suspected.

COBIT 1. Define a Strategic IT Plan	ITIL	Guidance	Objective
1.1. IT as Part of the Organization's Long- and Short-Range Plan	The Business Perspective, 4. Business/IS Alignment ICT Infrastructure Management, Design and Planning, 2.5 The process and deliverables of strategic planning	As part of the overall business objectives IT issues and opportunities are assessed and captured in the overall business objectives. If your IT Organization currently has no input into planning process this may be an opportunity to correct that issue.	Repositioning
1.2. IT Long-Range Plan	The Business Perspective, Business/IS Alignment, 4.3 The management governance framework	IT long-Range plans should be developed with input from business process owners. Regardless of the process you will need to ensure that your IT plans are link to the business process owners.	SOX & Repositioning
1.5. Short-Range Planning for the IT Function	ICT Infrastructure Management, Appendix C.1, General Planning Procedures and Preparation	IT Short-Range plans should be a subset of the IT Long-Range plans. Again, the business process owners should be in agreement with the objectives, timing and results.	SOX & Repositioning

Continued

COBIT 1. Define a Strategic IT Plan	ITIL	Guidance	Objective
1.6. Communication of IT Plans	ICT Infrastructure Management, Annex 3C, Example of a Communications Plan	Although COBIT states that Management is responsible for communicating IT long-range and short-range plans to business process owners, by assuming this role you can capitalize on another opportunity to change the image of your IT Organization.	Repositioning
1.7. Monitoring and Evaluating of IT Plans	ICT Infrastructure Management, The Management Processes Involved, 2.4.5 Reviewing and evaluating progress of the plan	This is control work in conjunction with 1.7, except now you are measuring your organization against what you communicated you'd do and reporting the results back to the business process owners.	Repositioning
1.8. Assessment of Existing Systems	Service Support, Planning the Implementation of Service Management, 11.3 Assessing the current situation ICT Infrastructure Management, Design and Planning, 2.8 The planning and implementation of new technology and services	Whether for SOX, COBIT or budget planning the first step in moving in any given direction is to determine where you are.	SOX & Repositioning

Section 2 deals with the storage of financial data such as databases, spreadsheets, etc.

COBIT 2. Define the Information Architecture	ITIL	Guidance	Objective
2.3. Data Classification Scheme	No ITIL correlation this is an example of where you would need to develop a compliance component for use with ITIL or augment ITIL with the COBIT control.	Establish a structure for the classification of data i.e. categories, security levels, etc. Although this should be done, depending on your company it might be a monumental task. So keep in mind the scope of what you need to accomplish for SOX – systems that are materials in financial reporting process.	SOX
2.4. Security Levels	Applications Management, The Application Management Lifecycle, 5.2 Requirements	Security levels should be defined and maintained for any security access above general access such as systems that are material in financial reporting process.	SOX

Section 3 deals with standardization and the technology considerations of your strategic plan. Whereas items 3.2 and 3.3 would be nice to have for an IT strategic plan, they should not be critical for SOX compliance.

COBIT 3. Determine Technological Direction	ITIL	Guidance	Objective
3.1. Technological Infrastructure Planning	Applications Management, The Application Management Lifecycle, 5.5 Deploy ICT Infrastructure Management, Design and Planning, 2.5 The processes and deliverables of strategic planning	This can be incorporated into planning	Repositioning
3.5. Technology Standards	ICT Infrastructure Management, Design and Planning, 2.8 The planning and implementation ICT Infrastructure Management Technical Support, 5.4 The technical support processes	IT management should define technology guidelines in order facilitate standardization. Most IT Organization have long been pushing standardization and have met with various levels of success, if yours is one that has met with resistance, this is an opportunity.	Repositioning

Section 4 deals with organizational issues, and we have eliminated the obvious choices for our fictitious company since they do not uses contracted staff related to financial operations. Please make particular note of item 4.1, as it is very important to note that it stipulates IT Planning **OR** Steering Committee and not IT Steering Committee.

COBIT 4. Define the IT Organization and Relationships	ITIL	Guidance	Objective
4.1. IT Planning or Steering Committee	Business Perspective, Business/IS Alignment, 4.3.5 The IS steering group	There is no need to have a Planning or Steering Committee, but you should have a defined planning process.	SOX & Repositioning
4.2. Organizational Placement of the IT Function	ICT Infrastructure Management, ICT Infrastructure Management Overview, 1.6.1 Organizational structure	There is no requirement of SOX concerning IT and Organizational placement; however, COBIT does recommend that the IT Organization be place to ensure authority, critical mass and independence from user departments.	Repositioning
4.3. Review of Organizational Achievements	Service Delivery, Service Level Management, 4.5 The ongoing Process	Process to ensure the IT organization continues to meet the needs of the company. This process should be tied into Executive Management.	SOX & Repositioning
4.6. Responsibility for Logical and Physical Security	Security Management, Guidelines for Implementing Security Management, 5.2.1 The role of	There is no requirement to have a dedicated person in this role but there will be a need to define will the function resides – could be defined in job descriptions.	SOX

Continued

COBIT 4. Define the IT Organization and Relationships	ITIL	Guidance	Objective
4.10. Segregation of Duties	No ITIL correlation this is an example of where you would need to develop a compliance component for use with ITIL or augment ITIL with the COBIT control.	From a SOX perspective you should not have the approver of an action be the same person as the implementer. Depending on the size of your IT Organization you may need to add head count to address this issue.	SOX & Repositioning
4.12. Job or Position Descriptions for IT Staff	ICT Infrastructure Management, Annex 2A, ICT Planner and Designer Roles	Job descriptions should already be a normal part of your structure but if they are not you should develop them be able to better articulate the function with IT and they will also be need as part of SOX.	SOX & Repositioning
4.14. Contracted Staff Policies and Procedures	ICT Infrastructure Management, Annex 2A, ICT Planner and Designer Roles	Hopefully, there are standard processes in place at your company to protect its IP (Intellectual Property) if this is the case you should be able to extend it to meet this control - as it relates to the financial reporting process.	SOX

It is not our intent to suggest by the elimination of "Cost and Benefit Monitoring" and Cost Benefit Justification" that these are not important items or that an effective IT Organization would not need to be cost effective. But rather that in the context of what we are attempting to demonstrate that these items can be deferred and/or address as part of another project.

COBIT 5. Manage the IT Investment	ITIL	Guidance	Objective
5.1. Annual IT Operating Budget	Service Delivery, Financial Management for IT Services5.2 Budgeting	This control can be met by a standard company yearly budget planning process.	SOX

As stated in Chapter 4, SOX compliance requires two types of control, "Entity Controls and "Control Objectives". The significance of this as it relates to Communicate Management Aims and Direction is that with the exception of Control Objective 6.3. The remaining item should be addressed at the Entity level.

COBIT 6. Communicate Management Aims and Direction	ITIL	Guidance	Objective
6.3. Communication of Organization Policies	The Business Perspective, The Value of IT, 3.4 Establishing a value culture	Simply put you will need a mechanism for conveying organizational policies to the employees. This process can be accomplished via an Intranet site or a printed document. Whatever the method, it would be advantageous to capture the employees acknowledgement	SOX

Continued

COBIT 6. Communicate Management Aims and Direction	ITIL	Guidance	Objective
6.8. Communication of IT Security Awareness	No ITIL correlation this is an example of where you would need to develop a compliance component for use with ITIL or augment ITIL with the COBIT control.	Same as the above but as it relates to IT Security	SOX

Personnel issues have been covered in previous chapters however we will formalize some of these items for SOX compliance.

COBIT 7. Manage Human Resources	ITIL	Guidance	Objective
7.3. Roles and Responsibilities	No ITIL correlation this is an example of where you would need to develop a compliance component for use with ITIL or augment ITIL with the COBIT control.	Covered as part of job descriptions	Repositioning
7.5. Cross-Training or Staff Back-up	No ITIL correlation this is an example of where you would need to develop a compliance component for use with ITIL or augment ITIL with the COBIT control.	If yours is a typical IT Organization in a small company, cross-training is like documentation – something you will get to. Although not require by SOX cross-training is a control for COBIT.	Repositioning

8. Ensure Compliance with External Requirements

Although this section contains important business objectives if your organization must meet compliance requirements such as HIPPA or OSHA standards, there are really no items that require our examination for SOX compliance, so in this case we can safely eliminate this section for most businesses, but of course there will be exceptions. No ITIL correlation this is an example of where you would need to develop a compliance component for use with ITIL or augment ITIL with the COBIT control.

9. Assess Risks

Risk assessment from an IT perspective is also an important subject to undertake as a normal course of capacity planning and disaster recovery, however from a strictly SOX perspective these activities are actually covered elsewhere and do not particularly serve a purpose in the current discussion, so again we can safely skip this section for now.

As we have stated, Good sound Project Management will be crucial the success of your Sarbanes-Oxley compliance process. This statement might appear to be an oxymoron when you review this section however you must keep in mind that from a SOX perspective the removed items will not be germane.

COBIT 10. Manage Projects	ITIL	Guidance	Objective
10.1. Project Management Framework	ICT Infrastructure Management, Design and Planning, 2.4.4 Design and implementing a plan	Throughout this book good project management skills and methodologies have been touted. This where you would develop the methodologies surround this discipline within your IT Organization.	Repositioning
10.2. User Department Participation in Project Initiation	No ITIL correlation this is an example of where you would need to develop a compliance component for use with ITIL or augment ITIL with the COBIT control.	As with the planning process, user involvement will be critical. This should be defined as part of your methodology.	SOX & Repositioning

Continued

COBIT 10. Manage Projects	ITIL	Guidance	Objective
10.4. Project Definition	ICT Infrastructure Management, Design and Planning, 2.4.4 Design and implementing a plan	This should be defined as part of your methodology.	SOX & Repositioning
10.5. Project Approval	ICT Infrastructure Management, Annex 3B, Running a Deployment Project	This should be defined as part of your methodology.	SOX & Repositioning
10.11. Test Plan	No ITIL correlation this is an example of where you would need to develop a compliance component for use with ITIL or augment ITIL with the COBIT control.	Although the specific of your test plan should be incorporated into a project plan the overall methodology should follow your change management process.	SOX
10.13. Post-Implementation Review Plan	No ITIL correlation this is an example of where you would need to develop a compliance component for use with ITIL or augment ITIL with the COBIT control.	Should be defined as part of your project methodology and should be closed loop by inclusion of the requester as part of the review – formal sign-off would be recommended.	SOX & Repositioning

11. Manage Quality

Of course quality management is an important part of IT planning since it is the bar by which you validate your delivery goals of products and services to an organization. In the interest of keeping SOX control objects to a manageable list however there are no specific items in terms of compliance that are applicable in this section, with the possible exception of the testing of your environment and controls.

The Transparency Test

The CFO Perspective

As with anything you do, planning is the key to success. Planning becomes more critical the more complex an interdependant activity becomes. Given the stakes, time pressures and interdependancy of SOX compliance, it is clear that the first step is to properly analyze and plan your activities. As a small company, you will be tempted to short-cut this step – don't! I would argue that given the limited resources and high time pressure of implementing SOX in a small company, planning is even more critical than at a larger organization, which can afford to catch up with added resources. As the author points out, good practices are good practices no matter what name they are given and no matter what size your organization.

–Steve Lanza

Working The List

If you are a little confused why some items were eliminated and other weren't, please keep in mind that the COBIT and ITIL frameworks and SOX compliance are not synonymous. COBIT and ITIL Guidelines were developed prior to SOX and were intended to provide IT organizations with guidelines for "Best Known Practices", and the focus of SOX compliance is to ensure the accuracy of financial reporting data and/or the systems that support this data.

If you recall there was a note at the bottom of control objective four that pointed out a subtlety of item 4.1, which stipulates IT Planning OR Steering Committee and not IT Steering Committee. Although various sources may tell you that you have to have an IT Steering committee in place in order to comply with the Sarbanes-Oxley Act, this is not true. If your organization already has one in place, or if you want to put one in place as part of Sarbanes-Oxley compliance this is fine. If not, it merely means you will need to do a little more work to effectively address this area.

For example, since NuStuff Electronics does not have the type of organizational structure that lends itself to having an IT Steering Committee, this is where we will start. The first thing you will want to do is document your current process for capturing projects. If your IT Organization is like most, then project activity usually comes in via two mechanisms Management and Departmental groups. If this information has been captured and is readily available, then this is fantastic news. If not, you will need to gather as much information from Management and the department heads as possible so you can develop your overall IT plan. Once you have this information for your organization you will need to synthesize it and present it back to the originator for concurrence. As with any Quality Process and as stated above this process should be document as well and be a closed loop process. When you have accomplished the above it will be necessary for you to get buy-in on your IT strategy from Executive Management. This process can take the form of a formal presentation to a one-on-one meeting with the CEO, CFO, etc... Whatever the mechanism, make sure you capture it in writing as part of your process.

Assuming that we have completed the aforementioned steps, now let see how this information may translate into form and processes for SOX. Figure 6.1 below represent a Single Page Strategy that in part fulfills SOX requirements for the items COBIT control objectives 1, 3, 4, 5, 10 and the corresponding ITIL. Figure 6.2 below represent the NuStuff Electronics IT Roadmap that in conjunction with the Single Page Strategy should fulfill the remaining SOX requirements for the items in control objectives one, three, five and ten.

Figure 6.1 NuStuff Electronics Example Single Page Strategy

NuStuff Electronics Inc.

Information Technology

Single Page Strategy

Program	Server Capacity
User Champion	Big Boss
Owner:	Sam Customer
Program Manager	Joe IT
Project Priority	1

Purpose and Scope

The existing Engineering server environment may not adequately support the needs of the Engineering Group during the next cycle of product engineering, particularly if multiple product lines are under development during the same time. Subsequently, it would be advantageous for NuStuff Electronics to allocate the necessary funds to increase the existing server compute farm environment.

Benefits

Increased server should improve product development cycle time, as contention for resources will be reduced.

Additional benefits would be:

New H/W may allow NuStuff to leverage 64-bit processing where appropriate in the compute farm environment.

Impact

If NuStuff elects not allocated the appropriate budget to perform the server capacity increase product development cycle times may be artificially elongated. The elongated cycle times may result in a negative impact to Engineering's schedule and delivery dates.

Financial Summary

	Headcount	Expense	Capital
Sustaining	N/A	0	N/A
Server Capacity Increase	0	0	$50,000

Figure 6.2 NuStuff Electronics Example IT Roadmap

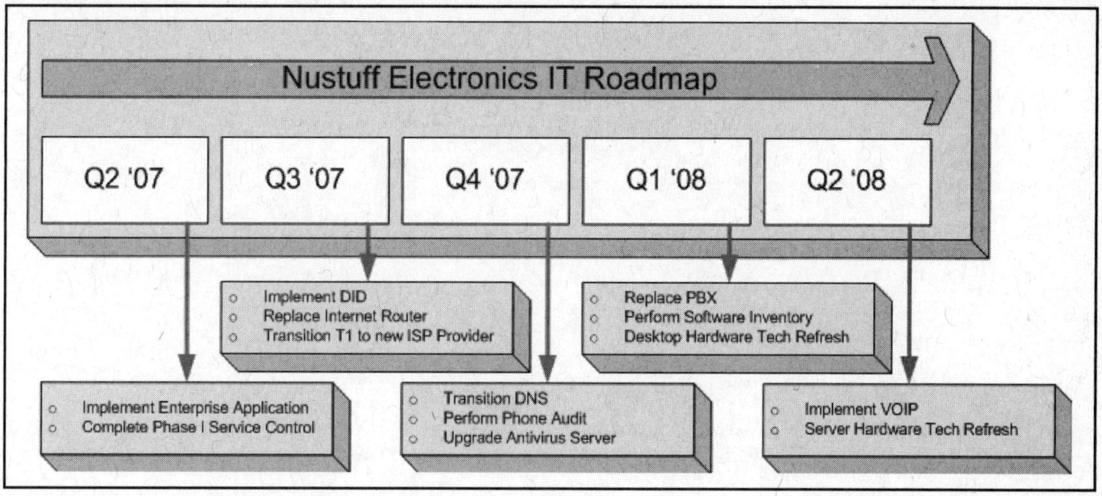

Believe it or not with the identification of two more documents we will have completed the remaining items in the COBIT "Planning & Organization Domain" and the ITIL Service Delivery. As a general rule, perhaps outdated, most companies will have organizational chart for the various departments. If yours is such a company, then you will merely need to update what you have, if not then you will need to created two new documents - functional & reporting organizational charts. Figure 6.3 below is a standard organizational chart based on reporting line structure. Figure 6.4 below may be a new one to some but it is a simple functional chart broken down by function verse people as with a standard organizational chart. With the development of the charts in Figures 6.3 and 6.4 we have now addressed the remaining organization item in number 4, which believe it or not completes the first COBIT domain.

Figure 6.3 NuStuff Electronics Example IT Organizational Chart

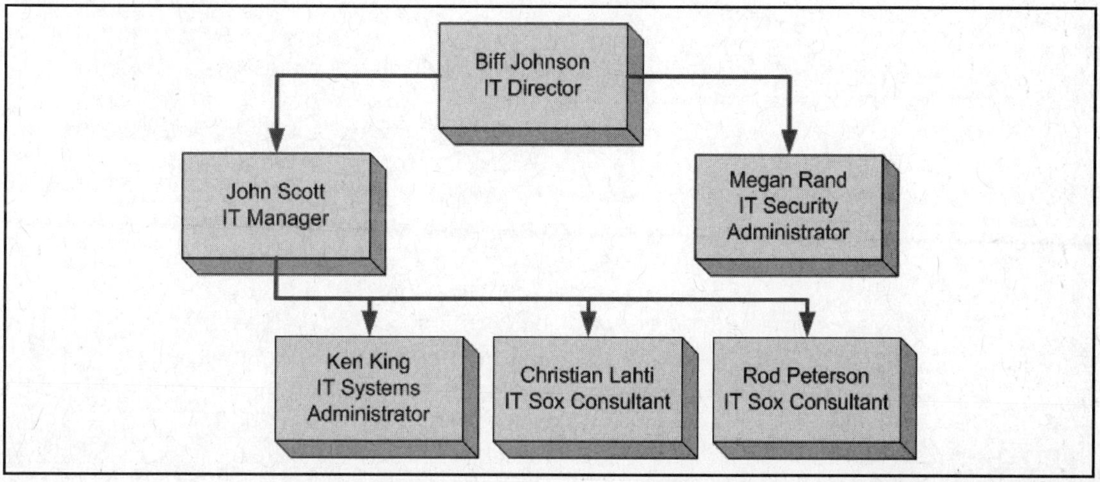

Figure 6.4 NuStuff Electronics Example IT Functional Chart

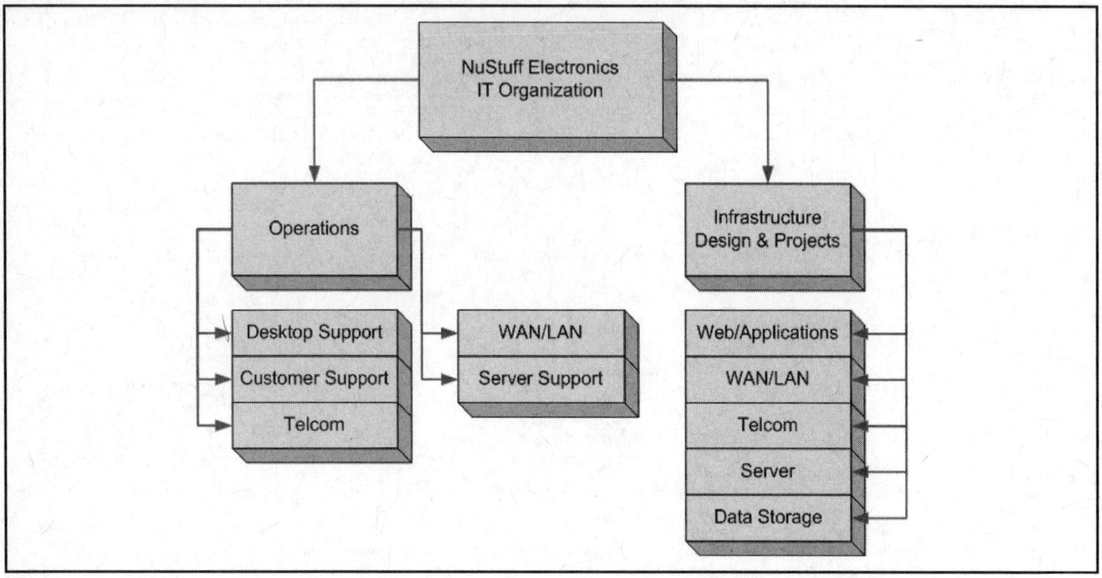

Now that we have discussed the basic process and forms needed for the first COBIT domain and the ITIL Service Delivery we now need to encapsulate them into an open source automated tool.

Lessons Learned

"I Have to do WHAT?"

During the SOX discovery process, my anxiety level reached heights not seen since those heady days when the company was started. But if it is not clear by now, this one fact should be. SOX compliance is all about ensuring financial reporting is accurate and reliable. If you have implemented your IT infrastructure well according to normal industry standards you should be ahead curve. However, there is a fair amount of work yet to be done to get through the audit. You will need to formalize and document processes and IT artifacts, but arguably that may be good thing in the long run. In fact, if you read between the lines here you'll realize the authors are presenting material for a foundation of good practices and open source tools to implement them with. This is great information, but what really eases my mind at this point is that they can be implemented over time. At least with the material presented here I know what is really required to pass the pending audit and what I can incorporate over time as part of an ongoing process improvement program.

–Matt Evans

Policy Definition and Management

https://itsox2.nustuff.com/egroupware/wiki/index.php

The first step in defining the scope of work for your SOX compliance is to define a set of policies as a company in support of the business objectives. While we could spend several chapters creating an exhaustive list of policies we have narrowed the list down to a few examples from which we will derive procedures, strategies and project plans for implementation throughout the book. We chose to utilize the Wiki eGroupware application to define the policies since each change has full document history, "Wiki-fied" HTML is easily learned and the very nature of a web application makes for an ideal distributed collaboration environment. You can apply these principles to your environment accordingly.

TIP

The Wiki supports access control and all documents in the Wiki application are readable by all persons in the system. A group called the "Policy Editors" has been created to restrict access to the editing of existing policies, consisting of those individuals who would be responsible for defining and refining the policy documents. The initial page does not have this restriction in place, so currently anyone can define a new Wiki page, but the editing of subsequent versions can and should be restricted. In your environment you should consider creating groups for every role that you might use Wiki for.

NuStuff Corporate Policy Documents

NuStuff has a much more complex environment; with multiple operating systems and platforms their policy documents are necessarily going to be more complex in some areas. You will notice however that the basic overall SOX compliance requirements are similar for both companies and as we have mentioned previously the scale is not linear. Here are the same policies for NuStuff that cover the same business processes; however several are expanded or subtly different in other ways.

Administrative Access Control Policy

The administrative access policy covers the access to servers, data and applications. This policy takes into account that NuStuff Electronic has a multi-platform environment.

Change Management Policy

The Change Management policy is one of the most complicated, since the scope of this document covers many practical implementation details. You will notice quite a bit of verbiage related to the procedural execution of the policy statements. While you might think that this should be left to the implementation pieces later on, we discovered that the auditors were happier with this level of detail when the procedures defined or clarified the policy statements.

Data Backup and Restore Policy

The backup policy defines standards for what gets backed up when, tape rotation, offsite storage, and restore procedures. The IT Manager John Scott was the person who drafted the policy, with Nancy Green then providing further feedback.

Firewall and Intrusion Detection Policy

Perimeter defense is an important ability to demonstrate from a SOX perspective since they represent entry points into the network and are prone to attack from external sources. A solid policy for dealing with protection and attack mitigation is one of the major issues you will deal with. A multiple layered approach of protection, detection and escalation works best.

Malicious Software Policy

This policy falls under the same category as the Server Environmental policy, ensuring uninterrupted and secure access to the company's financial data and other related systems. This policy is an example of one that was required by the AS2 standard, but as a result of the new AS5 standard it may be possible to negotiate with your auditor. Please keep in mind that we are not suggesting that operationally this policy should not be in place and/or implemented. But rather that as a result of the new AS5 standard your auditor may view it differently as it applies to Sarbanes-Oxley compliance.

Network Device Configuration Backup Policy

This policy is an example of a sub-policy related to change management. You will be required to demonstrate the ability to control change in your IT environment and quantify how, when and why any change occurred.

Network Security Monitoring and Controls Policy

This policy covers both security and availability of the financial systems and servers. Several sub-policies can be derived from this parent class.

Oracle New User Account Creation and Maintenance Policy

This policy governs the creation and maintenance of Oracle user account for NuStuff Electronics. As you can imagine since the potential for impact to the financial reporting process is high in regards to access to NuStuff Electronics Oracle system, the auditors scrutinized this policy very closely.

Oracle New User Password Policy

Since NuStuff Electronics has elected to outsource their Oracle environment they need to develop a separate policy for Oracle. However, for consistency the Oracle New User Password Policy follows the Password Control Policy closely.

Password Control Policy

This policy establishes the requirements for password management. This is an area where systems, automation, configuration files and logs will go a long with in demonstrating your controls to your auditor.

Physical Building Access and Budging Policy

This policy is an example of a subject that is related to IT in an indirect way; IT is not explicitly responsible for controlling access to the building, however can be impacted from a security standpoint if the building is breached. For this reason it falls under the scope of IT SOX compliance. This policy is an example of one that was required by the AS2 standard, but as a result of the new AS5 standard it may be possible to negotiate with your auditor. Please keep in mind that we are not suggesting that operationally this policy should not be in place and/or implemented. But rather that as a result of the new AS5 standard your auditor may view it differently as it applies to Sarbanes-Oxley compliance.

Server Room Access Policy

This policy is another example of the indirect relationship between operational responsibilities versus the operational consequences of a policy violation. You should look to other areas where IT may not own the process but might be adversely affected by violations to stated business policies. This policy is an example of one that was required by the AS2 standard, but as a result of the new AS5 standard it may be possible to negotiate with your auditor. Please keep in mind that we are not suggesting that operationally this policy should not be in place and/or implemented. But rather that as a result of the new AS5 standard your auditor may view it differently as it applies to Sarbanes-Oxley compliance.

Server Room Environmental Policy

This is an example policy for that relates to the reliability and availability of the critical systems, a core component of the IT SOX compliance mission.

System Security Policy

Locking down systems is important so that a baseline of operations can be established. By defining the approved system services change can be controlled and quantified, and hacking opportunities are diminished as much as possible.

Generic Template

Figure 6.5 contains a generic template example of an actual policy document used by NuStuff electronics. If as part of tour planning process you determine that you will need to develop a format for your policies you can you the below template as a guideline in your process. To this extent we have included a brief definition as to each element in the template.

Figure 6.5 Generic Template

NuStuff Electronics Company	
Information Technology	
Section	Depending upon the size of the manual section may or may not be of value
Policy Title	Name of policy i.e. Password Policy
Approved By:	Title and person who approved policy
Policy Number	Manual version of policy (can use revision control application)
Issue Date	Date the policy was issued
Supercedes	Policy name and version replaced by this policy
1.0 Overview	
General overview of the problem policy should address	
2.0 Purpose	
3.0 Scope	
The scope of policy i.e. could be based on department, geography, etc..	
4.0 Policy	
General overview of the problem policy should address	
4.1 General	
General statement of what is to be done and what is impacted	
5.0 Review	
Who reviews the policy and frequency of review	
6.0 Enforcement	
What happens if the policy is not adhered	

As part of the VM you will have not only an opportunity to see examples of the specific NuStuff Electronics policies as defined above but you will also have an opportunity and the ability to create your own policies based on these documents.

Spotlight: KnowledgeTree Document Management

KnowledgeTree™ Document Management Made Simple http://www.knowledgetree.com/

In the course of gathering information in the planning of your SOX audit all the way through to sustaining the momentum once you have achieved that goal you will inevitability amass a mountain of various documents. It is early in the process that you need to be considering how to deal with the long-term storage and retrieval of this information, so this is where some type of document management plays an important role. In all actuality you should consider a formal document management system (DMS) for the administration of all sensitive documents, for example financial information, company trade secrets, engineering data and customer documentation. This is where KnowledgeTree can be of tremendous value to your organization.

KnowledgeTree is a full featured, easy-to-use, and production-ready enterprise document management system which allows you to share, track, secure, and manage the documents and records your organization depends on. KnowledgeTree has a very active Open Source community its architecture allows you to easily customize and integrate this with any existing infrastructure, providing a flexible, cost-effective alternative to proprietary applications. Like eGroupware, KnowledgeTree is written in PHP and is available in three editions:

- Open Source Edition
- SMB Edition
- Enterprise Edition

KnowledgeTree SMB and Enterprise Edition are commercial editions of the Open Source project and are sold with commercial support and extended functionality that provides access to the KnowledgeTree document repository from Microsoft Windows Explorer and the Microsoft Office application suite. The Open Source edition is not lacking at all in features however, here is a short list:

- A central document repository with audited and version controlled document content
- Flexible document metadata definition, management and versioning
- Document authoring management and workflow
- Full-text indexing which allows you to search within document contents
- Security group and role-based security model and integration with enterprise directory servers such as Fedora Directory Server (the spotlight subject of the previous chapter).
- Plugin support for extending functionality

KnowledgeTree provides excellent documentation for the installation, deployment and use of all of its many features; we will take a somewhat brief look at some of major functionality in the next few sections. Current documentation is always available online at http://docs.knowledgetree.com/manuals/ug/index.html.

KnowledgeTree Web Interface

To bring up the web interface in the ITSox2 Toolkit VM simply launch a browser and click on the KnowledgeTree DMS link. You should be presented with the login screen as in

Figure 6.6. You can login as any of the "cast of characters" from our case study NuStuff, you might want to start with Biff Johnson as he has been granted administrative rights in the application (bjohnson/letmein!1).

Figure 6.6 KnowledgeTree Login Screen

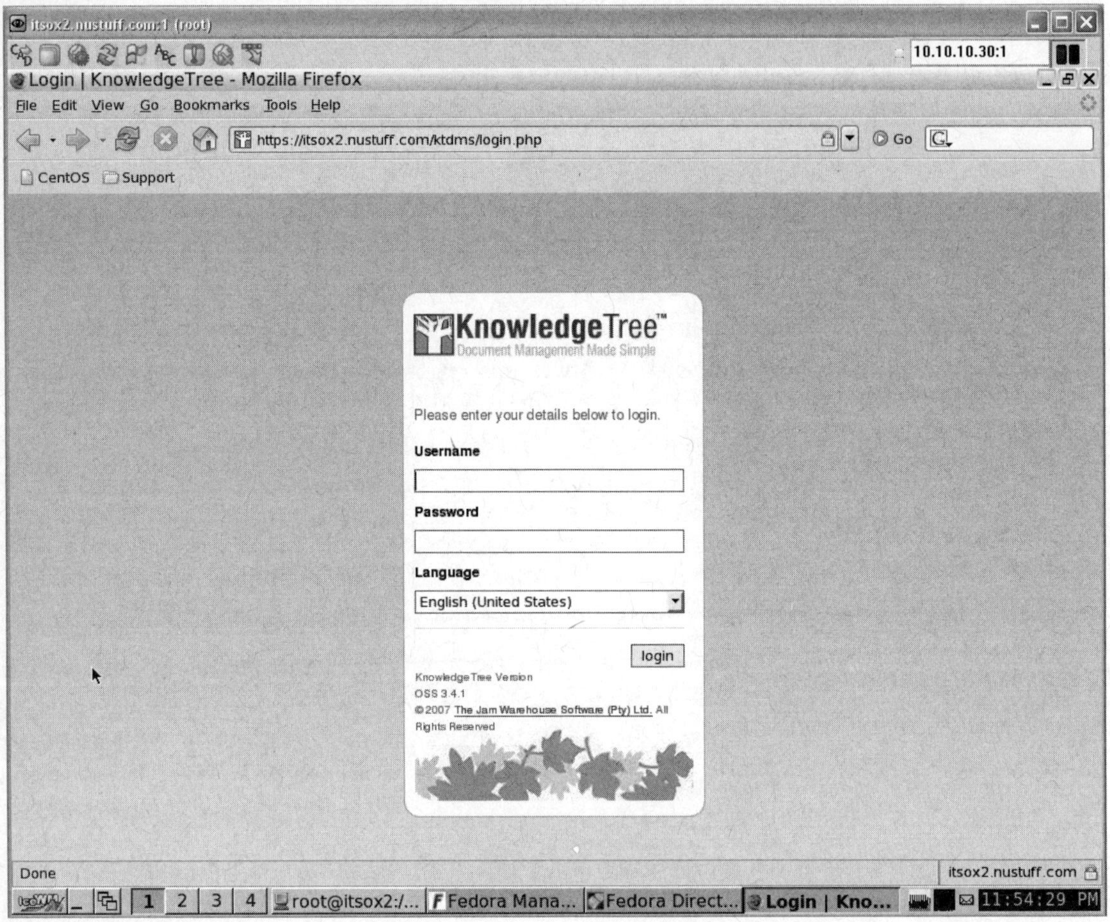

Once logged in there are three main views to the DMS, the Dashboard, Browse Documents, and Administration. You can also change your preferences here such as email notification options. With every page you are given a "breadcrumbs" navigation view located just below the toolbar so you can visually see where you are at any time and jump to a parent easily with one click.

The Dashboard View

The view you are initially presented after logging into the application is the Dashboard with gives you a comprehensive view of items related to your account (Figure 6.7).

Figure 6.7 KnowledgeTree Dashboard View

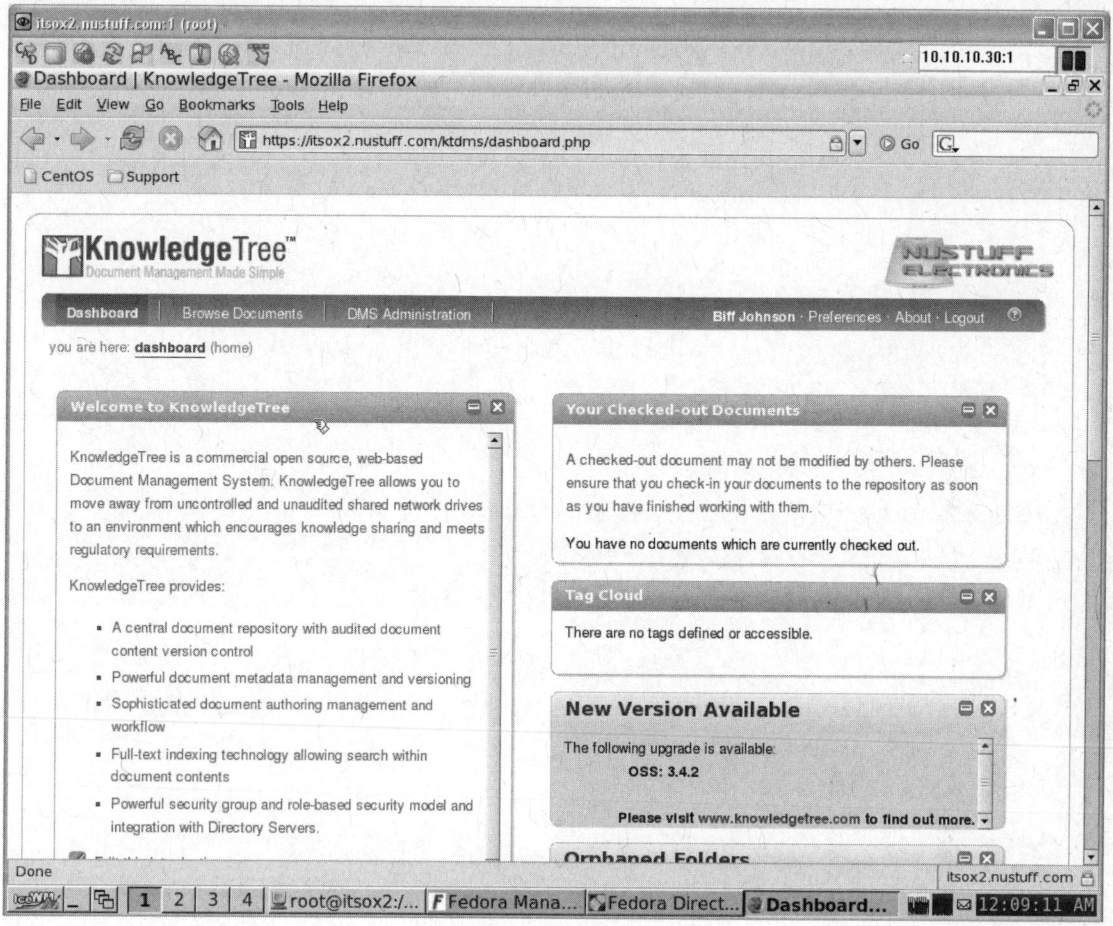

The Dashboard is divided into sections that you can rearrange via drag and drop or remove entirely depending on your personal preferences. Here is a brief description of each section:

- Welcome – this text appears for all users, and if you have administrative rights you can edit this to suit your organization's needs such as providing specific instructions and/or announcements.
- Your Checked Out Documents – if you have "checked out" any documents for editing (see later in the chapter) then a list of those documents will appear in this section.
- Tag Cloud – this is a new feature where you given a visual clue of the relevance of documents according to the frequency of tags (metadata) that are associated with documents.
- Orphan Folders – this provides a direct link to sub folders in the repository where you have appropriate access, however you *do not* have permissions to one or more

of the parent folders in the path to the sub folder, thus making an "orphan". In the Browse Documents view you would not be able to access this folder by drilling down since you need to navigate each parent, so this provides a convenient way to access the sub folder directly.

- General Metadata Search – provides a mechanism to search the repository.
- RSS Feeds – provides a "news" style interface for document activity such as check-ins and checkouts as well as new versions.

DMS Administration View

We are going to skip over the Browse Documents view temporarily to discuss the administrative features first. Figure 6.8 gives you a synopsis of the major administration categories for KnowledgeTree.

Figure 6.8 KnowledgeTree Administration View

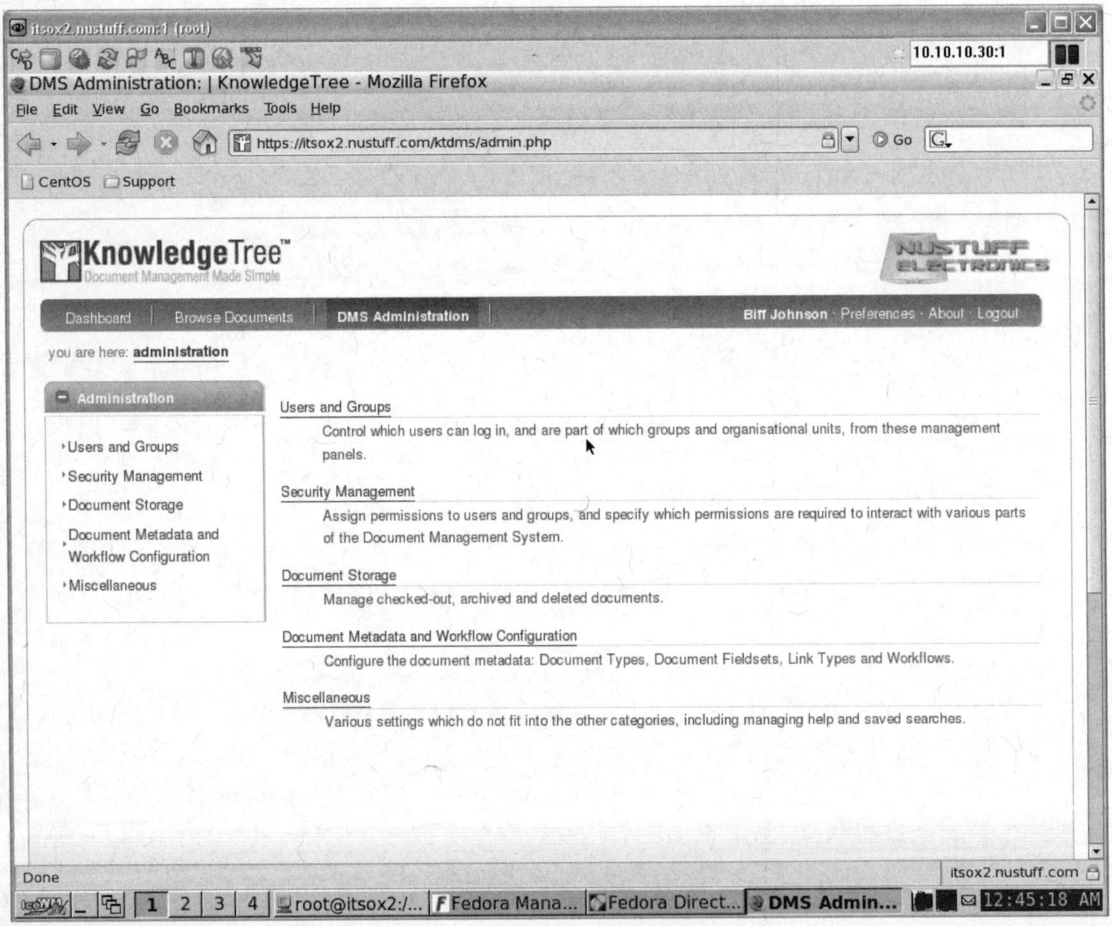

Each section has several administrative functions that you can manipulate.

Users and Groups

- Authentication – This is where you define both internal and external sources for authentication, including LDAP. We have "hooked up" KnowledgeTree to the running Fedora Directory server as the main authenticator for this application.
- Manage Users – this is where you either defined local accounts or pull in accounts from an external source. If you use LDAP for example, a user is created in KnowledgeTree however when they login the credentials are passed to the LDAP server for authentication.
- Manage Groups – is where you either defined local accounts or pull in accounts from an external source. If you use LDAP for example, existing group memberships are maintained. Keep in mind there is no problem adding LDAP users to locally defined groups. Groups may also be nested into sub-groups.
- Control Units – You can divide your document repository in to sections called Units, departments for example, and assign a user or group of users as having administrative rights to that individual unit. In this way you can delegate responsibility for administrative overhead to the people who should manage their own group's documents.

Security Management

- Manage Permissions – In KnowledgeTree you can define a set of permissions that can be assigned to a document folder in the repository. The application comes with standard permissions such as read, write, add folder, etc. but you can define more if your business situation calls for it. This is mainly for special permissions you might need to define for a plugin for example.
- Roles – these are used when defining workflows for documents. They differ from groups much like the roles in eGroupware's workflow application in that a role is a type of person or entity that needs to perform a function and a group is a list of actual people who fulfill that role. Three default roles exist: Creator, Publisher, and Reviewer.
- Dynamic Conditions – give the administrator the ability to define what permissions are applied to a folder based on a set of rules, which can be based on metadata, document contents, or transactional information. Dynamic Conditions are applied from the document folder's permissions section.

Document Storage

- Checked Out Document Control – this is where you may administratively override the "checked out" status of a document. This functionality is useful for example when a user goes on vacation but forgets to release the edit lock on a document.

- Archived Document Restoration – this is where you can restore previously archived documents.
- Restore or Expunge Deleted Documents – when a document is deleted within KnowledgeTree this in not a permanent action until an administrator comes here to permanently expunge them.
- Verify Document Storage – performs a check of the repository to make sure the information stored in the database matches the repository on disk, and reports any errors found.

Document Metadata and Workflow Configuration

- Document Types – here you define classes of documents that you want to capture distinct metadata for. The Document Types can be specified as generic, where they are applied to all documents added to the repository or individually selected when adding a document. Document Types contain one or more associated sets of Fieldsets.
- Document Fieldsets – these are the individual fields of information you want to capture as metadata. You can define fields as optional or required, and further specify the format as free form text, a list of items, or a tree of items to choose from.
- Workflows – a Workflow is the electronic representation of a document's lifecycle. It is made up of workflow states that describe where in the lifecycle the document is, and workflow transitions that describe the next steps within the lifecycle of the document. As we outlined in chapter 3 when discussing workflow as a concept, KnowledgeTree differs from eGroupware in that this is an Entity based workflow where eGroupware is an Activity based. There are a couple of generic workflows predefined in KnowledgeTree:
    1. Generate Document – defines a simple workflow that goes from draft, to final to publication.
    2. Review Process – defines a workflow that goes from draft, to approval, to publication.

- Link Type Management – it is possible for users to create links between related documents. Link types may include constructs such as "associated with" and "duplicated by" for example, and this facility is used to define and manage those relational types.
- Automatic Workflow Assignments – when using workflows this is where you configure how documents are allocated to workflows automatically. Assignment may occur on a per-folder or per-document type basis and only one mode may be selected for the system, we chose to activate the folder level for ITSox2 Toolkit.

Miscellaneous

- Edit Help files – you can customize your own context sensitive help screens that are presented to users.
- Saved searches – you can define a set of specific search criteria and save this for later use.
- Manage saved searches – these are the Saved Searches that you can make available by default to all users.
- Manage plugins – KnowledgeTree has nice plugin architecture and this is where you manage the registration of new plugins and activate or deactivate existing plugins. There are several Open Source plugins available developed by the community, see http://forge.knowledgetree.com/ for more information.
- Support and System information – this gives you information about this system and how to get support.
- Manage views – this allows you to specify the columns that are to be used by a particular view (e.g. Browse documents, Search).
- Edit Disclaimers – you may change disclaimers displayed on login and at the bottom of each page.

DMS Administration View

The Browse Documents view is where all document activity takes place. You are presented with a list of folders that you have permissions to in a drill down style as shown in Figure 6.9. Menu items are added or removed as applicable depending on where you are in the tree at the given moment.

Figure 6.9 KnowledgeTree Browse Documents View

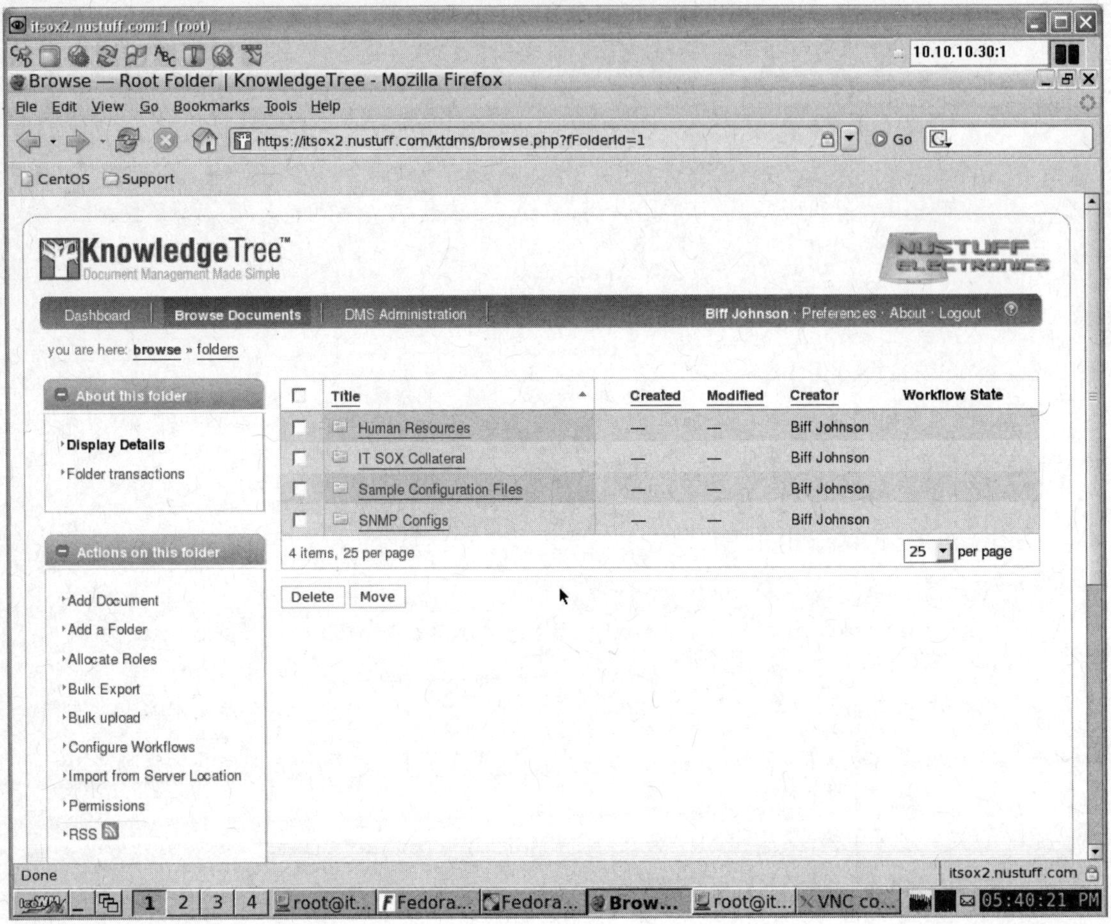

Folder Details and Actions

- Display Details – this displays the documents in the currently selected folder as well as any sub folders.
- Folder transactions – this displays the history of actions that have occurred on this folder.
- Add Document – add a document to the currently selected folder.
- Add a Folder – add a sub folder to the currently selected folder.
- Allocate Roles – this is where you would assign roles to be associated with this folder's documents. If any workflows exist that use this role then the role members will be notified when the workflow reaches the state in which they need to take action.

- Bulk Export – export an entire folder of documents, the server compresses the entire folder and its contents, including any subfolders as a zip archive and presents this to the browser for downloading.
- Bulk Upload – import a zip archive of files and folders into the system. All files that are placed into the repository with share the same metadata information you attach to the zip file when importing.
- Configure Workflows – this is where you assign automatic workflows to a folder. For example if you assign the reviewer workflow then any document added to this folder will automatically be in draft state and transition to the "approver" role. Members assigned to the approver role will be notified to approve the document for publishing or not.
- Import from Server Location – this is similar to the Bulk Upload feature in that it is designed to place multiple files into the repository at the same time, however this accepts a path rather than a zip archive. The entire tree of files starting with the base path will be imported and share the same metadata information you attach.
- Permissions – this is where you assign the various permissions for the folder access. Note that you can assign permissions to either the folder or each individual document. The default behavior is for files and folders to inherit the permissions of the parent. Default assignable permissions are Read, Write, Add Folder, Manage Security, Delete, Manage Workflow, Folder Details, and Rename Folder.
- RSS – you can copy the provided link into any RSS aggregator to create a feed to the selected folder.
- Rename – this allows you to rename the folder.

Document Information and Actions

- Display Details – this displays the details of the document in the repository, including all document version, metadata information that has been attached, creation and modification dates, owner, and tag cloud.
- Download – allows you to download the document.
- Permissions – this is where you assign the various permissions for the folder access. Note that you can assign permissions to either the folder or each individual document. The default behavior is for files and folders to inherit the permissions of the parent. Default assignable permissions are Read, Write, Add Folder, Manage Security, Delete, Manage Workflow, Folder Details, and Rename Folder.
- Transaction History – this displays the history of actions that have occurred on this document.

- Version History – this displays the check-in history of the document. KnowledgeTree by default internally assigns a version number to each document and then increments that number as new versions of the document are added to the repository, starting with 0.1 and then 0.2 and so on.
- View Roles – this allows you to view the roles currently associated with the folder this document lives in.
- Archive – this function allows you to move the document out of the production repository and into a special archive section, thus retiring documents that are no longer needed. The document is still visible in its original location by administrators who can subsequently choose to restore this back to production at any time.
- Change Document Ownership – this will reassign the current owner of the document. Only administrators and current owners have access to this function, and by default the owner of the document is the person that created the initial copy.
- Checkout – when working with documents it is sometimes desirable to prevent the document from being updated by other users. This gives you the ability to place an edit lock on a file until you check a new version in. Users may still view and download the document (subject to permissions) however they will not be allowed to update the document in the repository. An administrator can release locks on files.
- Copy – this allows you to make a copy of the selected document to any location in the repository (subject to permissions).
- Delete – this allows you to delete the document, which moves the document including all history to a special holding repository. An administrator may then remove the file permanently or restore this back to production at any time. Once a document is deleted from the holding repository all information including the document and its versions are permanently removed from the system.
- Discussion – this allows a threaded discussion to occur on each document by users.
- Edit Metadata – this allows you to update, add or remove metadata information and document types.
- Email – this allows you to email the document to one or more persons or groups.
- Links – this allows you to associate this document with other documents in the system.
- Make immutable – this locks the document from further updates, making the current version the only version available to non-administrative users. From this point forward only administrators are allowed to modify the metadata or remove the document from the repository.
- Move – this allows you to move the document to any other location in the repository (subject to permissions).

- RSS – you can copy the provided link into any RSS aggregator to create a feed to the selected document.
- Rename – this allows you to rename the file.
- Request Assistance – this allows you to notify the administrators of the Unit you are in that you require some kind of assistance, for example to change a metadata item on a file that has been marked immutable.
- Workflow – this is where you begin or continue a workflow process on a document.

Other Actions

- Administrator mode – There are instances where systems administrators may be denied access to certain documents in the repository, if the owner has explicitly omitted them from the assigned permissions. Sometimes however these documents require administrative manipulation so this gives administrators the ability to grant themselves temporary access to the file. Of course all actions are logged in the audit log.
- Search – the search functions of KnowledgeTree are very robust. By default you can do a simple metadata search or you can perform full text searching of document contents. The advanced view gives you the ability to specify multiple criteria for your search. Administrators can create special saved searches that are made available to users for commonly used criteria.
- Browse by… – this allows you to change the default behavior of the browse view, reorganizing the data based on:
    1. Folder (this is the default)
    2. Document Type
    3. Lookup Value (a specific metadata value from a lookup field)
- Subscriptions – this allows you to receive email notifications when a document version is updated in the repository.

A Document Class Example

To demonstrate the use of KnowledgeTree we have wish to a folder for sample Linux configuration files. The first thing we do is to decide the kind of metadata we want to capture for these documents, keeping in mind that generic Document Types already capture the Document Author, Category, Media Type and Tag Cloud. We added a new Document Type called Linux Configuration File and then associated a Fieldset that captures the following metadata fields:

- Use – what the configuration file is used for
- Default Location – where in the file system the configuration file should be placed

- File Permissions – a lookup list of the permissions that the file should be set to, choices are:
  1. 600 owner(read:write)
  2. 640 owner(read:write) group(read)
  3. 644 owner(read:write) group:world(read)
  4. 660 owner:group(read:write)
  5. 666 owner:group:world(read:write)
  6. 700 owner(read:write:exec)
  7. 750 owner(read:write:exec) group(read:write)
  8. 755 owner(read:write:exec) group:world(read/write)
  9. 770 owner:group(read:write:exec)
  10. 777 owner:group:world(read:write:exec)
- File Owner – the ownership the file should be set to (owner:group)

Once this is complete we are now able to add documents to the repository. The first thing we do is change the default permissions of the "root" folder to allow users at least read permissions since by default System Administrators are the only persons with initial access to this folder. Figure 6.10 shows the permissions for the root folder.

Figure 6.10 "root" Folder Permissions

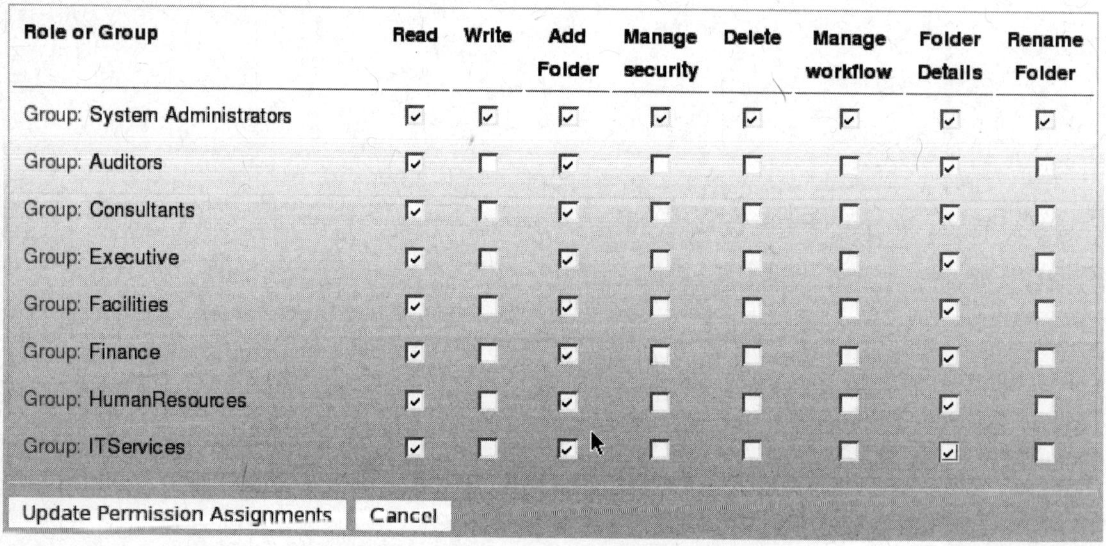

We then create a folder at the root called "Sample Configuration Files" and override the root folder permissions to assign the ITServices group with appropriate privileges to this folder and deny access to everyone else (Figure 6.11).

Figure 6.11 "Sample Configuration Files" Folder Permissions

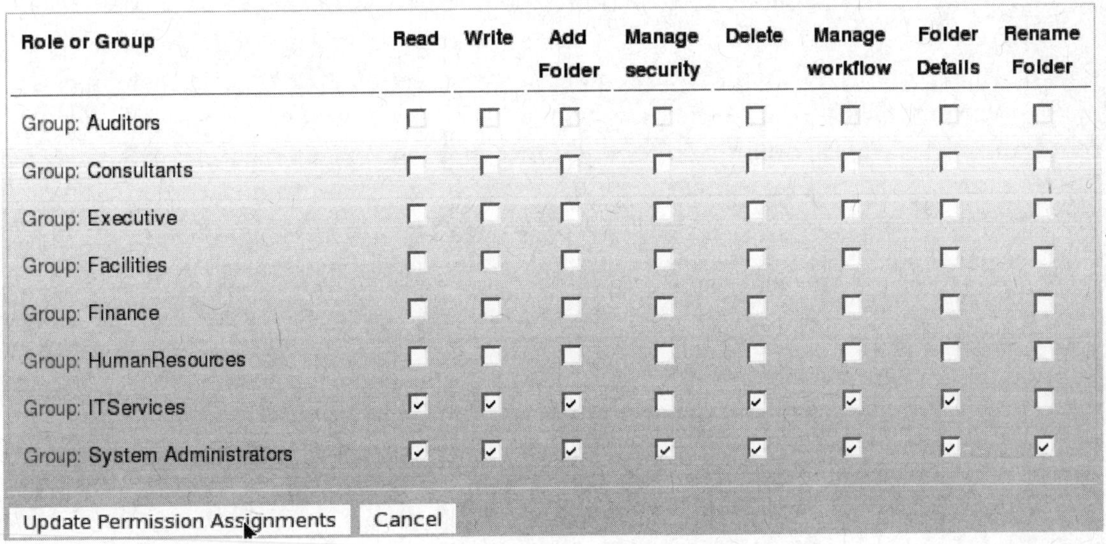

Role or Group	Read	Write	Add Folder	Manage security	Delete	Manage workflow	Folder Details	Rename Folder
Group: Auditors	☐	☐	☐	☐	☐	☐	☐	☐
Group: Consultants	☐	☐	☐	☐	☐	☐	☐	☐
Group: Executive	☐	☐	☐	☐	☐	☐	☐	☐
Group: Facilities	☐	☐	☐	☐	☐	☐	☐	☐
Group: Finance	☐	☐	☐	☐	☐	☐	☐	☐
Group: HumanResources	☐	☐	☐	☐	☐	☐	☐	☐
Group: ITServices	☑	☑	☑	☐	☑	☑	☑	☐
Group: System Administrators	☑	☑	☑	☑	☑	☑	☑	☑

Note that we do not grant "manage security" rights to the ITServices group as we want all documents in this folder to inherit the parent permissions, unless of course an administrator needs to override this default behavior. Once this is complete we now add subfolders for the various configuration files we wish to include in the repository, for example have a look at LDAP configuration files. We create a sub folder called LDAP and allow the permissions to remain inherited from the parent, and then add various client configuration files that relate to LDAP. As files are added to the system we attach our custom document type to each file and complete the metadata fields as required. For all of these files we added the keyword LDAP in the Tag Cloud so we can perform a search to find all LDAP related files in the repository. That's it! We could have easily chosen to create a root folder called ITServices, then create a Unit called ITServices and assign Unit owners for administering this section of files; it is really up to your needs how you choose to setup your particular repository. As you can see, KnowledgeTree is a powerful yet easy to use DMS that is a fantastic addition to our SOX arsenal.

Case Study: NuStuff Electronics

As part of their planning phase for Sarbanes-Oxley compliance, NuStuff Electronics looked at what the already had in place that fit the requirements for compliance. What they determined as a result of this process was that they had various components necessary for compliance but that the components had no cohesion. Simply put, their single page strategy document was only used once a year to justify budget and had no ties to policies, or Service Level Agreements (SLA's). Even though the intent of Sarbanes-Oxley compliance had nothing to do with quality by its nature it does however require that a closed loop process is put in place that facilitates accountability. Despite the fact that we will go into more detail on SLA's and policies later on in the book in order to set the stage for where NuStuff Electronics was prior to their compliance activities it was necessary to touch upon SLAs and policies in this chapter.

One could ask, while this is all good stuff like mom and apple pie, what does budget, SLA's and processes have to do with intentional wrongdoing? The simple answer would be absolutely nothing. But what we must understand is that Sarbanes-Oxley compliance was developed not only to prevent willful wrongdoing but also to prevent the unintentional misstatement of a company's financials. To this extent budget, SLA's and processes play a key role in insuring that system are properly maintained, administration is done in a consistent manner and that end-users are not motivated to find works around for a poorly performing or consistently unavailable system.

While NuStuff Electronics elected to go with COBIT as a framework, they could have just as easily elected to go with ITIL or just about any other framework. That being said, and given the discussion we have already had regarding frameworks and quality, it would stand to reason that NuStuff Electronics have to weave their indivdaul components into a cohesive quality-based process. What better cohesive process to drive accountability then the Deming quality process?

To give a little more detail about NuStuff Electronics, as we stated NuStuff Electronics already had a budgeting process, while not in an SLA format they even had informal SLA's and they had various IT processes – documented and undocumented. As you can see when NuStuff Electronics started their compliance activities they were probably no different then your environment or just about any other IT Organization in a small to medium size company. If thus far no other point has been driven home we hope that we have driven home the point that the majority of the frameworks that are being implemented today have their roots in quality. Therefore, in an effort to illustrate this process we will borrow the "Quality Control Process Cycle" diagram from chapter 5, with a slightly modification. In looking at their existing documents, SLA's, processes and taking into account the Deming quality method and Sarbanes-Oxley compliance requirements NuStuff Electronics decided to bring cohesion to their components by using a quality cycle. Figure 6.12 shows this cycle where Strategy flow into Policy, Policy flow into Process, Process flow into Strategy and the cycle repeats.

Figure 6.12 "Strategy, Policy and Process

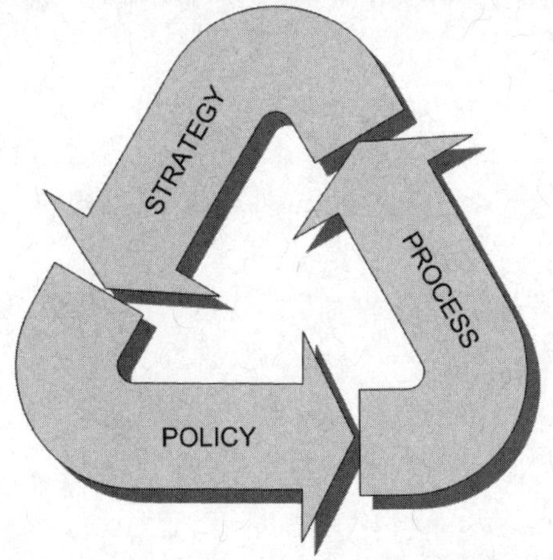

Whether, like NuStuff Electronics, your company employs a single page strategy or if your company requires the use of multiple documents, it is of no consequence to the overall process. What is important is that there is a process for capturing your IT strategies. However, as with NuStuff Electronics, there are some elements you should ensure are included as part of strategy document:

- Priority – This element should be incorporated to facilitate better resource management (Repositioning), so there is a clear understanding and agreement as to what items will be accomplish and in what order, based on resources.
- Purpose and Scope – This element should be incorporated so that clarity exists as to why the need, who benefits and to defines the overall size of project (group, department, geography, etc.).
- Benefit – This element should be included from not only the aspect of Repositioning but also to build accountability, and where applicable will be the measure by which SLA are based.
- Impact (Risk) – This element should address the risk to the company associated with implementing and/or not implementing a particular project. This element should take into consideration the overall Top Down Risk Assessment (TDRA) that we discussed earlier as part of Sarbanes-Oxley compliance.
- Budget (Financial) – This element should not only address the implementation costs but any sustaining costs as well, which should include personnel.

Beyond the elements list above the remaining elements on the Single Page Strategy document are mainly administrative and therefore we more then likely vary based on your company's particular nomenclature. But now that we have covered the elements of what a basic strategy should cover we need to have this strategy drive the necessary policy documents, as we identified previously and ultimately drive your processes (automate or manual) that we will discuss later in the chapter.

Defining your own policies

We have included a template for your convenience to define your own policies, which you can do by taking the following steps:

- Load the NuStuff Electronics portal and select Sample Configurations.
- Select the policy template; a standard html block listing will appear in your browser. Copy this to the clipboard and press the back button to return to the portal.
- You can now login as any "role" you desire and edit the first Wiki page to include a link to your new policy page. Even though the page does not yet exist, you should follow the same Wiki styles as the other policy definitions, a question mark will appear next to your entry once you save the page. This question mark means that Wiki understands a page is supposed to be linked there and is giving you the opportunity to define the page.
- Click the question mark link and you will be transported to a new blank page, select "source view" from the WYSWG toolbar and past the template into the code window.
- Press the preview button to revert the view back to the WYSWG editor. You can now modify the policy to your own liking, be sure to add a summary and optional searchable keywords in the category fields.
- Also pay attention to the Wiki page title. The squished together page name is standard Wiki nomenclature however you will most likely want to un-squish the title so it appears as a normal title. Wiki will work out the linking details as long as you leave the page name alone.

TIP

For complete Wiki documentation and numerous markup examples visit the WikiTikiTavi website at http://tavi.sourceforge.net.

Policy Approval Workflow

https://itsox2.nustuff.com/egroupware/wiki/index.php
https://itsox2.nustuff.com/egroupware/workflow/index.php

Once you have gone through the iterations of defining your policies, the next step in the chain is to submit them for approval between the appropriate parties who need to sign off on these items and make them "official". If there is one thing that brings a smile to an auditor is seeing a chain of approval attached to every control and business process in the organization, and the workflow application is ideally suited for this task. Not all Wiki pages in general need approval, so it is actually up to you to decide whether a page gets submitted or not. For the purposes of our examples here we are requiring approval of policies before they are placed into production. With this in mind any Wiki page can have any one of three states displayed in the upper left:

- Not Submitted – this is the initial default state of a Wiki page. A page may never need to be submitted, so this is what it will remain as. In this state a workflow link appears in the footer to allow you to initiate a workflow approval (see below)
- Pending – this is the state when a policy has been submitted for approval and is somewhere in that process. A Cancel workflow link is added to the footer to allow for a previous workflow request to be cancelled. Page editing and access to history are disabled in this state as well so that no changes can be made until the workflow is either cancelled or completed
- Closed – this is the state when a policy has completed a workflow approval cycle and was either ultimately accepted or rejected. You will not be able to submit this page for workflow approval again, but rather need to make an edit to force a new version which will revert back to the Not Submitted state.

For each of the policies above a single workflow has been defined called Corporate Policy Approval Workflow and Figure 6.13 shows how this workflow instance was derived.

Figure 6.13 Policy approval workflow diagram

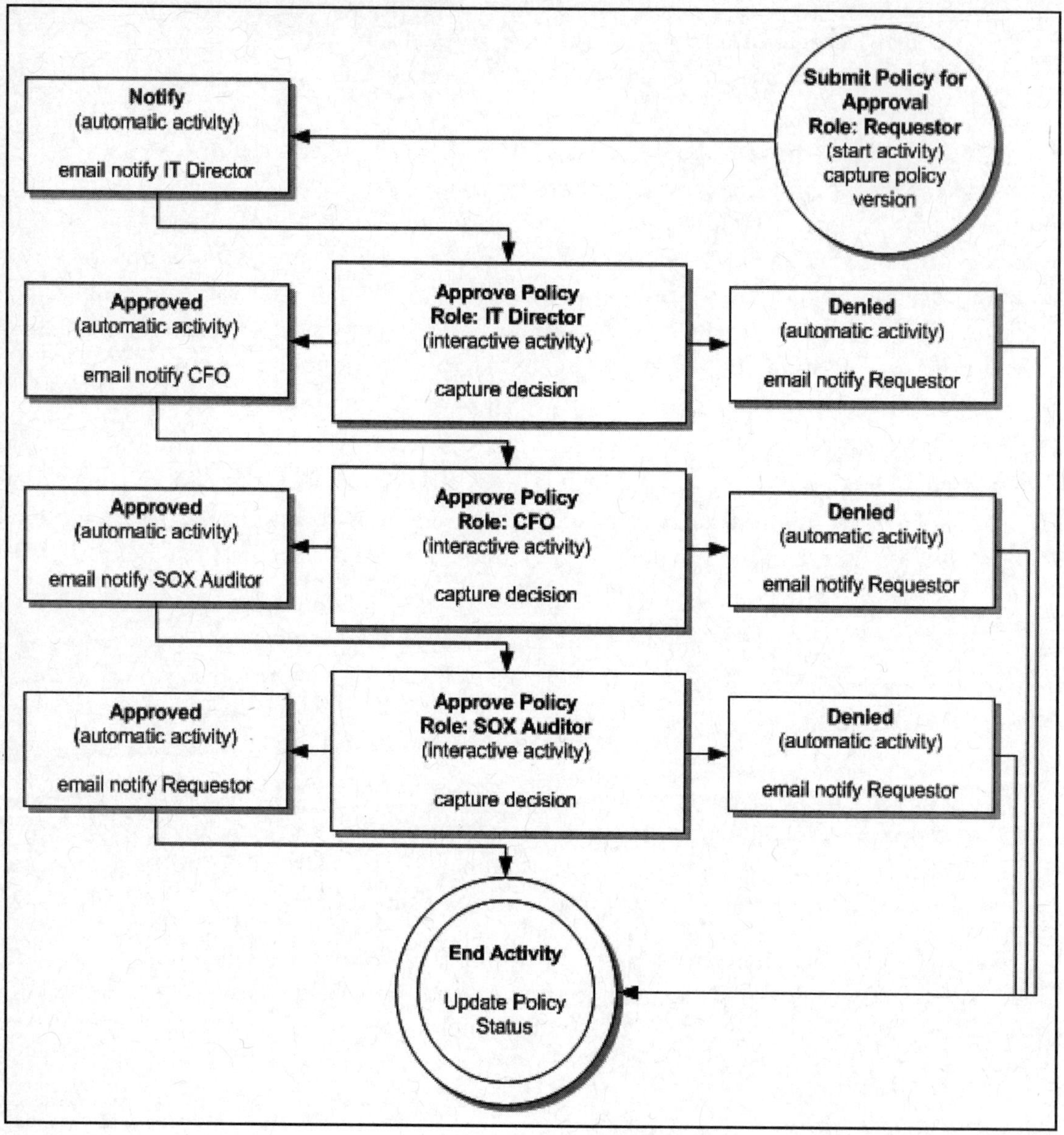

Workflow Roles

Each step in the activity that requires interaction with the system, i.e. the approval steps has an associated role. There are four roles for this workflow process, the Requestor, CFO IT Director, and SOX auditor. Role members are defined as:

- Requestor – This is the person who actually submits the policy for approval and is dynamically assigned at runtime to the currently logged in ID. Typically for policies the IT manager or person who last updated the policy will be the Requestor for submission
- CFO – this would be the company CFO, namely Joe Smith. He is the default person assigned to this role however Mary Wright is also assigned to the CFO role for this policy as a backup for Joe. In Joe's absence Mary should fill this role
- IT Director – again this is Biff Johnson who is the IT director. The IT Manager John Scott serves as the backup for this role.
- SOX Auditor – the role members for this are both Molly Fairbanks and Charles Morrison. Since a default person has not been assigned to this role, both Molly and Charles will receive notifications and either one can "grab" an existing policy request and approve or deny.

Workflow Activities

In addition to the defined roles there are several types of activities at various points in the workflow process:

- Start/End activity – this is the point where a new instance of the workflow is begun or completed.
- Interactive activity – this is any activity that requires human intervention, such as the capture of information or a decision to be made. In this example all of the approval activities are interactive.
- Automatic activity – this is any activity that has no human requirement. All of the notify activities are automatic and these send out email notifications to the appropriate roles associated with the next step in the process

Defining your own policy approval workflows

Chapter eight has a comprehensive "how-to" on defining a workflow from start to finish but we did want to touch on the subject here to get you considering your own approval chain. The workflow application is a fairly involved process to define from scratch, so really the place to start is to identify the activities and states your workflow process would encompass. It is helpful to use a diagramming application, and Dia is an open source diagram creation program released under the GPL license included on the VM. We will be seeing many different types of workflow over the next several chapters and approval routing is probably the most simple of the bunch, the process consisting mainly of:

1. Defining approval levels (called activities), such as IT Approval, Finance Approval, etc.
2. Defining the code that captures the decisions and routing activities based on those actions

3. Defining the roles for the levels of approval, such as IT Director and Controller
4. Adding persons to those roles
5. Assigning the appropriate roles to their corresponding approval levels
6. Optionally assigning a default person for the role
7. Defining your workflow as a target for Wiki submissions

By far the easiest approach initially is to modify one of the existing workflows in our example, placing your own roles and role members, perhaps adding an approval level once you have run through the tutorial in chapter eight.

Summary

In this chapter we've discussed the difference between COBIT, ITIL and Sarbanes-Oxley compliance. We have also discussed a basic process for developing IT strategic plans in an effort to implement the necessary COBIT and ITIL components to achieve Sarbanes-Oxley compliance. We also looked at some real world examples of forms and processes actually used to comply with the Sarbanes-Oxley Act.

Also in the chapter we explored the definition and approval routing of your IT business policies. These policies are what forms the core of your IT strategy and is the basis from which all procedures grow. Several policies are outlined as a representative set of items you will need to consider for SOX compliance. You can define or modify your own policies; we give you the details on how to accomplish this. You can also define or modify the policy approval workflow process if it does not suit your needs.

But in summarizing this chapter we would say there are three fundamental things you should take away with you:

- Based on your TDRA let your unique organizational structure drive the applicable domain items.
- When developing processes ensure that they follow a good Quality methodology such as, PDCA (Plan, Do, Check and Act).
- Above all, if you have existing processes that are good sounds processes and are already ingrained within the organization, by all means customize and modify them to using COBIT or ITIL for your Sarbanes-Oxley compliance.

Our VM Spotlight illustrated the major features of KnowledgeTree Document Management System, a full featured document control system with a feature rich web interface. KnowledgeTree serves as an excellent repository for all of the documentation bits and pieces you acquire in the discovery phase of your information gathering requirements. Although we chose to use the Wiki module for our policy examples you could easily choose to keep these in document form and use the workflow features of KnowledgeTree to route these for approvals. The choice is yours to decide which technology and methodology work best for you and your environment. Either way KnowledgeTree gives you full text searching, version control, and robust access permissions and serves as an extremely useful tool for all your document control needs.

Solutions Fast Track

The Work Starts Here

- ☑ The nature of IT lends itself to firefighting.
- ☑ Planning tends to be a back burner activity for most IT organizations.

- ☑ In order to comply with Sarbanes-Oxley IT, organizations will not only need to learn the discipline of planning, they will also need force themselves to adhere to this new discipline.
- ☑ The nature of IT lends itself to firefighting.
- ☑ COBIT is the defacto standard, which the majority of the audit firms have adopted.
- ☑ With the appropriate processes and documentation a small to medium size company can effectively implement the relevant guidelines for compliance.
- ☑ If implemented correctly compliance controls do not have to be so burdensome where paperwork becomes the main focus on an IT Organization.

Planning and Organization

- ☑ COBIT and ITIL guidelines can be tailored to better fit with a small to medium size company's structure.
- ☑ There is no one-size-fits-all template; COBIT and ITIL guidelines can and should be customized to a company's particular IT Organization.

Working The List

- ☑ COBIT, ITIL and SOX compliance are not synonymous.
- ☑ COBIT and ITIL guidelines were developed to facilitate IT Organization in deploying "Best Known Practices".
- ☑ The Sarbanes-Oxley Act was drafted to ensure the accuracy of financial reporting data and/or the systems that support this data.

VM Spotlight: KnowledgeTree Document Management

- ☑ KnowledgeTree is a formal Document Management System (DMS) that can be used for the administration of all sensitive documents, for example financial information, company trade secrets, engineering data and customer documentation.
- ☑ KnowledgeTree is a full featured, easy-to-use, and production-ready enterprise document management system, which allows you to share, track, secure, and manage the documents, and records your organization.
- ☑ KnowledgeTree is written in PHP and is available in three editions:
    a. Open Source Edition
    b. SMB Edition
    c. Enterprise Edition

- ☑ KnowledgeTree SMB and Enterprise Edition are commercial editions of the Open Source project and are sold with commercial support and extended functionality that provides access to the KnowledgeTree document repository from Microsoft Windows Explorer and the Microsoft Office application suite.
- ☑ KnowledgeTree has an abundance of features:
    d. A central document repository with audited and version controlled document content.
    e. Flexible document metadata definition, management and versioning.
    f. Document authoring management and workflow.
    g. Full-text indexing which allows you to search within document contents.
    h. Security group and role-based security model and integration with enterprise directory servers such as Fedora Directory Server (the spotlight subject of the previous chapter).
    i. Plugin support for extending functionality.

Policy Definition and Management

- ☑ Many example IT policies are presented for NuStuff Electronics. These policies illustrate the baseline requirements for SOX.
- ☑ In our examples the final policy versions are submitted for approval to become officially "Live" In SOX compliance one very important aspect is the signoff from management on IT practices in support of the business goals, and the workflow application is discussed in detail for this purpose.

Case Study: NuStuff Electronics

- ☑ Sarbanes-Oxley compliance was developed not only to prevent willful wrongdoing but also to prevent the unintentional misstatement of a company's financials.
- ☑ Poorly maintained systems can motivate end-users to find works around for a poorly performing or consistently unavailable system.
- ☑ Activities associated with the planning phase should be cohesive process and drive accountability, such as the Deming quality process.
- ☑ Quality is at the root of most frameworks.
- ☑ Strategy flow into Policy, Policy flow into Process, Process flow into Strategy and the cycle repeats.

Basic elements to include as part of strategy document:

- ☑ Priority – This element should be incorporated to facilitate better resource management (Repositioning) and so there is a clear understanding and agreement as to what items will be accomplish and in what order, based on resources.
- ☑ Purpose and Scope – This element should be incorporated so that clarity exist as to why the need, who benefits and to defines the overall size of project (group, department, geography, etc.).
- ☑ Benefit – This element should be included from not only the aspect of Repositioning but also to build accountability, and where applicable will be the measure by which SLA are based.
- ☑ Impact (Risk) – This element should address the risk to the company associated with implementing and/or not implementing a particular project. This element should take into consideration the overall TDRA that we discussed earlier as part of Sarbanes-Oxley compliance.
- ☑ Budget (Financial) – This element should not only address the implementation costs but any sustaining costs as well, which should include personnel.

Frequently Asked Questions

Q: Can I really customize COBIT and ITIL?

A: Yes. But prior to formalize any processes or documentation based on your customizations, you will want to run it by your auditor.

Q: Can the customization really be that simple?

A: Yes, as long as you allow your TDRA and your environment to drive the process. But keep in mind you will need to justify your decisions.

Q: Can I use the example forms?

A: Yes, you can use any example forms but you should format it to work best in your environment. These examples were developed to cover many of the items in the "Planning & Organization Domain".

Q: Is there a particular type of environment that COBIT or ITIL works better in?

A: No. Both the COBIT and the ITIL guidelines are platform and environment agnostic.

Q: If COBIT and ITIL are so cumbersome why should I use them?

A: Because the guidelines are sound. However, we think you should also consider the application of another Quality principle when looking at COBIT or ITIL guidelines and that is the 80/20 rule.

Q: Can I use my existing policies?

A: Yes, you may have to make some slight modifications. The main thing to remember is that they need to be documented and support a control. By using a tool such as Wiki, the collaborative effort between you and your auditors becomes much easier to manage and track.

Chapter 7

What's Second

Solutions in this chapter:

- Definition of Information Requirements
- Evaluating Open Source In-House Expertise
- Automation is the Name of the Game
- Working The List
- VM Spotlight – Webmin
- Case Study: Automation and Workflow

☑ Summary

☑ Solutions Fast Track

☑ Frequently Asked Questions

Definition of Information Requirements

"To face tomorrow with the thought of using the methods of yesterday is to envision life at a stand-still. To keep ahead, each one of us, no matter what our task, must search for new and better methods – for even that which we now do well must be done better tomorrow."

– James F. Bell

At the risk of diminishing the above quote we will venture to relate it in the context of this chapter. If we assume you have followed a quality standards framework such as COBIT or ITIL as a guideline and implemented an conforming IT environment, given the changes and advances in technologies as well as the ever changing requirements of your particular business it would be fool hearted if not even negligent of organization to assume that the implemented methods, practices and infrastructure could persist without evolving as the requirements and needs change. To clearly state the intent of this chapter, it is to setup guidelines for processes, practices and policies to ensure that a company's IT Organization continues to be able to meet the needs of the company by evolving with those needs with an eye to achieving and sustaining necessary compliance objectives.

In this chapter, we will look at the numerous control objectives in the COBIT "Acquisition & Implementation" domain and ITIL materials that relate to change management. Based on our experience, we offer suggestions on how a small to medium-size company might be able to reduce these to a manageable process. In Chapters 4 and 5, we established a high-level definition for the COBIT and ITIL frameworks. To review, the COBIT definition was "Once the plans are developed and approved, there may be a need to obtain new applications and acquire or develop a new staff skill-set to execute the plans. Upon completion of the Acquisition phase, the plans need to be enacted in the implementation phase and should include maintenance, testing, certifying, and identification of any changes needed to ensure continued availability of existing systems as well as the implementation and support of new systems. Now as part of Chapter 7, we will look at the specifics of this and attempt to summarize and distill the various control objectives to lend themselves more to the structure of a small to medium size company.

In Chapter 6 we cautioned you to minimizing overlap as much as possible in the applications that you will select for your environment. With this we would like to add to another caution: given that there are little or no acquisition costs associated with Open Source tools the tendency might be to over implement. Our point is that if you implement Open Source applications based on a specific requirement within your environment versus the looking at the totality of your infrastructure needs you will more then likely discover that you may have over implemented, and thereby in essence increased your environment resource and skill set requirements rather then reduced them. If the proper steps have been taken to view your needs holistically, while at the same time using a quality framework as a reference, you should be able to reduce this risk. We refer again to our case study company

NuStuff Electronics for our examples, and as we look objectively at the frameworks in total that apply to infrastructure acquisition and implementation there truthfully are not a significant number of control objectives that would apply to this fictitious company. So what we will do is identify the objectives that might not necessarily pertain to these companies, objectives that you may want to consider from a different aspect, and/or objectives that will feed into other processes that you might also need to give some thought. One example of this pruning process might be in-house development. If your organization does not perform in-house development then the framework items that relate to this activity can safely be skipped. However, even if you do not perform in-house development activities, you will want to be able to not only articulate this but also demonstrate it to your auditor. The point is that your audit team will be looking at a set of objectives that they feel need to be addressed, and it is up to you to point out where their ideas of your environment and the realities may differ.

Evaluating Open Source In-House Expertise

In an ideal world, every IT organization would have the expertise to execute and implement the necessary Open Source or proprietary tools that have been identified as necessary to cultivate its infrastructure into one that lends itself to sustained compliance. Indeed, we have identified many tools in this book as ones that can be used by a company in their quest to obtain compliance with the Sarbanes-Oxley act. However, as we all know, having all these skills may not inherently be the case. As a general rule, an IT organization's skill sets and expertise are usually related to their current environment and not necessarily to other IT disciplines not contained within that environment, or more importantly introduced by Sarbanes-Oxley. Since the focus of this book is how to effectively use Open Source to obtain compliance with the Sarbanes-Oxley act, and since the Open Source tools you choose to implement may run on mixed platforms such as Linux and UNIX-based operating systems, any organization that would like to avail themselves of these Open Source tools should have minimum level of expertise in the management of systems and platforms on which they run. If your company has already deployed Open Source in your organization to any significant degree, then you have probably already have the necessary expertise to evaluate and element the necessary Open Source tools to assist you with your compliance efforts. If your organization does not already have the aforementioned discipline, or you are uncertain about your organization's expertise level, it does not necessarily prevent you from taking advantage of the Open Source tools listed in this book, running in the ITSox2 VM Toolkit or other Open Source in general. However, what it does mean is prior to committing to Open Source as part of your Sarbanes-Oxley compliance efforts, or to assist in fulfilling some of the COBIT and/or ITIL guidelines you will first need to determine what skill sets you have in your organization and whether those skill sets are adequate, and address any deficiencies found. When evaluating the necessary skill set you will most likely need to look at two general aspects:

- Linux & Unix experience
- Open Source experience

It may appear somewhat peculiar to some of you that we separated Linux and UNIX experience from Open Source experience, as Linux is considered to be Open Source. But in the context of what we are discussing when we say Open Source we are referring to applications versus operating systems, where the expertise needed for each category may be slightly different. Additionally, many major Open Source applications run on a Windows platform, so the operating system does not necessarily determine the entire necessary skill set required for deployment and maintenance of these applications. It can be a major consideration, however, if you are a completely Windows-based infrastructure and you desire to run an Open Source tool that requires or "best runs" on a Linux platform. We do not wish to deter you from using Open Source by any means if you have not already deployed this but rather to leave you with the knowledge that you might have to consider your options for in-house staff and outsourcing.

Deployment and Support Proficiency

Although many Open Source applications run perfectly well in a Windows environment, it is useful to discuss Open Source expertise in the context of Linux since there is a good chance that this is the preferred platform for many applications you might consider for deployment. We won't spend too much time discussing Windows expertise and instead focus somewhat on Linux since there is a good chance your staff already possesses the necessary skills for the "incumbent" operating system, and regardless of platform all of the application considerations exist for Open Source regardless of this fact. In order to properly evaluate the expertise needed to support the Open Source you must examine your existing infrastructure and determine the extent in which Open Source will play a part in your overall strategy. One of the advantages of Linux for example is that if deployed for a specialized purpose, such as web servers or databases there tends to be very good setup and support options for these applications. On the other hand if Linux is going to be deployed on your desktops the skill sets required to maintain adequate levels of support might be quite different. Universally it is desirable to have a staff knowledgeable in both operating systems and the applications you plan to deploy and support in your organization, but since this may not be realistic or even necessary there are a few things to keep in mind. Figure 7.1 depicts the external support options typically available to Open Source.

Figure 7.1 Open Source Support Stack

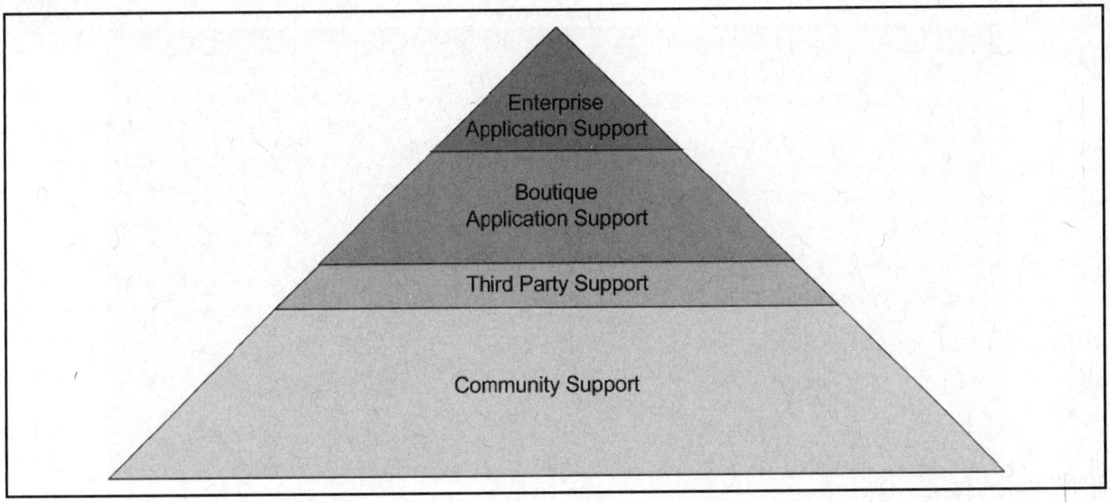

At the top of the support chain are large enterprise vendors such Red Hat and Oracle. If your intention is to deploy these distributions and applications the importance of in-house expertise might not be as critical due to the availability of vendor deployment expertise and support contracts. As we covered in Chapter 1, often the business model of organizations who develop and release Open Source applications or operating system distributions offer paid-for installation and support; a couple of good examples being SugarCRM (a customer relations management suite) and MySQL (an Open Source database application). Additionally there are companies that specialize in Open Source support, providing consulting for a wide variety of applications that you might wish to deploy into your organization, however your selection may be limited depending on your geographical location since these companies tend to be local or regional. By far, the most common avenues for support are community driven via documentation (some being better than others), mailing lists, forums and discussion lists. From a SOX perspective a majority the applications that you might consider, such as eGroupware, Zabbix, Fedora Directory, and Webmin, which are all examples that we spotlight in the ITSox2 VM Toolkit, all have very good to excellent documentation and support channels. Depending on your deployment goals the general skills you will be looking for are:

- First and foremost you must have solid ability to roll your sleeves up and dig in to solve problems that arise. While the mature projects in Open Source have good documentation and best known deployment practices, much of Linux and Open Source can involve trial and error to adapt to your specific needs, which is why the need for an adequate test environment is important and discussed later in this chapter.

- A good understanding of TCP/IP, networking, and the Linux Standards Base (http://www.linuxbase.org/) which outlines the general layout of a Linux system is also important so you know how and where configuration files and data are generally stored.
- Solid research skills might be required to navigate forums, mailing lists, and internet resources are an important addition to any good system administrator, regardless of platform and Open Source is no different.
- From a SOX perspective you should have a very good understanding of security in general, OS hardening, PAM/NSS, file and application security so that you ensure that your deployments are rolled out in a secure fashion, we cover this more in Chapter 8.

TIP

A quick way to evaluate someone's Linux and/or Open Source skills is giving them a standardized test such as the A.P. Lawrence's Linux Skills Test which can be found at http://aplawrence.com/Tests/Linux/.

Addressing Deficiencies

If after evaluating your in-house expertise you determine that any deficiencies exist there are several ways that you can effectively address this, each approach having its pro and cons. If time and resources permit you can elect to train existing staff, which could possibly be the best long-term solution but will yield the least short-term result as the learning curve will be steep and mistakes could potentially arise. If your budget and management will allow, you can also elect to hire consultants. This solution would address your short-term requirements as well as have the potential to address your long-term requirements. The caveats to using consultants effectively are:

- You will need to ensure that you source qualified consultants.
- The consultant you elect to use have a methodology and planning for transitioning the new environment back to your in-house staff.

If your budget and management will allow, you can elect to hire additional resources to satisfy this requirement. However, hiring an unknown entity under these circumstances may add more risk of failure to your process. The reality is that combination of any of the above is what most organizations end up doing in order to accomplish their compliance objective.

Automation is the Name of the Game

In implementing the COBIT and ITIL framework guidelines, complying with Sarbanes-Oxley, or installing a new application within your infrastructure, the ability to automate existing and potential new functions will be absolutely paramount. Not to lament the problems of a small to medium-size company, but if yours is typical then resources and activities are always at odds – too few resources and too many activities. That being said, one of the goals that should be specified as part of the acquisition and implementation of potential applications for your environment is the ability not only to automate new requirements, but also to automate some of your existing requirements. With the automating of new and existing functions you will be able to better leverage existing resource, and possibly stave off the necessity to add additional resources.

In addition to wanting to better be able to manage resources, another objective of the IT managers at case study company NuStuff Electronics is (as much as possible) to take the "People" out of the "Process". Although initially they decided to specify this as a requirement purely from the knowledge that people's behavior is difficult to change and even if you are able to effect behavioral change it is usually difficult and takes a considerable amount of time. Whereas, when you automate tasks via an application any required changes can be effected, if not in real-time, in a fraction of the time without automation. But as NuStuff progressed through their Sarbanes-Oxley compliance and automation efforts they quickly realized that this process yield additional benefits they hadn't initially determined. One of the areas that rapidly became apparent was that the automation process was forcing them to better define their processes by removing subjectivity from the processes. The other benefit they realized was that by eliminating subjectivity in their processes and removing the human element as much as possible that they could better demonstrate that the processes where being executed as defined. And from a Sarbanes-Oxley compliance perspective this ability will be crucial. Since you are reading this book we will assume that you are some where in the IT management chain and therefore you will either need to assert to your management or your auditor that the processes as they have been defined is how they are implemented. So, if you remember nothing else, remember this question – without some sort of automation how comfortable would you be in making this assertion?

The Transparency Test

The CFO Perspective

The authors once again emphasize the value of planning as it relates to a well-run project. With the enormity of Sarbanes-Oxley compliance and consequences of not meeting your timeline, this is especially pertinent. Given the focus of this book to small and mid size companies, resources will clearly be an issue all through the process. The correct use of automation can have a significant favorable impact on all aspects of the project size, timing and cost. As with most projects, you need to clearly define the needs and

Continued

> insure that project creep does not find its way in, thusly eliminating the benefits automation offers. Containing any project is not easy to do, the natural tendency will be to let the program continue to creep thusly adding some element of timing and cost risk. The project needs to keenly focus on Sarbanes-Oxley certification. Automation done properly can help achieve this end.
>
> – Steve Lanza

1. Identify Automated Solutions

This section as it pertains to COBIT deals with identification of Automated Solutions, and more specifically system that were developed in-house and not acquired from a third party. But you think you are in the clear because you have no development staff, you will need to consider any applications you'd had developed by consultants and contractors. For clarity in the section I will again, state that the scope of Sarbanes-Oxley is any system that is significant in the financial reporting process.

COBIT 1. Identify Automated Solutions	ITIL	Guidance	
1.1. Definition of Information Requirements	Application Management, The Application Management Life Cycle, 5.2 Requirements	The COBIT practice states that a methodology should exist to ensure that business requirements are met existing systems and by future development activities. Perhaps not as it relates to development but this is an area that you may want to look to as you prepare your plans and evaluate open source tools.	Repositioning
1.10. Audit Trails Design	No Correlation	The COBIT practice states that an organization's system development life cycle methodologies specify that systems have the ability to track activities via audit trails. Even if you do not any in-house developed or system provided by consultant or contractors you might still not be	SOX

COBIT 1. Identify Automated Solutions	ITIL	Guidance	
		in the clear. If you have systems that have been significantly customized or modified you will want to examine them.	
1.15 Third-Party Software Maintenance	Application Management, The Application Management Life Cycle, 5.6.2 Day-to-day maintenance activities to maintain service levels	Although by the very nature of open source it cannot be sold, there is however commercials companies that have embraced open source as part of their business model by providing third party support. As a good IT Organization application support should always be included in the overall plans.	Repositioning

2. Acquire and Maintain Application Software

This section as it pertains to COBIT deals with design elements of systems developed in-house. However, keep in mind that as with the previous section heavily customized or modified system may fall in this category form a SOX perspective.

COBIT 2. Acquire and Maintain Application Software	ITIL	Guidance	
2.2. Major Changes to Existing Systems	Application Management, Concluding Remarks, 8.3 Application evolution	COBIT states that major systems changes should follow a similar process as a systems development process. This should go without saying	SOX & Repositioning

Continued

COBIT 2. Acquire and Maintain Application Software	ITIL	Guidance	
		but this process should also contain Change Management.	
2.8. Definition of Interfaces	Application Management, Aligning the Delivery Strategy With Key Business Drivers and Organizational Capabilities, 4.3 Preparing to deliver the application	COBIT basically states the system development life cycle methodology should include the requirement that all system interfaces are properly specified, designed and documented. If you have system that are significant in your financial reporting process that feed into your financial system you will also need to include the interface as part of your testing and certification.	SOX
2.14. IT Integrity Provisions in Application Program Software	No Correlation	COBIT basically states the application should routinely verify the tasks performed by the software to help assure data integrity, and provide rollback capabilities. As this provides a "Prevent Control" it may be a requirement to use to evaluate future application and or interfaces you may need to implement.	SOX

COBIT 2. Acquire and Maintain Application Software	ITIL	Guidance	
2.15. Definition of Interfaces	Application Management, Control Methods and Techniques, 7.2 Understanding the characteristics of the application Application Management, The Application Management Life Cycle, 5.4 Build	COBIT defines that unit, application, integration, system, load and stress testing should be in accordance with project test plan. But form a SOX perspective what you should be cognizant of is that if you use production data as part of this process in essence you have placed your test environment in production and thereforeall SOX production environment criteria now applies to your test environment.	SOX

3. Acquire and Maintain Technology Infrastructure

This section as it pertains to COBIT as it deals with the acquisition and maintenance of systems related to your IT infrastructure. Depending on what processes you identified to implement from the previous domains, this could have the impact on your open source tools selection.

COBIT 3. Acquire and Maintain Technology Infrastructure	ITIL	Guidance	
3.3. System Software Security	No Correlation	COBIT states set-up of system software to be installed does not jeopardize the security of the data	SOX

Continued

COBIT 3. Acquire and Maintain Technology Infrastructure	ITIL	Guidance	
		and programs. This is true from a SOX perspective as well. If an infrastructure type system interfaces to your financial system i.e. monitoring etc. it too will need to follow the same guidelines as your financial system.	
3.5. System Software Maintenance	ICT Infrastructure Management, 4. Operations	This is in line with 1.15 of Identify Automated Solutions.	SOX & Repositioning
3.6. System Software Change Controls	ICT Infrastructure Management, Deployment, 3.3.3 External interfaces (change management)	As part of your Change Management Process procedures for system software changes should identified and documented.	SOX
3.7. Use and Monitoring of System Utilities	ICT Infrastructure Management, Design and Planning, 2.7.2 The tools ICT Infrastructure Management, Operations, 4.1.1 Managed objects	This particular control is not only in line with 3.3 but it expands it to include systems utilities.	SOX

4. Develop and Maintain Procedures

This section as it pertains to COBIT as it deals with the development and maintenance of procedures for system development activities. However, as we previously established, systems that have been customized or significantly modified will also fall under this umbrella.

COBIT 4. Develop and Maintain Procedures	ITIL	Guidance	
4.1. Operational Requirements and Service Levels	Service Delivery, Service Level Management, 4.4.1 Produce a service catalogue The Business Perspective, Managing the Provision of Service, 6.1.6 Service level management	COBIT states that the system development life cycle methodology should ensure the timely definition of operational requirements and service levels. When looking at open source systems and/or application acquisition you will want to also ascertain SLA methodology management and processes.	SOX
4.3. Operations Manual	No Correlation	COBIT states operational manuals should be prepared and kept up to-date as part of system development metho-dology. Again, remember that customized or significantly modified applications will also need to be taken into consideration.	

5. Install and Accredit Systems

This section as it pertains to COBIT as it deals with the installation and verification of systems pre and post migration to production. As in the case of NuStuff the control objectives in this section are not germane, but I would like to make a couple of general statements. At the risk of belaboring points already covered, keep in mind that:

- If required the focus of these activities should only pertain to systems/application significant in the financial reporting process.
- Customized or significantly modified applications will also need to be taken into consideration.

- Control Objectives listed in this section should be incorporated into your Change Management Procedure, whether it is specific to application/software develop or a general Change Management procedure.

For detailed specifics about these Control Objectives, please refer to Appendix A.

6. Manage Changes

This section as it pertains to COBIT deals with change and the management of that change. If you haven't noticed the various COBIT Control Objectives fundamentally exist in each Domain, but merely restated as it pertains to a particular Domain. So, in an effort not to be redundant we will site the previous Control Objectives as they apply to this section and guidelines for Change Management.

COBIT 6. Manage Changes	ITIL	Guidance	
6.1. Change Request Initiation and Control	Service Support, Change Management, 8.5 Activities	COBIT states that the system development life cycle methodology should ensure the timely definition of operational requirements and service levels. When looking at open source systems and/or application acquisition you will want to also ascertain SLA methodology management and processes.	SOX & Repositioning
6.2. Impact Assessment	Service Support, Change Management, 8.5.6 Impact and resource assessment	COBIT states procedure should be developed and put in place to assess the proposed change to the environment. At any point of impact assessment testing should be performed whether it be functional, or integration.	Repositioning
6.3. Control of Changes	Service Support, Configuration Management,	COBIT states that Change Management and software control and distribution	SOX & Repositioning

COBIT 6. Manage Changes	ITIL	Guidance	
	7.9 Relations to other processes	are integrated with a configuration managementsystem. The change control system should be automated.	
6.4. Emergency Changes	Service Support, Change Management, 8.2 Scope of change management	COBIT states that parameters should be established for defining emergency the procedures that changes and control these changes when circumstances require circumvention of normal processes. All and any changes should have prior approval of IT Management and be recorded. This should and could be part of your overall Change Management Process.	SOX & Repositioning
6.5. Documentation and Procedures	Service Support, Relationships Between Processes, 2.2 Change management Service Support, Change Management, 8.2 Scope of change management	COBIT states the change process should ensure the appropriate documentation and procedures are updated accordingly, when a change occurs to the system. This should and could be part of your overall Change Management Process.	SOX & Repositioning

Continued

COBIT 6. Manage Changes	ITIL	Guidance	
6.6. Authorized Maintenance	No Correlation	COBIT states procedures should be in place to ensure personnel with system access right work is monitored and that they do not perform unauthorized activities. This can be addressed as part of the access policies and procedures.	SOX

Working The List

At the risk of belaboring a point a point, I will restate, COBIT, ITIL and SOX compliance are distinctly different entities. This is a point I believe can't be stressed enough, because as you progress through your Acquisition and Implementation phase, the two can very easily and probably will become convoluted. It is at this point you will need to refocus yourself and activities to accomplish the primary goal of Sarbanes-Oxley compliance. Even though one of the stated goals of this book was to better enable you to reposition or position your IT Organization within your company via COBIT, the main objective of this book is Sarbanes-Oxley compliance utilizing open source and the COBIT or ITIL frameworks as vehicles, of which you may be able to derived additional benefits beyond compliance.

We will be using one of our fictitious companies, NuStuff to drive the process of customizing the control objectives in the COBIT Acquisition and Implementation Domain and correlating ITIL to the COBIT controls. The control objectives as in Chapter 6 listed will also be a combination of SOX and COBIT but each will be clearly identified. Again, the example is just that, your particular environment should drive your customization activities and you should work with your auditor prior to finalizing your efforts.

Project Management is Key

This should go without saying, but project management will be a key factor in the success or failure of your effort to obtain Sarbanes-Oxley compliance. Yes, as IT professionals or Executive Management, we have managed our share of projects and I will assume that at least some of those projects have been successful. I will also assume that few–if anyone–has ever been responsible for a project that not only can touch every aspect of their IT organization, as well as the company in general. That is precisely what you will have to contend with to obtain

Sarbanes-Oxley compliance, the scope of this project that is that wide reaching. Any project under which one of the COBIT domains would fall under normal circumstances could be considered a viable project by itself, but the combined total of all of these projects becomes the overarching task where, as of today, the consequences are so severe that failure cannot be an option. After recovering from what a former employee deemed as "Analysis Paralysis" I realize that the fundamentals of good project management still applied:

- Project Definition and Scope
- Shared Vision
- Role & Responsibilities
- Accountability
- Project Plans

Even though the task at hand may appear to be daunting if you keep the objective in mind and apply good sound project management your odds of success will be greater.

VM Spotlight – Webmin

Webmin
Web-based system administration

http://www.webmin.com/

Webmin is a web-based interface for system administration for Linux and UNIX. With Webmin you can setup and maintain virtually any aspect of your system remotely using any modern web browser. Webmin removes the need to manually edit UNIX configuration files and comes with over 100 standard modules in the base distribution. Some of the standard modules include:

- Apache Web Server – Configure almost all Apache directives and features.
- BIND DNS Server – Create and edit domains, DNS records, BIND options and views.
- Backup Configuration Files – Perform manual or scheduled backups and restores of configuration files managed by Webmin modules.
- Bandwidth Monitoring – View reports on bandwidth usage by host, port, protocol and time on a Linux system.
- Bootup and Shutdown – Setup scripts to be run at boot time from /etc/init.d or /etc/rc.local.
- Command Shell – Execute shell commands and view their output.
- Custom Commands – Create buttons to execute commonly used commands or edit files on your system.

- File Manager – View, edit and change permissions on files and directories on your system with a Windows-like file manager.
- Filesystem Backup – Backup and restore filesystems using the dump and restore family of commands.
- SSH Server – Setup the SSH server for remote secure logins.
- Samba Windows File Sharing – Create and edit samba file and print shares.
- Scheduled Commands – Schedule the execution of one-off commands or scripts.
- Scheduled Cron Jobs – Create, edit and delete Cron jobs.
- Software Packages – Manage software packages on your system, and install new packages.
- System Logs – Configure the syslog server on your system and view its log files.
- Upload and Download – Upload multiple files to the server, and download multiple URLs either immediately or in the background at a scheduled time.
- Webmin Actions Log – View detailed logs of actions by Webmin users.

The above is a very short abbreviated list of the tasks you can perform with Webmin. We consider this to be a "killer" application for systems administration from a SOX perspective because not only does this allow your administrators to be able to work with virtually any application or system task that has a corresponding module without necessarily having to know the "ins" and "outs" of these items, Webmin allows your administrators to perform functions that would otherwise require root access to the machine *without* giving them the root credentials. For new systems administrators or those who might not be skilled in Open Source or Linux this is an ideal tool. It is very simple to install and makes tasks that are often beyond newer administrators simple, if not trivial. In effect Webmin augments your staff's skills by providing an easy to use GUI based administration interface, and allows you to control this access with very fine grained detail that with normal root access would otherwise give them the keys to the kingdom for that particular system. We cannot overstate the value of these two points, we really do not know of any more useful tool for administering Linux when it comes to the SOX requirement of "segregation of duties". The only possible drawback of using Webmin might be that you do not see what the changes to the system or application are, so you cannot easily learn from the changes it makes. However, as your system administrators come up to speed with their skills in general, this becomes less of an issue. In the interim, Webmin gives them the ability to get something up and working in a short time, regardless of their current skill set. Even if you have more advanced administrators, Webmin is still very useful for access control, since for each module it allows different levels of access to different users so that you can delegate tasks.

TIP

Jamie Cameron, the main developer of Webmin, has released an excellent book that provides an in-depth look at Webmin and covers virtually all the possible situations you can come across using it in over 60 chapters. We highly recommend you obtain a copy of this for your own reading. *Managing Linux Systems with Webmin: System Administration and Module Development,* ISBN: 0131408828.

As you can see from the above list, Webmin can help in a number of ways other than configuring aspects of your system. You can automate processes such as backing up and logging, firewalling, as well as many "cluster aware" tasks. Webmin allows you to manage multiple machines from a single interface, providing additional modules to manage:

- Cluster Webmin Servers – Install and manage modules, themes, users, groups and access control settings across multiple Webmin servers.
- Cluster Change Passwords – Change passwords on multiple systems in a Webmin cluster at once.
- Cluster Copy Files – Schedule the transfer of files from this server to multiple servers in a Webmin cluster.
- Cluster Cron Jobs – Create scheduled Cron jobs that run on multiple servers simultaneously.
- Cluster Shell Commands – Run commands on multiple servers at once.
- Cluster Software Packages – Install RPMs, debian and solaris packages across multiple servers from one source.

Webmin is written in Perl and has a completely documented application programmer's interface for the Open Source community to write their own modules that are not included in the base distribution. Many hundreds of these modules are available as Open Source on the website http://webmin.com/third.html. One such example we alluded to in Chapter 5 is the FDS Toolkit http://fdstools.sourceforge.net. This is an application that provides user, group and host management for Fedora Directory Server written by the authors of this book and released under the terms of the GPL. In addition to the command line utilities, another feature of FDS Toolkit is a complete Webmin module to provide the command line functionality in a web browser GUI. We will use this module as our example a bit later in the chapter to further spotlight the design and capabilities of Webmin in general.

Webmin Users

When Webmin is first installed a "root" user is created by default that has full access to all installed modules. While this is necessary at first to get Webmin setup initially, we do recommend that you change the password from system authentication to being specifically set within Webmin and make this sufficiently complicated to prevent the Webmin root account from being improperly used as it is not necessary to login with the root account unless something is wrong. In the next section we will take a look at adding users to your Webmin installation however for now you can login as bjohnson, who has been granted access to all modules. Once logged in you can browse through the standard modules, the FDS Administration module is listed under the "Servers" tab as shown in Figure 7.2.

Figure 7.2 Servers Module Tab for bjohnson Account

This is sharply contrasted with the login for jscott, who has been granted more limited privileges, as depicted in Figure 7.3.

Figure 7.3 Servers Module Tab for jscott Account

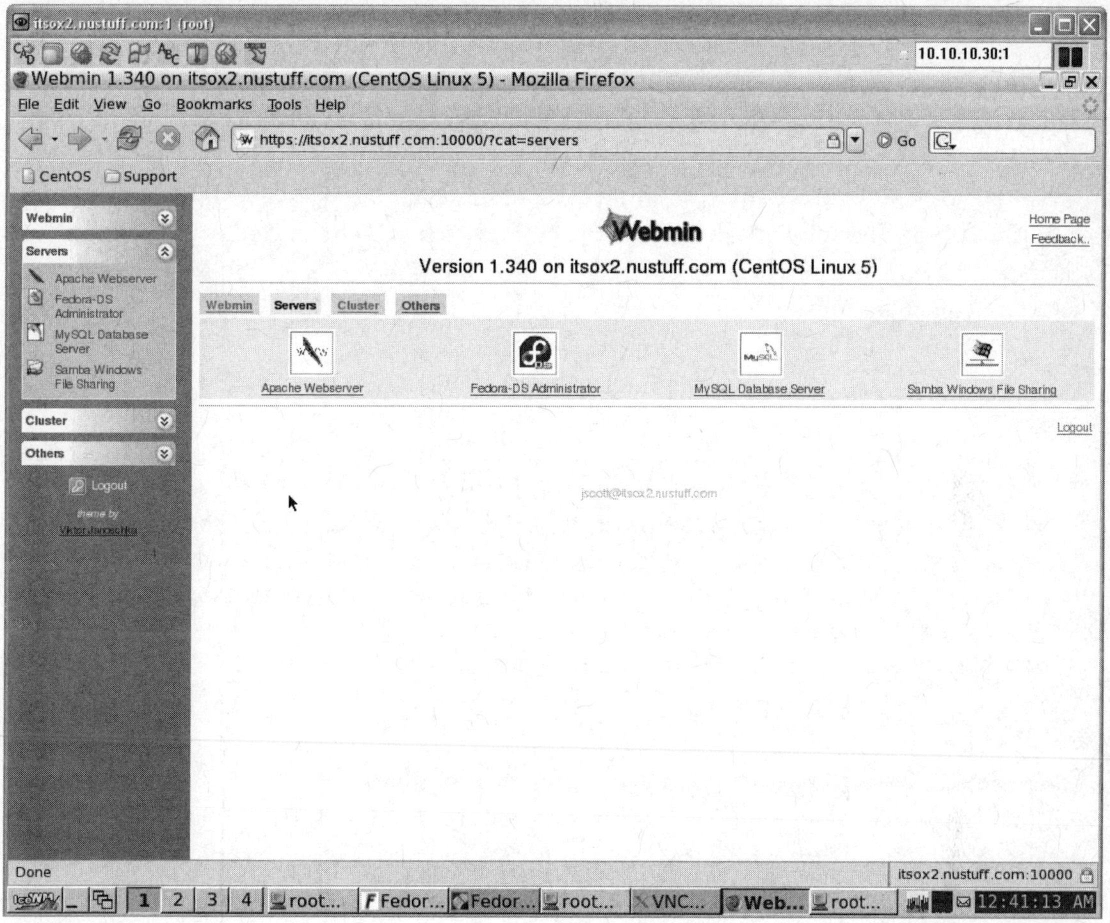

As you can see you can easily control access to virtually any administrative aspect of a Linux system with Webmin that would otherwise require the administrator to have root privileges on the system.

Adding Users

Webmin supports the ability to define the administrators that will be using the system, located under the section Webmin…Webmin Users. After a default installation you will notice the root user and all of the modules that have been granted by default. Webmin uses the Pluggable Authentication Mechanism (PAM) to link this with the system level credentials for the system root account. When defining your own users, you can choose to either set a password local to Webmin, or use the UNIX credentials. If the underlying PAM has been setup to authenticate against FDS for instance, your administrators will then be able to use their FDS password to login to Webmin. PAM is the commonly used authentication layer in Linux systems. It essentially sits between applications and authentication "backends," which store account and group information. With PAM, a developer can write an application using a well-defined API and not have to know about or deal with any specific implementation details associated with authentication and authorization. This means a PAM aware application such as Webmin can use any number of mechanisms to authenticate against, such as the /etc/password file, NIS, LDAP, Kerberos, or even Microsoft Active Directory.

TIP

Although you can and probably should use the default PAM system authentication settings for your Webmin users it is always a good idea to have at least one account (such as root) that uses a Webmin-locally set password for failsafe reasons. This is so that in the event the authentication provider (such as FDS LDAP) is unavailable or down you will still be able to login to the system to perform administrative tasks.

As you add users to Webmin, you can optionally control what IP addresses the users can login from, as well as place time restrictions on accounts by day and/or range of hours. You may also demand a client certificate for maximum security as well as specify an inactivity logout interval. This gives you complete flexibility and security for even the most sensitive security needs, such as your financials system. It is when adding/editing user accounts that you also select the default modules the account will have access to.

Applying Security Rights

Once you have added your accounts to Webmin you can optionally drill down into each individual module and set appropriate granular rights applicable to the application. For example in the Apache Webserver module you can choose to allow users to edit the global configuration file, restrict administration to specific virtual hosts, and even limit the directives the administrator can use when defining a virtual host. Most Webmin modules have application specific rights such as these to further refine the actions that can be performed. It is probably a good idea at this time to create Webmin groups, define general and specific module access details and then add members to these groups so that you can base your administrative tasks on roles rather than individuals. More time spent here will mean less administrative overhead in the future, and you will be able to show your SOX auditors exactly the access restrictions you have in place.

Webmin also provides a few additional useful features regarding users and security as well:

- Convert Unix to Webmin users – This feature allows you to automatically (and selectively) create Webmin users from the system user list. This includes and PAM backends that have been configured for use by either the system or Webmin specifically. Note you must have at least one group defined in order to use this feature as Webmin will use that group as their default security footprint.
- Unix User Synchronization – This feature allows you to determine whether a Webmin user is automatically created or deleted as system users are created or deleted. It is important to note that this only works when using the Webmin Users and Groups module to do the actually add./delete operation.
- Unix User Authentication – This facility fundamentally changes the way Webmin authenticates users. You can configure Webmin to validate login attempts against the system user list and PAM in general or restrict logins to the specific users defined in Webmin. If you have a large number of users who will be using Webmin, this can be useful to reduce ongoing maintenance, but we caution you that on your sensitive systems you should consider restricting logins to Webmin defined users.
- Current Login Sessions – This shows you the users currently logged in via Webmin, and gives you access to their specific action logs (see below for audit trail).

Fedora-DS Administrator, a Webmin Module

We introduced The FDS Toolkit in Chapter 5, which is a system for managing Fedora Directory Server users, groups, and hosts. In addition to the command line utilities this application also features a Webmin module to provide a GUI interface to manage those functions in a web browser. To begin using this you can navigate to the Fedora-DS Administrator by selecting the Servers section using either the bjohnson or jscott accounts to access the module.

Managing Users

The Accounts subsection provides the following functionality as depicted in Figure 7.4.

Figure 7.4 FDS Accounts

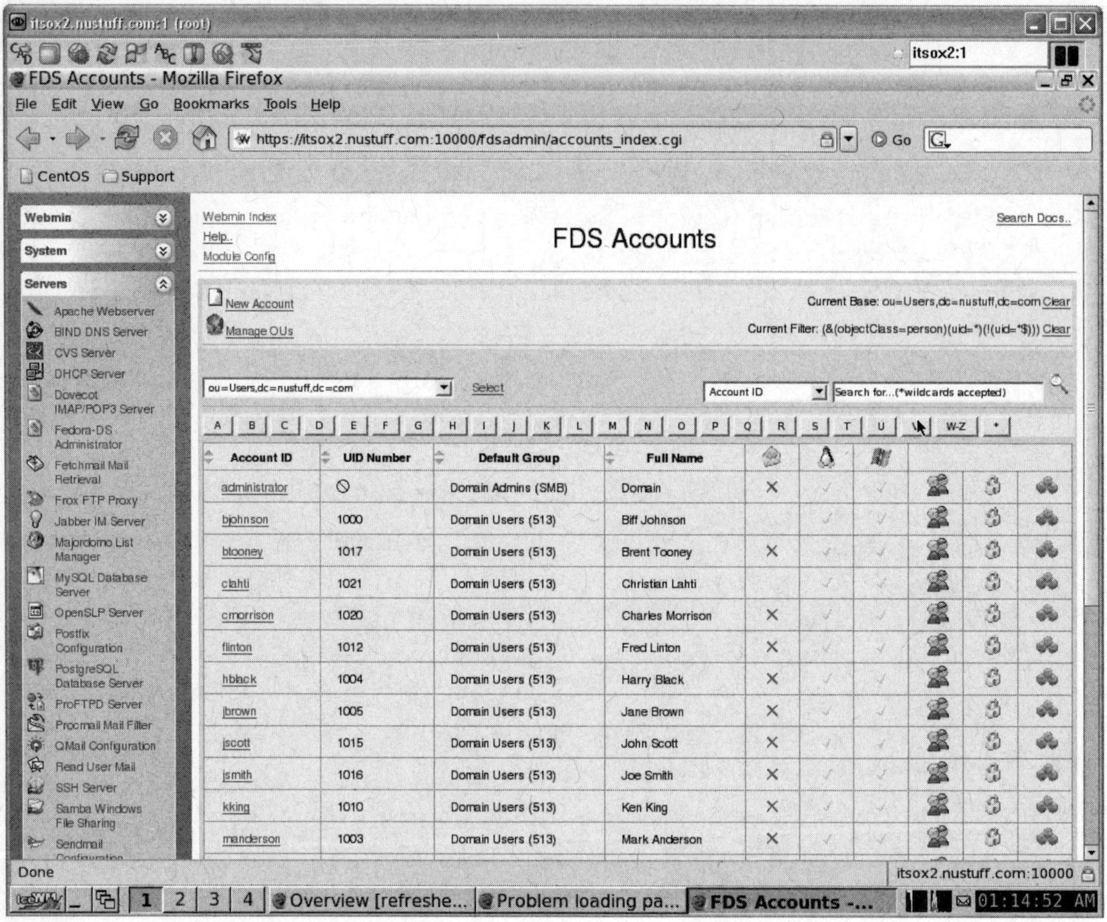

- User List – this provides a list view of accounts which fall under the selected Base and Search constraints you have defined (see below). The list can be sorted by any column where an arrow appears. In the module configuration you can optionally display Posix Status, Samba Status, and AIX Status icons for each account which gives a visual clue to whether the account has these properties. Additionally columns are provided to show whether the account is currently enabled, give access to group membership, password settings and special account handling such as moving the user in the LDAP tree.
- Current Base – this displays the current position in the LDAP tree for the accounts you wish to manage. A drop list is provided to change the current OU context, and if you select the top of the tree then all accounts are available for management. The module will "remember" on a per user basis where you last left off and will apply that view the next time you return to the screen.

- Manage OU's – this link will allow you to manage the tree of sub-OU's you would like to use for your particular installation under the configured user base (ou=Users, dc=nustuff,dc=com). You can add any number of branches under this base, but all users will be contained under this root suffix.
- Current Filter – this shows you the current search filter applied to the account list. When you first enter the module no filter is applied, and hence no records appear in the list. You have to supply a filter to begin managing user accounts. The "letter button" quick filters allow you to filter by a particular letter, if you select the (*) button then all accounts will be returned. Be careful when selecting this option on large directory instances since it can take quite some time to return say 10,000 records. Additionally you can filter by specific attributes or even build your own custom filter. An example might be 'firstname = Biff'. If using a custom filter you need to use complete LDAP filter syntax, so that example would actually be `(&(objectClass=person)(givenName=Biff)`. Again the module will "remember" the current filter and reapply the same on subsequent visits to this screen until you change it.

TIP

You can use wildcards in your search criteria by using an asterisk. For example:

`Biff*` will return all accounts beginning with Biff
`*Biff` will return all accounts ending with Biff
`*Biff*` will return all accounts containing Biff

- New Account – Depending on the module access permissions set you can add new accounts or edit existing accounts. If no rights exist, these links will not be present, which is useful for helpdesk personnel whose only role might be to reset passwords. To add or edit an account click on the appropriate link. When adding a new account the OU (organizational unit) Account UID, First Name, and Last Name fields are required. You can optionally set many other values as appropriate. There are various other required options depending on whether you add Posix attributes such as User Home Directory and Posix Login Shell. If you do choose to add Posix attributes Fedora-DS Administrator can create the user's home directory using configuration defaults stored in the /etc/fdstools/fds.conf configuration file. One more item worth mentioning, Fedora-DS Administrator will also track user UID numbers and generate these as necessary so you do not have to. This is the normal behavior and you should let the system choose the UID number unless there is a compelling reason to set this to a specific value initially. Just leave the field blank to accept the next generated number.

Managing Groups

The Groups subsection provides the following functionality as depicted in Figure 7.5.

Figure 7.5 FDS Groups

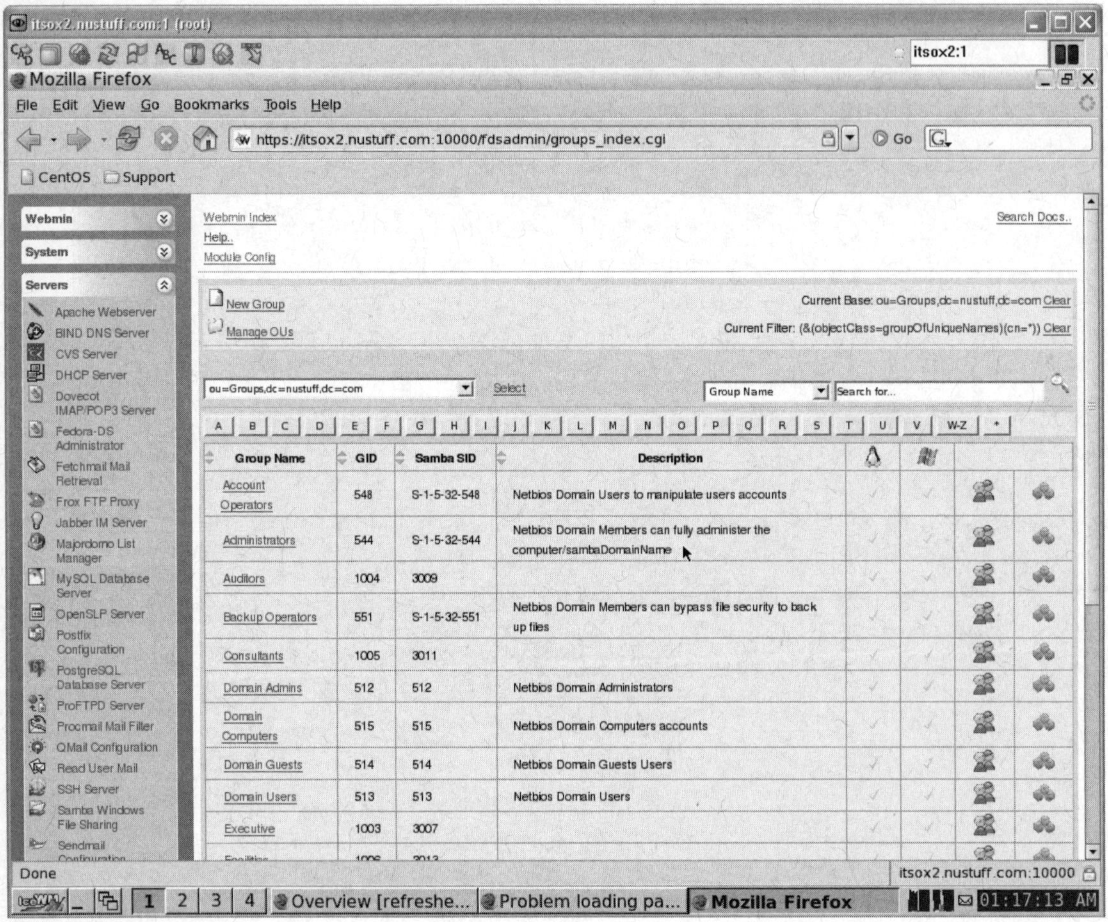

- Group List – this provides a list view of groups which fall under the selected Base and Search constraints you have defined. As with Users and Hosts the list can be sorted by any column where an arrow appears. In the module configuration you can optionally display Posix Status, Samba Status, and AIX Status icons for each group which gives a visual clue to whether the group has these properties. Additionally icons are provided to give access to group membership as well as special account handling such as moving the group in the LDAP tree.

- Current Base – this displays the current position in the LDAP tree for the groups you wish to manage. A drop list is provided to change the current OU context, and if you select the top of the tree then all groups are available for management. As with the Accounts and Hosts screens the module will "remember" on a per user basis where you last left off and will apply that view the next time you return to the screen.
- Manage OU's – this link will allow you to manage the tree of sub-OU's you would like to use for your particular installation under the configured group base (ou=Groups,dc=nustuff,dc=com). You can add any number of branches under this base, but all groups will be contained under this root suffix.
- Current Filter – this shows you the current search filter applied to the group list. When you first enter the module no filter is applied, and hence no records appear in the list. You have to supply a filter to begin managing groups. The same "letter button" quick filters and attribute filters are present so you can narrow your search criteria with virtually any valid filter criteria. Again the module will "remember" the current filter and reapply the same on subsequent visits to this screen until you change it.
- New Group – Depending on the module access permissions set you can add new groups or edit existing groups. If no rights exist again these links will not be present. To add or edit a group click on the appropriate link. When adding a new group the OU (organizational unit) Group Name, and Group Type fields are required. Optionally you can assign an Organization, Department and Group Owner for the group. As with user accounts you should leave the group GID Number blank to let the system generate the next valid number. Additionally you need to set the appropriate Group Type from the following choices:

    1. Group Of Unique Names (no Posix or Samba attributes, such as a mail list)
    2. Posix Only (this group will only be used for UNIX/Linux)
    3. Posix plus Samba Global Group (most common option, the group will be available for both Linux and Windows)
    4. Posix and Samba Local (not normally used, but sets the Samba attributes to a local group status)
    5. Posix and Samba Builtin (not normally used, but sets the Samba attributes to a builtin group status)

Managing Hosts

The Hosts subsection provides the following functionality as depicted in Figure 7.6.

Figure 7.6 FDS Hosts

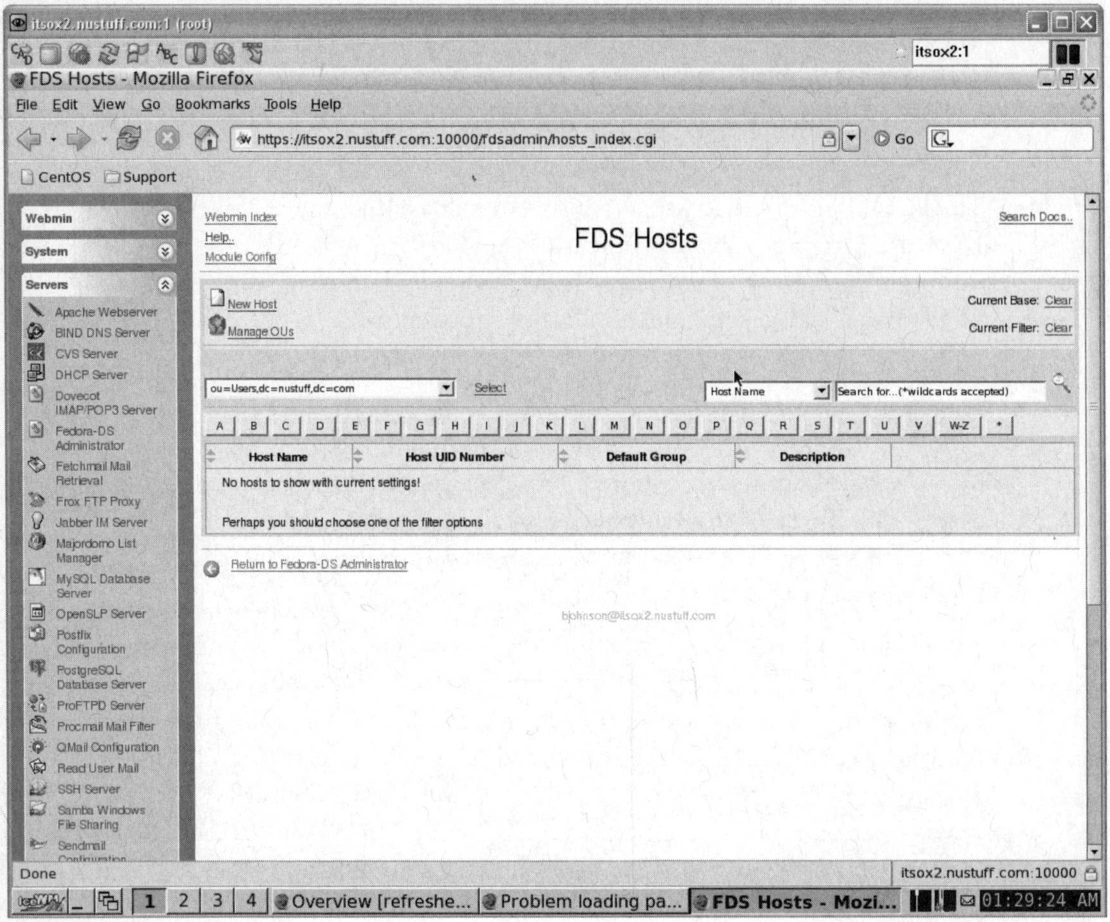

The FDS Hosts facility is only useful for managing Windows (Samba) Computer and Trust accounts. If you do not use your Fedora installation as a backend to a Samba Domain controller then you will generally have no need to use this section. Samba computer and trust accounts are specialized user accounts, the main difference is the addition of a "$" symbol at the end of the account name to differentiate this as a computer account. Fedora-DS Administrator automatically manages the proper naming so you do not need to supply the "$" when managing computer and trust accounts, and you will notice that these are not displayed in FDS Hosts. Trust accounts need a password assigned to them so when adding a host of this type a button will appear to allow you to do this. The same familiar Base, Search, and Filter options are available and operate identically to the Accounts and Groups views. Valid Host types are Workstation and Trust account.

Webmin Audit Trail

An important aspect of SOX is the ability to generate audit trails. Webmin thankfully has a facility to track all actions taken by any Webmin user. To view the Webmin Actions Log navigate to the Webmin Section. Filter options are:

- Search for any module
- Search for specific module
- Any time
- For Today only
- Specified Date/Time range

Case Study: Automation and Workflow

Automation is a key partner to being able to demonstrate compliance and to sustain this over time. As you have seen in the Chapter's 1 and 5, we have introduced the concept of the workflow automation to assist in achieving this goal and how it can be used to automate what would other wise be a paper or resource intensive process. The practical applications of this application are nearly limitless; any process that can be translated to an electronic equivalent is a good candidate for process automation via the workflow. To illustrate this constructed several example workflows for NuStuff Electronics that fall into the various categories, and these tie directly back to the policy documents and control objectives that were outlined in Chapter 5. Figure 7.7 depicts the categories of workflow automation that fall under the scope of SOX compliance and how they tie into the overall goals of your compliance objectives.

Figure 7.7 SOX Workflow Automation Categories

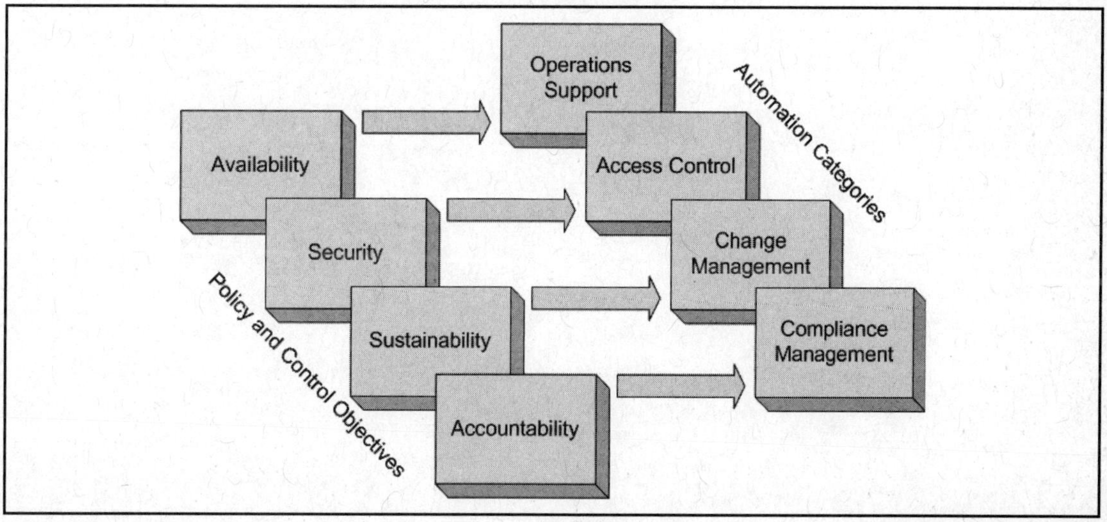

In the remainder of this chapter we will walk through a couple sample implementations for NuStuff Electronics that highlight some of the important considerations for introducing change to your organization, and how you can automate many aspects of this. Any change can and should be applied to the same automation process, which is important for SOX since your auditors will expect justification, signoff and documentation of any change in your environment that might affect your financial systems and/or reporting chain.

NuStuff Electronics Example Implementation: Intrusion Detection System

NuStuff Electronics has recently become aware of port scanning intrusion attempts on their main corporate firewall. By reviewing the firewall logs, they have determined that there have been almost daily attempts to find open ports to penetrate their network, but what is not known is if any of these penetration attempts have been successful. In order to improve the security and threat response of the IT organization, management has decided to deploy Snort Intrusion Detection Systems, which will not only tell them when port scans are attempted but if and when any known exploits have penetrated their firewall. Again, the main objectives that this project addresses are:

Availability and Security

Security is one of the most important aspects of SOX compliance. The consequences of successful penetration from an external hacker are significant depending on the extent of privileges an attacker is able to gain, such as theft, alteration, or destruction of your company's critical intellectual property and proprietary information. NuStuff has opted to implement a detect control in order to quantify the penetration attempts from the outside, as well as define a specific process to deal with specific threats that are discovered as part of the detection process.

TIP

There are many types of intrusion detection systems, including host, network and file based. This subject contains far more information than can be covered in this chapter however the SANS (SysAdmin, Audit, Network, Security) Institute, which is the largest source of information security training and certification available, has a comprehensive FAQ regarding IDS. http://www.sans.org/resources/idfaq/. We will cover more open source solutions in Chapter 8.

Sustainability and Accountability

Again since this is augmenting your security infrastructure your SOX auditor will want to see the control objectives, policies, procedures, and signoff of the key stakeholders for this project. In addition we cover the issues of security in more detail in Chapter 7.

Infrastructure Change Request Workflow

https://itsox2.nustuff.com/egroupware/workflow/index.php

The general process for the change management approval workflow is illustrated in Figure 7.8. While it may seem somewhat complicated at first glance, it is actually pretty straightforward. From a change management perspective this captures the information needed for SOX and follows the change through to production deployment. Since this is automated there is little effort necessary to capture and document the controls necessary for SOX compliance while still maintaining the good IT practices based on the COBIT framework.

Figure 7.8 Infrastructure Change Request Workflow

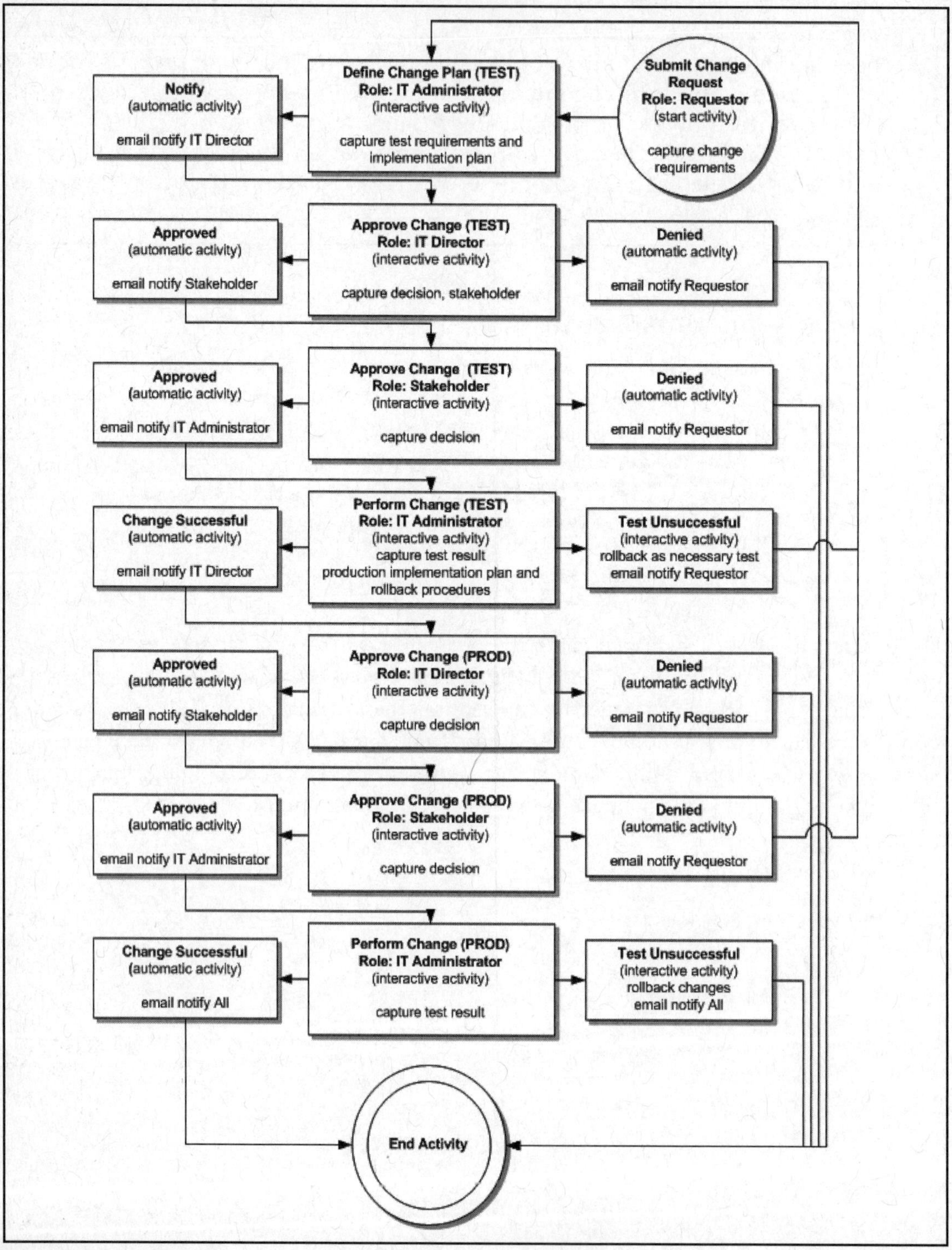

Workflow Roles

The Roles for this workflow and their members are as follows:

- Requestor – This is the person who submits the infrastructure change request. This is limited to the IT Staff group since this workflow is for IT specific changes.
- IT Administrator – The administrator is responsible for gathering the change requirements and procedures and performing the actual change. Again this role is assigned to the IT Staff group, with Ken King being the default assignee for this role, since he is the IT administrator.
- IT Director – Biff Johnson is the IT director, responsible for approving the original request and the transition into production. The IT Manager John Scott serves as the backup for this role.
- Stakeholder – This role can potentially be anyone in the company. This role is defined as the person who would own the process as a result of the change or the main beneficiary of the change itself. The IT Director role defines the Stakeholder in his approval process for routing purposes. For our examples the following persons are selected to fulfill this role:
  1. NuStuff Electronics – Although the intrusion detection system is purely an IT driven function, the goal is to improve security and the main beneficiaries are the business owners of the data being protected. Since this globally deployed the key stakeholder chosen for this process is the CEO Jane Brown, ensuring she is apprised of the security measures being implemented in the company. There is no backup for this role since it is dynamically chosen by the IT Director at runtime

Workflow Activities

Here are the activities associated with this workflow:

- Start/End activities – This is the point where a new instance of the workflow is begun or completed.
- Interactive activities – This workflow has several different types of interactive activities associated with it:
  - Definition Activities – The IT administrator fills in the requirements and procedures to implement the change in a test environment as well as defining the implementation and rollback plans for production.
  - Approval Activities – The IT Director approves the plan and assigns the Key stakeholder initially. Both sign off on both test and production stages.

- Implementation Activities – The IT Administrator performs the changes in test and production. Please note that the test implementation and the production definition activities are combined in the same step, since the production plan is only necessary if the test implementation plan is successful.
- Automatic activities – These activities automatically send out email notifications to the appropriate roles associated with the next step in the process following an approval, denial, or result of a change implementation, both test and production.

Implementation Planning

https://itsox2.nustuff.com/egroupware/index.php?menuaction=project manager.uiprojectmanager.index

The first step in implementing any change is to define what the change will be, how it will be implemented and what recovery procedures need to be taken if something goes wrong. Here are the implementation plans for our sample projects.

NuStuff Electronics Snort IDS

Again this is not designed to be a comprehensive Snort tutorial so we will cover the project from a high level overview as an example and to demonstrate the documentation capture and workflow requirements for SOX. In yet another example of the power of open source we will be using another Live CD distribution for our Snort IDS needs based on Knoppix derivative called Network Security Toolkit (NST). If you would like to actually perform a site security audit for this tool you can download the latest version from http://networksecuritytoolkit.org/nst/index.html.

Test Procedure

Again, we have a plan to evaluate the system for our needs prior to placing into production.

- Download and burn the latest NST Live CD http://www.networksecuritytoolkit.org/nst/index.html
- Boot Live CD on test network and set root password
- Use ifconfig to determine/set ip-address
- From test workstation login to NST web interface
- Select "Bleeding Snort" ruleset
- Start Snort IDS

- Use IDS Policy Manager For Windows to tune ruleset http://www.activeworx.org/programs/idspm/
- Use Metasploit on NST Live CD to test IDS

NOTE

Metasploit is a collection of tools, scripts and applications gleaned from known security exploits in the wild. The Metasploit framework provides and consistent and easy to use web interface for executing these tests, however they are *strictly* provided for legal penetration testing and research purposes only. Please be sure to get management signoff before using these on your own network, and of course, only your own network. See http://www.metasploit.com for more information.

Production Procedure

Once the testing phase is complete, our production procedure is nearly identical, we just now operate on the production network instead of the test network.

- Boot Live CD on test network and set root password
- Use ifconfig to determine/set ip-address
- From test workstation login to NST web interface
- Select "Bleeding Snort" ruleset
- Start Snort IDS
- Use IDS Policy Manager For Windows to tune ruleset http://www.activeworx.org/programs/idspm/

Rollback Procedure

When introducing new systems, services or applications it is sometimes as simple as turning things off if a rollback is needed. You need to be aware of any residual side affects as a result of the rollback and deal with them appropriately, but in this case it would not be necessary.

- Shutdown the NST Live CD system

Lessons Learned

Building an Effective Test Environment

Once your policies and control objectives have been identified and defined, the next step in the process is to identify application software that satisfies your objectives. In most cases you will not have every control objective necessary for compliance met at the beginning of your compliance cycle. Whether you choose to self appraise or have a company assist you in the preparation for your SOX audit, the initial assessment of your environment will yield areas that you need to address in any of the four main categories of availability, security, sustainability or accountability. You may have to deploy new software or make considerable changes to existing configurations in order to mold them into a SOX friendly mode. In the face of a major augmentation or change, this will rather obviously require planning and testing before deploying into your production environment. This being the case, to what point do you need to replicate your current environment, how close is close enough?

The unfortunate answer is… it depends. What you need to deploy into production might be very simple and superficial, or it might be a new and completely independent system from your existing infrastructure. In these cases your test environment would be tailored to reflect the needs of the application rather than any systems you might already have deployed. On the other hand if you are implementing something that will integrate with your present architecture to any degree you will need to simulate this environment accordingly in order to adequately test your integration efforts. For illustration, if we use the examples given in this chapter for our sample companies the security example would require the networking pieces to be in place, such as a firewall with a similar rule-set as well similar networking topology. To adequately test the web migration project you would need server hardware of similar specifications with the same software installed as your intended production target. The advantage of using open source is there will be no licensing issues from a software perspective for your lab systems.

Another important consideration for your test lab that might otherwise get overlooked is the subject of storage. It is good to remember that Production Data in a test environment = Production Data. A common example is the replication of a financials database to a test server in order to test a change or operation that would otherwise have an effect in your production server that would be difficult to rollback, such as closing a period. Whatever the need to have this type of data in your test environment, you must subject this to the same security and access controls that you would apply to production.

Implementation

https://itsox2.nustuff.com/egroupware/index.php?menuaction=projectmanager.uiprojectmanager.index

We will not spend too much time in this section other than to highlight the capabilities of some open source projects to assist in the project management, tracking, and deployment aspects of your implementation. As highlighted in the sidebar "Project Management is Key" the importance of keeping track of the various tasks and subtasks are important to ensure everything stays on track and focused on the end result. There are two main applications in the eGroupware suite to assist in the deployment stage, InfoLog and Projects. Depending on the complexity of your project, you may choose to utilize either the more simple InfoLog application to just assign tasks to users, or the full blown Projects module which is a complete project management application with milestones, jobs, time tracking, and Gantt charts. For our example implementations we have chosen to use the Projects module for NuStuff Electronics to give you an idea on how you might use these applications for your own needs.

Documentation

https://itsox2.nustuff.com/egroupware/phpbrain/index.php

Documentation is also an integral part of any ongoing IT function, and the Knowledgebase module is a very good tool for making this available to the right people. Documentation can be separated into categories and each category can have its own set of permissions. On the sample portal sites we have setup the following examples to give you an idea how this facility can be useful. As usual this is not exhaustive by any stretch and is just meant to give you and idea of what you might use the Knowledgebase application for.

- IT User Documentation – It is important to disseminate information to your users to help them "help themselves." Our simple examples include instructions on how to change their password and the corporate hardware standards for desktops and laptops.
- IT Administration – we have placed documents for administrators here, mostly based on the operational workflows discussed earlier in the chapter.
- HR Documentation – in an effort to illustrate that not only HR documentation can go here we included an HR section with a sample Employee Handbook. Other departments to consider would be finance, sales and marketing engineering, really anything and since you can define permissions on a per category group you can control access to sensitive documentation, such as monthly finance close procedures and budgets.

Other Change Management Workflow Examples

https://itsox2.nustuff.com/egroupware/workflow/index.php

Firewall Change Request

This workflow is an example where it is important to capture the approval of IT management, the test procedure and rollback plan. This differs from the General Infrastructure workflow since the key stakeholder is IT and this workflow is strictly designed with SOX documentation in mind. Figure 7.9 illustrates the simplified change request.

Figure 7.9 Firewall Change Request Workflow

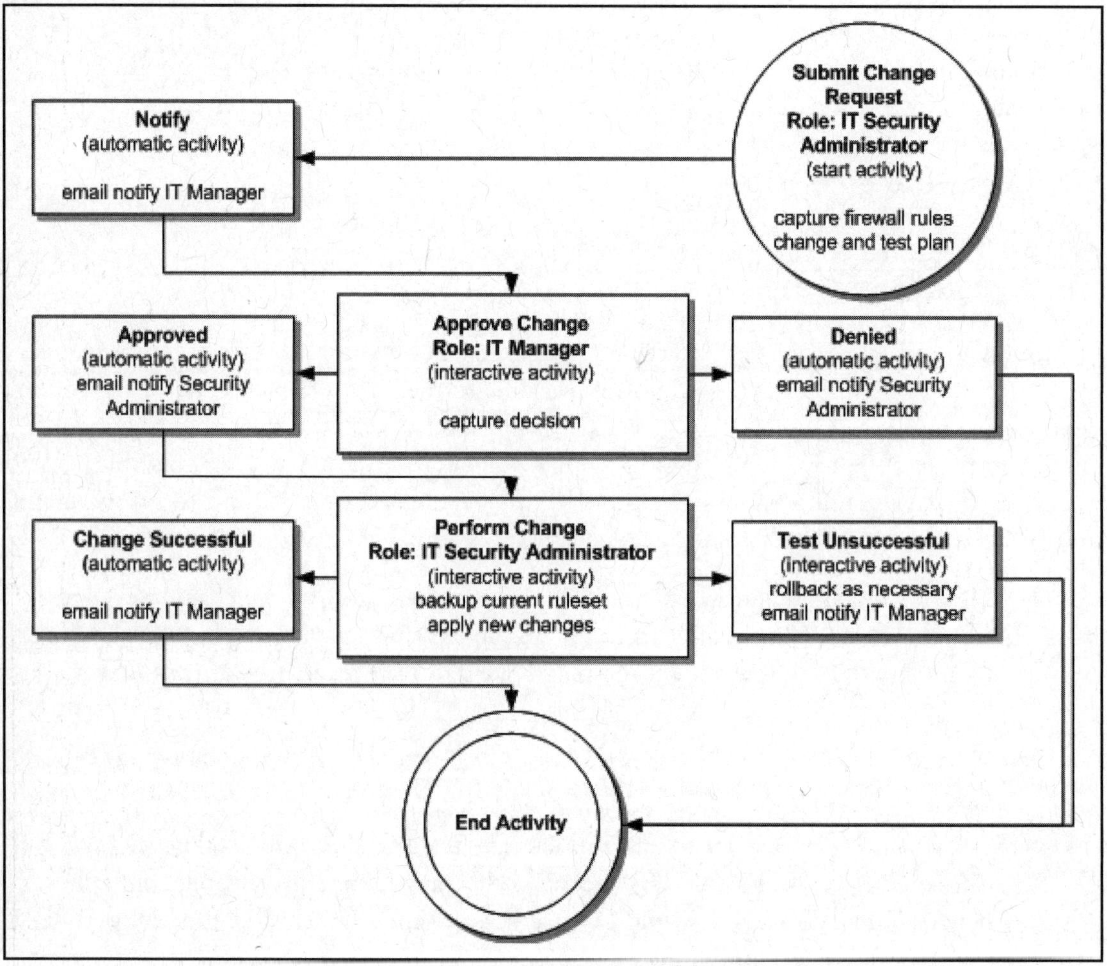

Workflow Roles and Activities

The roles associated with this workflow are the IT Security Administrator who is responsible for gathering the change requirements and procedures and performing the actual firewall change. The IT security administrator is the only person who can submit such change requests, and this role is assigned to Megan Rand, with Ken King serving as the backup for this role. The IT Manager role is responsible for approving the original request and since segregation of duties plays an important part of SOX compliance John cannot serve in the backup role of IT Security Administrator by definition. Biff Johnson serves as the backup for this role.

The workflow is fairly straightforward, the IT Security Administrator submits the workflow with the rules that need to be modified or added and the IT manager approves or denies the request. On approval the IT Security Administrator performs a backup of the current ruleset and applies changes. The test plan is executed and the new rules are either kept or the previous rules are reinstated if there is a problem. Implementing this workflow does not add to the workload of any IT staff and effectively captures change activity for later review.

Oracle Change Request

The Oracle Change Request is very close to the General IT Change request however there are subtle differences. Figure 7.10 shows the workflow activity.

Figure 7.10 Oracle Change Request Workflow

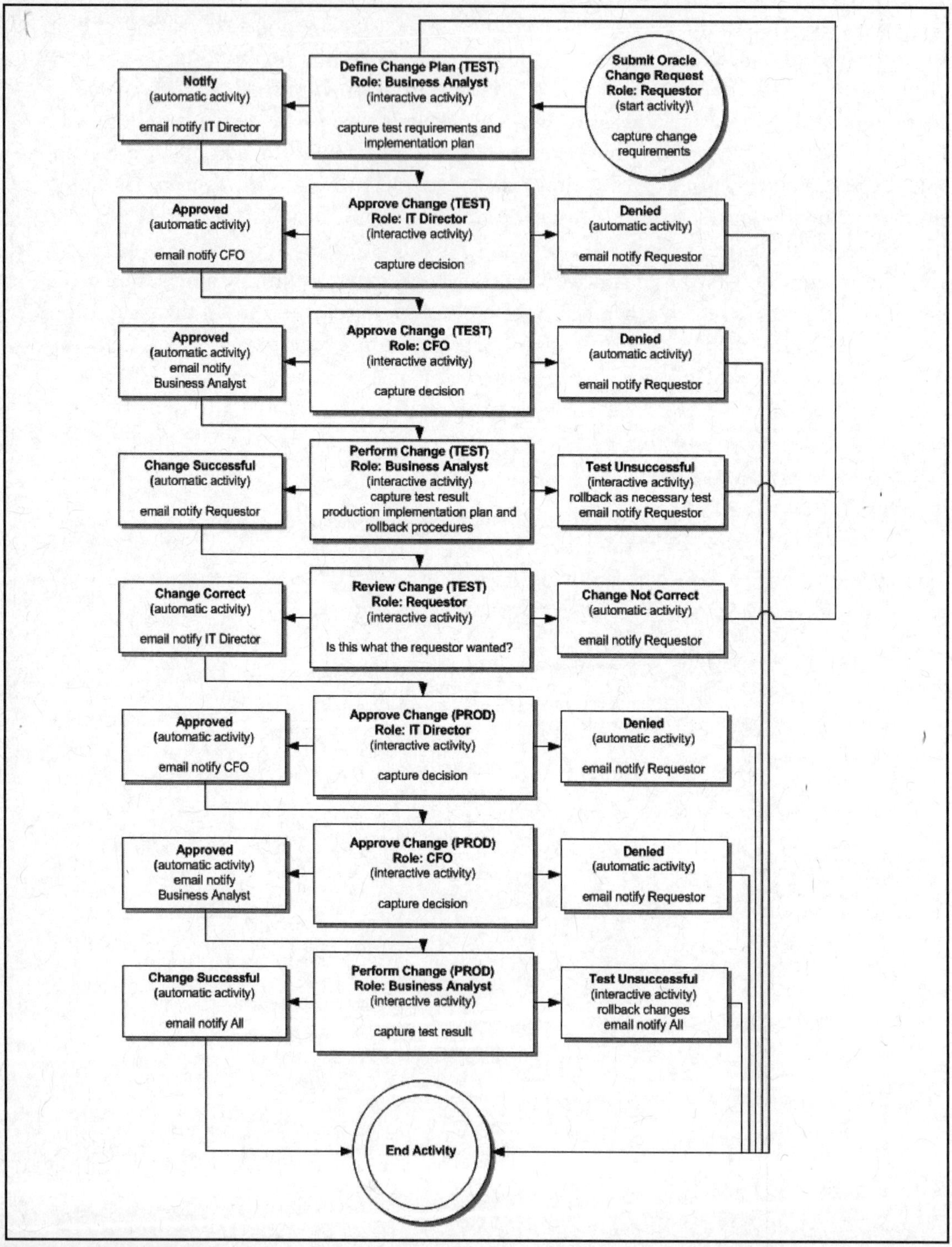

Workflow Roles and Activities

This workflow is designed to capture major changes to the financial system, including patches, upgrades, and interface or database changes. The requestor for this workflow is anyone in finance or the business analyst, so a group has been defined for the entire finance department. The business analyst is responsible for translating the user's requirement into a solution plan that can be submitted for approval. If the business analyst is also the requestor then he or she needs to play both roles. The IT Director is part of the approval chain and has first say for approving the intended change. Since the key stakeholder in this case is Finance, the CFO is the next stop in the process. Once approval has been granted the business analyst will perform the change in test. If the change is unsuccessful, the workflow loops back to the requirements gathering stage so that it can be revised and corrected.

TIP

In the event of an outsourced application, the business analyst role can be performed by someone outside the company. In this case you will need to provide a login to the eGroupware system for this person, and if you decide to make the website available outside your firewall in order to grant access to external partners you should ensure that the server is encrypted via SSL. For information on how to perform this type of installation refer to the Installation and Security Guide available on the eGroupware Sourceforge project page http://sourceforge.net/project/showfiles.php?group_id=135305

At this point an additional step has been added to the workflow to ensure that the requestor can review the change in the test environment to make sure that it fulfills his or her original requirements. Again if the requestor does not think the problem has been solved the workflow loops back to the requirements gathering stage so that it can be revised and corrected. The workflow then continues with approvals from the IT Director and CFO, and a step to implement the change in production.

Summary

In this chapter we have discuss automation and why it should be a key component of any small to medium sizes company's Sarbanes-Oxley compliance actives. We also developed guidelines by which you cannot only assess your in-house expertise as they relate to the skill sets that will be necessary to utilize open source tools. We have also provided action and alternative for acquiring the necessary skill set if it does not currently exist within your organization.

Also as part of the VM Spotlight we discussed how Webmin was a web-based inter-face for system administration for Linux and UNIX. How Webmin could be used to remotely setup and maintain system by use of a web browser, and that it removes the need to manually edit UNIX configuration files. We also identified some of the standard Webmin modules:

- Apache Web Server – Configure almost all Apache directives and features.
- BIND DNS Server – Create and edit domains, DNS records, BIND options and views.
- Backup Configuration Files – Perform manual or scheduled backups and restores of configuration files managed by Webmin modules.
- Bandwidth Monitoring – View reports on bandwidth usage by host, port, protocol and time on a Linux system.
- Bootup and Shutdown – Setup scripts to be run at boot time from /etc/init.d or /etc/rc.local.
- Command Shell – Execute shell commands and view their output.
- Custom Commands – Create buttons to execute commonly used commands or edit files on your system.
- File Manager – View, edit and change permissions on files and directories on your system with a Windows-like file manager.
- Filesystem Backup – Backup and restore filesystems using the dump and restore family of commands.
- SSH Server – Setup the SSH server for remote secure logins.
- Samba Windows File Sharing – Create and edit samba file and print shares.
- Scheduled Commands – Schedule the execution of one-off commands or scripts.
- Scheduled Cron Jobs – Create, edit and delete Cron jobs.
- Software Packages – Manage software packages on your system, and install new packages.

- System Logs – Configure the syslog server on your system and view its log files.
- Upload and Download – Upload multiple files to the server, and download multiple URLs either immediately or in the background at a scheduled time.
- Webmin Actions Log – View detailed logs of actions by Webmin users.

In concluding our discussion on Webmin we identified it as an Open Source tool that could assist a small to medium size company with potential "segregation of duties" issues.

Additionally, we've looked at the various control objectives of the "Acquisition and Implementation Domain", correlated them to ITIL and identified ones that relate specifically to Sarbanes-Oxley compliance, using our fictitious company as an example. But in summarizing this chapter there are three fundamental things you should take away with you:

- Let your unique organizational structure and your specific TDRA drive the applicable domain items
- Automation will be critical but you will need to avoid the urge to over implement
- Utilization of good project management methodologies will better position your compliance effort for success

The remainder of the chapter shows examples of automation, project planning and tracking for the sample companies. NuStuff Electronics on the other hand has opted to augment their security infrastructure with an Intrusion Detection System. Snort has been selected as a network based detection system since it is the leading open source solution and there is copious documentation and several books written for its deployment. In order to leverage the hard work of the open source community the NST Live Security CD will be deployed since it contains both the Snort IDS and a testing framework. These project examples are then put through the COBIT framework for approval, planning and implementation, using the example workflow, project management, and documentation modules on the ITSox2VM Toolkit. The chapter closes with a discussion of additional example change management workflows where there might be special considerations such as the need for additional activities on one side of the spectrum and the desire to simplify the generic change management workflow on the other, which ultimately demonstrates the flexibility of the system.

Solutions Fast Track

Definition of Information Requirements

☑ Implemented methods, practices and infrastructure should evolve as the requirements and needs change.

☑ There could be a need to obtain new applications and acquire or develop a new staff skill-set to execute new plans.

- ☑ Plans enacted in the implementation phase should include maintenance, testing, certifying, and identification of any changes needed to ensure continued availability of existing systems as well as the implementation and support of new systems.
- ☑ Given that there are little or no acquisition costs associated with Open Source tools the tendency might be to over implement. If the proper steps have been taken to view your needs holistically, while at the same time using a quality framework as a reference, you should be able to reduce this risk.

Evaluating Open Source In-House Expertise

- ☑ An IT organizations skill sets and expertise are usually related to their current environment and not necessarily to other IT disciplines not contained within that environment, or more importantly introduced by Sarbanes-Oxley.
- ☑ Any organization that would like to avail themselves of Open Source tools for Sarbanes-Oxley compliance should have minimum level of expertise with Open Source tools.
- ☑ When evaluating the necessary skill set you will most likely need to look at two general aspects:
  - Linux & Unix experience
  - Open Source experience
- ☑ Linux & UNIX experience is different from Open Source experience.

Automation is the Name of the Game

- ☑ Resources and activities are always at odds in small to medium size companies – too few resources and too many activities.
- ☑ The acquisition and implementation of potential applications for your environment should be the ability not only to automate new requirements but also to automate some of your existing requirements.
- ☑ The goal of automating of new and existing functions should be able to better leverage existing resource, and possibly stave off the necessity to add additional resources.
- ☑ The automation process will force better-defined processes and remove subjectivity from the processes.
- ☑ Eliminating subjectivity from processes facilitates the ability demonstrate that processes are executed as defined.

Working The List

- ☑ COBIT, ITIL and SOX compliance are distinctly different entities.
- ☑ You may be able to derived additional benefits from implementing COBIT or ITIL components beyond compliance.

VM Spotlight: Webmin

- ☑ Webmin is a web-based interface for system administration for Linux and UNIX.
- ☑ Webmin removes the need to manually edit UNIX configuration files and comes with over 100 standard modules in the base distribution.
- ☑ Webmin allows your administrators to perform functions that would otherwise require root access to the machine without giving them the root credentials.
- ☑ Webmin augments your staff's skills by providing an easy to use GUI based administration interface.
- ☑ Webmin could be a useful tool for administering Linux when it comes to the SOX requirement of "segregation of duties."
- ☑ Each Webmin module allows different levels of access to different users so that you can delegate tasks.

Webmin can help in a number of ways other than configuring aspects of your system. You can automate processes such as backing up and logging, firewalling, as well as many "cluster aware" tasks. Webmin allows you to manage multiple machines from a single interface, providing additional modules to manage:

- Cluster Webmin Servers – Install and manage modules, themes, users, groups and access control settings across multiple Webmin servers.
- Cluster Change Passwords – Change passwords on multiple systems in a Webmin cluster at once.
- Cluster Copy Files – Schedule the transfer of files from this server to multiple servers in a Webmin cluster.
- Cluster Cron Jobs – Create scheduled Cron jobs that run on multiple servers simultaneously.
- Cluster Shell Commands – Run commands on multiple servers at once.
- Cluster Software Packages – Install RPMs, debian and solaris packages across multiple servers from one source.

Webmin is written in Perl and has a completely documented application programmer's interface for the Open Source community to write their own modules that are not included in the base distribution. Many hundreds of these modules are available as Open Source on the website http://webmin.com/third.html. One such example that we alluded to in Chapter 5 is the FDS Toolkit http://fdstools.sourceforge.net. This is an application that provides user, group and host management for Fedora Directory Server written by the authors of this book and released under the terms of the GPL. In addition to the command line utilities, another feature of FDS Toolkit is a complete Webmin module to provide the command line functionality in a web browser GUI. We will use this module as our example a bit later in the chapter to further spotlight the design and capabilities of Webmin in general.

Case Study: Automation

- ☑ There are four main categories of automation that talk directly to your control objectives:
  a. Operations Support sustains availability objectives
  b. Access Controls reinforce security objectives
  c. Change Management supports sustainability objectives
  d. Compliance Management sustains accountability objectives
- ☑ Changes introduced into your IT environment must have the proper approval and acknowledgement from the key stakeholders. Change management is a key SOX compliance objective, and this chapter shows demonstrates that automation workflows can be leveraged to accomplish this.
- ☑ An effective test environment will ensure your projects successful rollout and integration.
- ☑ Production data in a test environment = production data, and should be subjected to the same security and access controls.
- ☑ Implementation can be streamlined and managed effectively through the use of project management and tracking tools.

Frequently Asked Questions

Q: Can I automation really minimize resource requirements?

A: Yes, as long as you don't over implement

Q: Is over implementation really an issue?

A: Yes, if you lose sight of the business objectives and become engrossed in the open source applications and capabilities.

Q: Why shouldn't I use the template in the book?

A: You probably could if your environment mirrors NuStuff. But I would recommend that you use the process to develop your own template rather then use the ones in the book, but they can serve as a good place to start.

Q: Are there any open source tools that can assist me with Project Management?

A: Yes, some of these tools are included on the ITSox2 VM Toolkit. Additionally you might want to check out:

- GanttProject http://ganttproject.sf.net/
- PHProjekt http://www.phprojekt.com/
- dotProject http://www.dotproject.net/
- Planner http://developer.imendio.com/wiki/Planner

Q: Can I fully take the People out of the Process?

A: No, but as in NuStuff you can wrap automated processes around the functions but you will not be able to completely take the People out of the Process.

Q: Can tools that are not open source be used to assist with compliance?

A: Of course, but since this book is about open source we will focus our discussion there.

Q: This workflow stuff seems complicated. Does it add more time and effort in performing day-to-day operations?

A: As with any new procedure, you might experience some initial resistance to using the system, however once this is ingrained into your internal procedures you will find that it actually helps your efficiency as an IT group and gives you the ability to quantify your deliverables to the organization, not to mention the fact that your SOX auditors will be happy.

Chapter 8

Are We There Yet?

Solutions in this chapter:

- All About Service
- Delivery & Support
- Working The List
- Service Level Agreements
- Managing The Infrastructure
- VM Spotlight – Subversion
- Case Study: NuStuff Electronics Segregation of Duties

☑ Summary

☑ Solutions Fast Track

☑ Frequently Asked Questions

All About Service

"The first mark of good business is the ability to deliver. To deliver its product or service on time and in the condition which the client was led to expect. This dedication to provision and quality gives rise to corporate reliability. It makes friends and, in the end, is the reason why solvent companies remain solvent."

–Michel Ferguson on the delivery of quality service

While Mr. Ferguson may not have been speaking of IT services specifically, we found the above quote not to only be a very powerful statement but one that struck at the very heart of what we believe to be the essence of what IT should mean to your organization. Assuming this is the premise for what IT is all about and also assuming that as the responsible CFO, CIO or IT Director you have been confronted with the questions of "what do those IT people do"? What value does the IT Organization provide? Let's not forget the granddaddy of them all – why shouldn't we outsource the IT Organization?

In thinking about these questions, as people responsible for an IT Organization we can probably come up with several reasons why these questions continue to reoccur and never go away. Although the reasons we would identify might be valid, we propose that the main reason these question continue to confront IT Organization is because, whether perceived or not, the company sees no value from their IT Organization. We are clearly cannot judge the effectiveness of your IT Organization but what we can do is make the point that if a company sees a return for its IT expenditures and sees value from its IT Organization then these are usually readily dismissed by the company. So you might be saying to yourself – that's all fine and dandy but how does an IT Organization add value? There is no simply answer to this question, as each organization is different. But if we refer back to the above quote and make a slight modification – "To deliver IT product or service on time and in the condition which the client was led to expect."

Whether you subscribe to the notion or not that in essence the people and departments that you support are your clients. What we hope you do realize by now is that where any IT Organization brings value to their company is by as the quote states – providing – product or service on time and in the condition which the client was led to expect. The inference of on time is pretty self-explanatory but from a SOX perspective we would like to add some criteria to "in the condition which the client was led to expect", like within budget, within requirements, etc. Whether you're main objective is to merely comply with Sarbanes-Oxley or whether you have elected to use SOX compliance as an opportunity to reposition your IT Organization, this is where the rubber meets the road.

The Transparency Test

The Manager's Perspective

Recently I had the opportunity to speak with a former employee. He had recently changed jobs and had been correlating the similarities between his new boss and myself. He was explaining to me how his new boss and I supported many of the same concepts as they relate to being customer focused, understanding the customer's business requirements and the delivery of IT services in support of stated customer's requirements. He further explained how his new manager had given him two books to read, "Crucial Conversations, Tools for Talking When the Stakes are High" and "Managing IT as a Business, A Survival Guide for CEOs. The rationale his new boss had for giving him these books to read was because he has a keen focus on customer communications. He further explained that he requires his managers to work with stakeholders, set and manage expectations for deliverables, timelines and budgets. Any and all new capabilities must meet the stated needs of the organization and be completed within the work plans objectives and timelines. To get a gauge of this each year, their IT publishes the results from their customer satisfaction survey. These results and actions to address them are published and distributed on a tri-fold 8 1/2 × 11 report card.

In thinking about our conversation I thought this strategy would make for a good example of how if IT organizations are going to be successful and support their company they must embrace change. The days of IT organizations building cases to justify technical expenditures then delivering to their customers what they think meets their requirements, or worse yet what they want to give their customers are quickly coming to an end. The new model of the successful IT organization will be one where IT partners with their customers, one where customers' requirements and SLA's are drafted to manage, measure, report project results. Although SOX compliance may have been the impetus for you to read this book, as the authors have pointed out, you should use this as an opportunity to review your own IT organization and start the transformation into the organization of the future.

–Bill Haag

In this chapter we will look at the day to day delivery of IT services, and based on our experience offer suggestions on how a small to medium size company might not only be able to scale this to a manageable process, suggest possible Open Source tools that can help along the way and contribute to a sound organizational processes that facilitate compliance. We begin the chapter with a look at Service Level Agreements.

Delivery & Support

The COBIT Delivery & Support phase ensures that not only do systems perform as expected upon implementation but they continue to perform in accordance with expectations over time usually managed via SLAs". Now as part of Chapter 8 we will look at the specifics of each of the control objectives of the "Delivery & Support Domain", summarize and distill the various control objectives to lend themselves more to the structure of a small to medium size company and correlate them as appropriate to ITIL.

Not to minimize the significance and/or effort required for any of the previous COBIT Domains and ITIL components, but without a doubt Delivery & Support is probably the one that will cause the most concern in a small to medium size company. The concern will more then likely rise out of two concerns 1) given the number of IT resources can these activities be sustained and 2) is there really a need for all this bureaucracy. If we have accomplished nothing else hopefully by now you have come to realization that the answer to the first concerns is yes. Although the answer to the second concern is yes as well, if the right Open Source tools are implemented the gains in efficiencies, security, user satisfaction and environmental stability will more then sufficiently mitigate the introduction of any bureaucracy. Similar to the "personal processes" process we discussed in Chapter 6 and the need to understand your company's culture, there is also a need to understand the normal perception about and/or within a small to medium size company. Two of these perceptions are:

- It's is easier and quicker to relay requests via phone or in person
- People don't have time for processes and filling out forms

While it might be interesting to debate whether these perceptions are valid for a small to medium size company the reality is that Sarbanes-Oxley has made this debate a moot point. We cannot emphasize this point enough as it relates to this chapter because not only will you need to get Executive Management to understand it but your customers as well.

As we stated the COBIT "Delivery & Support Domain and ITIL components covered in this chapter is where an IT Organization can gain the most benefit from a repositioning perspective and therefore the control objective identified for our fictitious company have been weighted to accomplish that goal.

1. Define and Manage Service Levels

This section as it pertains to COBIT deals with Service Level Agreement and the various element that should be taken into consideration and the processes surrounding these agreement.

COBIT **1. Define and Manage Service Levels**	ITIL	Guidance	Objective
1.2. Aspects of Service Level Agreements	Service Delivery, Service Level Management, 4.6 SLA contents and key targets	COBIT basically states, that explicit agreement should be reached with regards to the elements of SLAs The SLAs should cover areas such as availability, reliability, performance, capacity, levels of support provided to users, acceptable level of satisfactorily delivered system, etc…	SOX & Repositioning
1.3. Monitoring and Reporting	Service Delivery, Service Level Management, 4.7 KPI and metrics for SLM efficiency and effectiveness	COBIT basically states, a service level manager should be put in place that is responsible for monitoring and reporting on the achievement of the specified service performance criteria and all problems encountered during processing. This is an area where if automation is leverage and Quality process deployed cost associated with an additional	Repositioning

Continued

COBIT 1. Define and Manage Service Levels	ITIL	Guidance	Objective
		H/C or the burdening of an already taxed resource can be avoided.	
1.4. Review of Service Level Agreements and Contracts	Service Delivery, Service Level Management, 4.3.2 Plan monitoring capabilities	COBIT basically states, a regular review process for service level agreements and contracts with third-party service providers should be performed. The management of this activity can be automated through eGroupware via Work-Flow Process.	Repositioning

2. Manage Third-Party Services

This section as it pertains to COBIT deals Third-Party qualifications and management. This is an area that is of particular note as it relates to our fictitious companies since they have elected to outsource their Oracle requirements.

COBIT 2. Manage Third-Party Services	ITIL	Guidance	Objective
2.4. Third-Party Qualifications	No ITIL correl ation this is an example of where you would need to develop a com-pliance compo-nent for use with ITIL or	COBIT basically states, potential third parties need to be assessed as to their capabilities and competence to deliver the required services. Since our fictitious companies elected to outsource their Oracle	SOX & Repositioning

COBIT 2. Manage Third-Party Services	ITIL	Guidance	Objective
	augment ITIL with the COBIT control.	requirements they ensure that the selected outsource provide had the required SASE 70 certification for SOX compliance.	
2.7. Security Relationships	No ITIL correlation this is an example of where you would need to develop a compliance component for use with ITIL or augment ITIL with the COBIT control.	COBIT basically states that processes and procedures need to be put in place and followed to ensure that proper security practices have been put in place to govern third-party service providers. i.e. non-disclosure agreements, contract, etc...	SOX
2.8. Monitoring	Service Delivery, Service Level Management, 4.4.7 Establish monitoring capabilities The Business Perspective, Supplier Relationship Management, 7.4 Contract management	COBIT basically states, that processes should be put in place to monitor the delivery of services by third-party providers. These processes should measure delivery of service against the contractual agreement. This is an area that should be included as part of your SLA and managed via eGroupware.	Repositioning

3. Manage Performance and Capacity

This section as it pertains to COBIT drives to the heart of SLAs by way of Performance and Capacity management. Please bear in mind that areas identified as SOX pertain only to system germane in the process of reporting financial results.

COBIT 3. Manage Performance and Capacity	ITIL	Guidance	Objective
3.2. Availability Plan	Service Delivery, Availability management, 8.5 Availability planning	COBIT basically states, processes and procedures should be established and implemented to en sure the availability of information services. Again, this can be part of SLA requirements and documented via eGroupware.	Repositioning
3.3. Monitoring and Reporting	*Service Delivery*, Availability Management, 8.7 Availability measurement and reporting *Service Delivery,* Capacity Mana-gement, 6.3.1 Monitoring	COBIT basically states, a process should be established to ensure that the performance of IT resources is continuously monitored. The process should have an exceptions are reporting element to ensure timely resolution of issues. The key com ponents for this objective could be Zabbix, Quality Process and eGroupware.	SOX & Repositioning
3.4. Proactive Performance Management	Service Delivery, Availability Management, 8.5.3 Designing for availability Service Delivery, Capacity Management, 6.3.7 Modeling	COBIT basically states, the performance management process should include fore-casting capabilities to allow for proactive problem resolution. This is why the establish-ment of SLA monitoring criteria and the setting of your monitoring tool (Zabbix) threshold will be important. You will	SOX & Repositioning

COBIT 3. Manage Performance and Capacity	ITIL	Guidance	Objective
		want to set your threshold in a manner in which you are afforded the opportunity to correct problems.	
3.8. Resources Availability	Service Delivery, Availability Management, 8.3 The avail ability management process	COBIT basically states, availability of systems has been identified step should be taken to ensure the system(s) availability. This step can be in the form of a fault tolerant architecture. Although there is no stipulation to implement fault tolerance for SOX you will need to be able to rationalize your architecture and plans if an outage occurs during a critical period such as Finance cycle close periods.	SOX & Repositioning

4. Ensure Continuous Service

This section as it pertains to COBIT deals with an organizations ability to provide Continuous Service to their supported customer base. Although for the purpose SOX the only COBIT control that has been listed is 4.12.Off-site Back-up Storage (SOX).

COBIT 4. Ensure Continuous Service	ITIL	Guidance	Objective
4.12. Off-site Back-up Storage	Service Delivery IT Service Continuity Management, 7.3.2 Requirements analysis and strategy definition	COBIT basically states, off-site storage for critical back-up media, documentation and other IT resources should be established to support recovery and business continuity plans. As described later in this chapter, business process owner should be involved in establishing criteria for off-site storage. This Control Objective is also one where certification will be beneficial.	SOX

5. Ensure Systems Security

This section as it pertains to COBIT deals logical security of systems and system resources. In reviewing the Control Objectives in this Domain we have elected to leave several in that don't necessarily pertain to SOX, or that will be visible from a customer perspective and therefore can't be considered as "Repositioning" but nonetheless should prove of value.

COBIT 5. Ensure Systems Security	ITIL	Guidance	Objective
5.2. Identification, Authentication and Access	Security Management, Security Management Measures, 4.2.4 Access control	COBIT basically states, that procedures should be implemented and update to ensure that logical access to and use of IT computing resources has adequate control to ensure that users of the resource has been granted proper approval. Based on your own unique environment this might be accomplished via LDAP, Active Directory, etc... It is also noteworthy to mention that these controls or similar one should be extended to remote access.	SOX
5.3. Security of Online Access to Data	Security Management, Security Management Measures 4.2.2 Access control	COBIT basically states, policies and procedures should be develop that access security control based on the individual's demonstrated need to view, add, change or delete data. This particular control can be considered as working in concert with the above Control Objective.	SOX
5.4. User Account Management	Security Management, Security Management Measures, 4.2 Implement	COBIT basically states, procedures should be established to ensure timely response for the requesting, establishing, issuing, suspending and closing of user accounts.	SOX

Continued

COBIT 5. Ensure Systems Security	ITIL	Guidance	Objective
		This control should have a formal approval procedure outlining the data or system owner granting the access.	
5.5. Management Review of User Accounts	Security Management, Security Management Measures, 4.3 Audit and evaluate security reviews of IT systems	COBIT basically states, a control should be established governing the periodic review of access rights and a mechanism of reconciliation. The reconciliation process can be as simple as comparing your authentication layer log against your user access request process logs.	SOX
5.8. Data Classification	Security Management, Security Management Measures, 4.2 Implementation	COBIT basically states, all data and access should be classified in terms of sensitivity via a formalize process with the data owner. This is definitely a good practice but it may be a daunting one. So, it may be advantageous to limit this activity to financial data.	SOX
5.9. Central Identification and Access Rights Management	Security Management, Security Management Measures 4.2 Implementation	COBIT basically states, that the identification and access rights of users as well as the identity of system and data ownership are established and managed from a centralized point to ensure consistency and efficiency of global access control. Although this	SOX

COBIT 5. Ensure Systems Security	ITIL	Guidance	Objective
		control makes administration of access/account more efficient it is not a SOX requirement. But you will need to be able to demonstrate control over financial systems access.	
5.10. Violation and Security Activity Reports	Security Management, Security Management Measures, 4.5 Report	COBIT basically states, security activity and any violation are logged, reported, reviewed and when appropriate escalated to identify and resolve incidents involving unauthorized activity. As it relates to SOX this capability and any documented policies are important to be able to demonstrate to auditors. It would be advisable to enable logging and audit trails if the capability is present in your systems.	SOX
5.19. Malicious Software Prevention, Detection and Correction	Security Management, Security Management Measures, 4.2 Implementation	COBIT basically states, processes and procedures should be established for the prevention, detection, correction and response to computer viruses or trojan horses. The aforementioned processes and procedures should encompass clients and servers. Capabilities to consider when looking at virus detection software are:	SOX

Continued

COBIT 5. Ensure Systems Security	ITIL	Guidance	Objective
		1. Centralized management 2. Ability to record client virus activity 3. Centralized and automated virus definition file distribution	
5.20. Firewall Architectures and Connections with Public Networks	Security Management, Security Management Measures, 4.2 Implementation	COBIT basically states, adequate firewalls should be in place to protect against denial of services and unauthorized access to the internal. With regards to SOX it will be important to ensure that your firewalls have logging capabilities and that there is a tracking method in place to capture any occurrences.	SOX

6. Identify and Allocate Costs

Although managing costs and allocating budgets is an important IT related activity it does not have a direct bearing on our SOX compliance efforts, so we can skip this section.

7. Educate and Train Users

This section as it pertains to COBIT still deals security of systems and system resources. No matter how well security measures have been applied if the end-user is not aware of what constitutes good security practices, the security measures are likely to fail.

COBIT 7. Educate and Train Users	ITIL	Guidance	Objective
7.3. Security Principles and Awareness Training	Security Management, Security Management Measures, 4.2 Implementation	COBIT basically states, personnel must be trained and educated in system security principles. This education process should include: ethical conduct of the IT function, security practices to protect against harm from failures affecting availability, confidentiality, integrity and performance of duties in a secure manner. The aforementioned training can be done via policy statements on an intranet site or distribution of these policies via employee handbooks.	SOX

8. Assist and Advise Customers

This section as it pertains to COBIT still deals strictly with the customer experience in dealing with your IT organization whether you have a dedicated Helpdesk or merely process of contacting IT personnel directly. The listed control objectives are fairly standard and should for the most part be common sense.

COBIT 8. Assist and Advise Customers	ITIL	Guidance	Objective
8.2. Registration of Customer Queries	Service Support, Incident Management, 5.6.2 Classification and initial support	COBIT basically states, procedures should be in place to capture all customer calls. SOX does not require you to implement a call systems. This objective can be accomplished via eGroupware (Workflow Process) or by placing a process in place to receive requests via email.	SOX & Repositioning
8.3. Customer Query Escalation	No ITIL correlation this is an example of where you would need to develop a compliance component for use with ITIL or augment ITIL with the COBIT control.	COBIT basically states, customer requests not immediately involved should be appropriately escalated. This too can be accomplished through the use eGroupware in conjunction with Zabbix. We cover Zabbix in detail in Chapter nine	SOX & Repositioning
8.4. Monitoring of Clearance	Service Support, Incident Management, 5.4.3 Investigation and diagnosis Service Support, Incident Management,	COBIT basically states, procedures should establish to monitor the timeliness of customer requests resolution. The same process as described above	SOX & Repositioning

COBIT 8. Assist and Advise Customers	ITIL	Guidance	Objective
	5.6.6 Ownership, monitoring, training and communication	in "Customer Query Escalation" can be utilized for this Control Objective.	
8.5. Trend Analysis and Reporting	Service Support, Service Desk, 4.9 Incident reporting and review Service Support, Problem Management, 6.8 Proactive Problem management	COBIT basically states, Procedures should be put in place that ensure, sufficient reporting of customer queries and resolution, response times and trend identification. If you have follow our suggestion with regards to Quality the aforementioned elements ill be part of any processes and/or procedures.	SOX & Repositioning

9. Manage the Configuration

This section as it pertains to COBIT still deals with server and client configuration management and ensuring that process exist to ensure standardized configuration as well as to ensure that implemented configuration are not changed without going through the proper change process.

COBIT 9. Manage the Configuration	ITIL	Guidance	Objective
9.4. Configuration Control	Service Support, Configuration Management, 7.6.3 Control of CIs	COBIT basically states, procedures should be in place to ensure that the deployed configuration remain the same, useless changes have been approved and have gone through the change process. There are many tools that provide this functionality for switches, routers and firewalls but the one we found to be particularly good is Kiwi Cattools – it's not Open Source but it is relatively inexpensive.	SOX
9.5. Unauthorized Software	Service Support, Release Management, 9.3.6 Definitive software library	COBIT basically states, policies restricting the use of personal and unlicensed software should be developed and implemented. Again, the key is that all changes should follow the approved process and procedures should be in place to detect deviations.	SOX
9.7. Configuration Management Procedures	Service Support, Configuration Management, 7.11.1 Level of control Service Delivery, Availability Management, 8.9.1 Component failure impact analysis	COBIT basically states, procedures should be in place to ensure that critical components have been appropriately identified and are maintained. You can satisfy to this control, by merely establishing processes that dictate where configuration file should be store and incorporating the backup of this location.	SOX

10. Manage Problems and Incidents

This section as it pertains to COBIT still deals with server and client configuration management and ensuring that process exist to ensure standardized configuration as well as to ensure that implemented configuration are not changed without going through the proper change process.

COBIT 10. Manage Problems and Incidents	ITIL	Guidance	Objective
10.1. Problem Management System	Service Support, 4. The Service Desk Service Support, 5. Incident 8.1 Operational Management Service Support, 6. Problem Management Service Support, Incident Management, 5.3.5 Relationship between incidents, problems and known errors	COBIT basically states, that a problem management should be implemented to capture all operational events which are not part of the standard operation (incidents, problems and errors) are recorded, analyzed and resolved in a timely manner. With the correct implementation of eGroupware and Zabbix this Control Objective should be easily met.	SOX
10.2. Problem Escalation	Service Support, The Service Desk, 4.4.4 Escalation management Service Support, Incident Management, 5.3.3 Functional vs. hierarchical escalation Service Support, Incident Management, 5.6.6 Ownership, monitoring, tracking and communication	COBIT basically states, problem escalation procedures should be defined and implemented to ensure that problems are efficiently resolved and in a timely manner. Again, with the correct implementation of eGroupware and Zabbix this Control Objective should be easily met.	SOX

11. Manage Data

This section as it pertains to COBIT deals with a very important aspect of what your SOX audit will be and that is the ability to backup and restore data. This is also an area where you my want to ensure that the focus is kept on systems that are material in the reporting of financial data.

COBIT 11. Manage Data	ITIL	Guidance	Objective
11.23. Back-up and Restoration	ICT Infrastructure Management, Annex 4D, Back-up and Storage	COBIT basically states, processes and procedure should be developed and implemented to ensure that the strategy for back-up and restoration satisfies the business requirements and business process owners.	SOX
11.24. Back-up Jobs	ICT Infrastructure Management, Annex 4D, Back-up and Storage	COBIT basically states, procedures implemented to ensure back-ups performed in accordance with the defined back-up strategy and periodic restores are performed and verified. This particular Control Objective is an ideal candidate to be automated via eGroupware. What is important will be the capturing and maintaining of verification for this activity.	SOX

12. Manage Facilities

This section as it pertains to COBIT still deals with server and client configuration management and ensuring that process exist to ensure standardized configuration as well as to ensure that implemented configuration are not changed without going through the proper change process.

COBIT 12. Manage Facilities	ITIL	Guidance	Objective
12.1. Physical Security	No ITIL correlation this is an example of where you would need to develop a compliance component for use with ITIL or augment ITIL with the COBIT control.	COBIT basically addresses, the physical security and access control associated with securing IT related hardware, system and applications. Believe it or not with a little creativity this particular Control Objective can be met with Open Source – Nagios to be precise. You will need to acquire an additional component. The component you will need is the EM01B which will give you the capability of monitoring temperature, relative humidity, illumination and DC voltage monitoring to detect contact open/ closure. If your facility does not have card key control for doors the above can still be utilized using a manual key process.	SOX
12.3. Visitor Escort	No ITIL correlation this is an example of where you would need to develop a compliance component for use with ITIL or augment ITIL with the COBIT control.	COBIT basically states, that procedures need to be in place to ensure that all Non-IT staff and or non-approved employees are only given access to computer facilities in the accompaniment of an escort. This Control Objective can be accomplished via a visitor's log and the reviews of the logs managed and captured by eGroupware	SOX

Continued

COBIT 12. Manage Facilities	ITIL	Guidance	Objective
12.5. Protection Against Environmental Factors	No ITIL correlation this is an example of where you would need to develop a compliance component for use with ITIL or augment ITIL with the COBIT control.	COBIT basically states, processes should be implemented and maintained to ensure that critical computer equipment is protected from environmental factors (e.g., fire, dust, power, excessive heat and humidity). In conjunction with some additional safeguards (fire suppressants, extinguishers, etc.) this Control Objective can be met with the implementation of the Nagios EM01B component.	SOX
12.6. Uninterruptible Power Supply	Service Delivery, Availability Management, 8.3 The availability management process	COBIT basically states, measures should be taken to protect critical computer equipment against power failures and fluctuations. This Control Objective is traditionally part of the standard IT audits; the main difference for SOX is the necessity of periodic testing and the capturing of the test result.	SOX

13. Manage Operations

We cover operational workflows as part of our discussion later in this chapter however from a SOX compliance perspective this section deals with general operations that are not specifically applicable to the discussion.

Working The List

"Working The List" for the COBIT Domain III Delivery & Support and the respective ITIL components follows the same process as the previous chapters utilizing our fictitious companies as our environments. However, although not the primary objective of this book, we did commit to raising note to areas will Sarbanes-Oxley compliance could be used to address previous IT issues and/or assist in the repositioning of IT within an organization. Well, this chapter and the COBIT Domain of Delivery & Support and ITIL components definitely present that opportunity. It is our expectation that after reviewing the Control Objectives for our example that there was an observation that the majority of the Control Objectives that were selected and were defined are related to Service Level Agreements (SLAs). The most important advice we can convey as it relates to this Domain and the establishment of SLAs is that if they are stated and defines a part of your processes and procedures and pertain to your financial systems your auditor will expect you to demonstrate or show evidence to such. As the other focus of this book is SOX compliance utilizing Open Source later in this chapter we will provide you with more specific tools that can be used to assist in this domain.

The control objectives in Chapter 8 will also be a combination of SOX and COBIT and each will be clearly identified. Again, the example is just that, your particular environment should drive your customization activities and you should work with your auditor prior to finalizing your efforts. As you have probably surmised COBIT requires a lot of documentation and although your Sarbanes-Oxley effort should not require as much documentation a fair amount of documentation will be required. To compound this even more, FDIC's regulations require that the audit firm rendering the attestation be independent of the compliance effort. What this means is that not only will your audit firm have there own requirements for documentation, they will be able to give you little guidance as to what documentation you will need. Now with all these requirements for documentation one might think that there is nothing that could have been overlooked as part of this process. Well, they would be wrong. Although the intent of the majority of the efforts listed in this book are focused on "taking the people out of the process", don't forget to update the IT Organization Role and Responsibilities and Job Descriptions base on how the organization will look and what will need to perform after all of the changes.

Service Level Agreements

While you might associate Service Level Agreements with a service you might provide to customers, you can also consider your IT group as a service organization to the company in general, end users of systems and applications being your "customers". The combination of Policies and Service Level Agreements is what defines and articulates the IT goals and the expected results of the delivery of those goals to the company. From the top down executive view you should ask the question "what am I paying for and what can I expect in return?" From the bottom up user perspective these define how they can expect to get their work done, and finally from a SOX perspective you now have a basis in which to establish how you will test your environment and demonstrate compliance. This will ultimately become a key area of focus when you actually undergo your audit. In the case study for NuStuff Electronics we provide some sample Service Level Agreements for you to consider, and while these examples are simple and will not completely satisfy the individual needs of your organization they do illustrate the sections that should address when writing your own SLAs.

TIP

The sample SLAs provide in part on the CD are based on the standard templates available from NextSLM.org, which provides information concerning the strategies and practices surrounding IT service level management. In addition to templates they provide tips, recommendations and guidelines for managing IT as a service organization. For more information please visit http://www.nextslm.org/

Service Level Agreements are probably nothing new to you or your organization and we are sure that either you have implemented some SLAs in your environment, or that from time to time you may have tried to implement SLAs. But if yours is like most IT Organization in a small to medium size company theses effort have probably yielded minimal if any results. This is where you can leverage COBIT as part of your Sarbanes-Oxley compliance efforts. Fortunately or unfortunately, whether strictly for Sarbanes-Oxley compliance or for COBIT, there is no possible way to succeed if a clear and agreed upon definition of success has not been developed – success criteria is your Service Level Agreement. Hopefully there is no necessity articulate the value of Service Level Agreements beyond the establishment of success criteria so that will not be our focus but rather the processes and elements that derive good SLAs.

What is a Service Level Agreement?

In general, a Service Level Agreement defines what services and service levels an organization or company will provide. As mentioned above if defined and implemented correctly SLAs will provide you with an effective method for measuring current performance levels, as well as a way to anticipate future needs. Keep in mind an SLA is basically a contract and therefore should not only address what the services and/or performance to be delivered but also resources and funding necessary to meet the requirements. The addressing of resources and funding requirements and the agreement from your management and customers will be a key strategy for establishing meaningful SLAs. This is an area that may have not been successful or over looked in the past but it is critical that SLAs are integral to budget process. Thereby, there needs to be the understanding that if budget or resources change the SLA will be affected and therefore will need to be renegotiated. The key elements of an SLA are:

- SLA metrics levels should be driven by business objectives and meet user requirements, be agreed upon by the parties involved, and be attainable.
- Executive Management should understand the correlation between IT funding and the ability to deliver agreed upon services and service levels.

- SLA matrices should have performance buffer to allow for the recovery from breaches.
- To avoid user dissatisfaction it is essential the service levels defined are achievable and measurable.
- Service level commitments should be monitored, managed and measured on a continual basis. Monitoring and alerting should be done in a proactive manner and should contain a performance buffer stated above.
- Document, document, document – any/all performance matrices should be included in the appropriate documentation, and if feasible contain sign-offs.
- Communicate, communicate, communicate – communication is essential. If a problem arise or an SLA can't be met, proactive communication on the nature of the problem and plan of action will go a long way in establishing creditability for your organization.

The following is a generic template of a service level agreement that has been annotated for your consideration:

Template: Internal Service Level Agreement

1. Statement of Intent
   This section defines exactly what you are agreeing to provide to the organization. This is the subject of what you intend to do and perhaps why but does not address how you will go about meeting your promises.
   1.1. Approvals
   This section traditionally defines the key stakeholders who have participated and signed off on this agreement, but in our case we apply a Service Level Agreement Workflow to address this need. See the next section for information on this.
   1.2. Review Dates
   The dates and versions that the SLA stakeholders reviewed the SLA. This information is also captured via the workflow automation process.
   1.3. Time and Percent Conventions
   This section defines the conventions used in the document for describing service availability in terms of percentages, business hours, and time zones. This is typically standardized across all SLAs.
2. About The Service
   This section defines what aspects of the service are to be considered for measurement and how it is to be measured. It is important to keep your commitments realistic and to set expectations that are achievable. Although in an ideal environment your

SLA will be defined before any infrastructure has been implemented, but this is not usually the case so it is important to consider known limitations of any existing systems and applications when writing your SLAs since you don't want your SLAs to be ineffective.

2.1. Description
This section provides a detailed description of the application or service and defines who will be providing support for the service.

2.2. User Environment
This section defines the environmental characteristics of the application or service. This can be in terms of number of users, geographical locations and supported platforms.

3. Service Availability
This section defines the normal availability of the application or service.

3.1. Normal Service Availability Schedule
This section describes the schedule that the application or service will be available under normal circumstances in terms of hours or days.

3.2. Scheduled Events That Impact Service Availability
This section is for normally scheduled downtime for the application or service, or events and conditions that need to be present for this to be available.

3.3. Change Management Process
This section is for managing change to the application or service, such as enhancement requests. This includes the procedures for requesting change, workflow approvals, implementation scheduling, and notification.

4. Roles and Responsibilities
This section defines who is responsible for providing support for the application or service and how this support will be provided. In addition this section addresses how new users will be added, the request mechanism, who can make the request, who needs to approve the request, and the timeframe in which new users will be added.

5. Service Measures
This section defines what will be monitored and what thresholds of performance are acceptable. For example, this SLA might describe storage availability in terms of percentage of disk space consumed before action is taken to either reduce current footprint or acquire more capacity. We will cover service monitoring in Chapter 9.

Signoff and Approval

As with the requirements for Sarbanes-Oxley compliance, the approval process for SLAs will vary from company to company. But what we would like to do is offer a few suggestions on how better to capture and manage the process:

- Whether you use KnowledgeTree, Wiki or some other application it would be of benefit to capture the history of your SLA so you can demonstrate what was changed and when it was changed.
- Just as important as knowing what was changed and when it was changed, is knowing who changed it and who authorized the change. To this goal we would recommend that you use Workflow and that SLA changes follow the same process as any other change in your environment.

Managing The Infrastructure

In day's gone-by only large companies with various international locations required that their IT department and infrastructure be available 7/24. So if you were in IT and didn't want to or for some reason couldn't be accessible 7/24, you naturally went to work for a small to medium size company. This model stayed pretty much consistent and for the most part work well until the late 20th century and/or early 21st century when Globalization started to occur. Now with Globalization if a small to medium size company wanted to be competitive they need to have their IT staff and systems available 7/24, or did they really (we'll answer this question later). Anyway, to address the need to have their IT staff and systems available 7/24 most companies small, medium and large implemented the dreaded on-call. If you worked for a large company on-call was shared among several staff members and although not the thing you most looked forward to it was tolerable. However, if you work for a small to medium size company where there were only two IT staff or in some cases where you were they only IT staff, the requirement for on-call lead to a poor quality of life. While on-call gave the perception of addressing the requirements of the business, in essence it didn't – systems still failed and with this lost time and business opportunities continued. Even though the on-call structure gave someone to call in an event of a failure it did nothing to prevent failures. As you can see on-call as the single solution to Globalization was not beneficial for the IT personnel or their company. Keeping this in mind, you can see there was a need, especially at small to medium size company to do something different. We can only speculate that in order to develop a more effective solution to the effects of Globalization companies and their IT Organization needed to better identify the real IT requirements. Again, we are going to speculate that in order to help better define their requirements they deployed a quality process similar to the "Five Whys". The process of the "Five Whys" is pretty simple, you repeatedly ask the question "Why" (sometimes it can be less then five and sometimes more) until you get past the symptoms of a problem and are left with the root cause. If you apply the concept of the "Five Whys" hopefully you will see that small to medium size company didn't necessarily need or want IT personnel to be on call but rather to have the IT systems available.

As we had indicated as one of our goals early, in situation where Sarbanes-Oxley compliance, repositioning and enhancing your IT infrastructure coalesced, we would illustrate these opportunities. Some of he items we are about to discuss fall into this category, and although we briefly cover some of them as part of a control, in these instances we felt they warranted covering in more detail.

Performance, Capacity and Continuity

The subject of making your IT environment "available" is a large enough topic to justify its own book, so we will cover what we reasonably can in order to get you pointed in the right direction. Part of the 404 section requirements is that your financial systems be available to the end users and that an explicit recovery plan is documented and executed in the event of a failure. There are several methodologies and open source technologies that implement the ideas that fall into this category, here are a few things you might consider as part of your strategy for both implementation and day to day management of the systems that make up your financial reporting chain.

Service and System Virtualization

It seems the application "du jour" today in the datacenter is virtualization, and for some very good reasons. You have been experiencing virtualization with the ITSox2 VM Toolkit running on the VMware platform. While VMware is not Open Source it is free to use and provides many benefits that you might consider for deployment in your own environment. We gave a brief overview of virtualization as a concept in chapter 8 so we won't go over that information here, however we do want to point out a couple of other virtualization products that are also free and/or Open Source:

Xen Virtual Machine

Xen is the premier Open Source product for server virtualization for the Linux platform. The Open Source version allows you to create Linux and NetBSD, providing the fastest and most secure virtualization software available for these architectures. XenSource and other vendors offer a Windows version and formal support. Xen allows you to increase your server utilization and lower your TCO by consolidating multiple virtual servers on a smaller number of physical systems, each with resource guarantees to ensure that its application layer performance is met, hence enabling you to meet your SLAs With Xen virtualization, a thin software layer known as the Xen hypervisor is inserted between the server's hardware and the operating system. This thin software layer provides an abstraction layer that allows each physical server to run one or more "virtual servers," effectively decoupling the operating system and its applications from the underlying physical server.

Once a virtual server image has been created it can run on any server that supports Xen. Some of the key features of Xen include:

- Support for up to 32-way SMP guest.
- Intel® VT-x and AMD Pacifica hardware virtualization support.
- PAE support for 32 bit servers with over 4 GB memory.
- ×86/64 support for both AMD64 and EM64T.
- Extreme compactness – less than 50,000 lines of code. That translates to extremely low overhead and near-native performance for guests.
- Live relocation capability – the ability to move VM's to any machine brings the benefits of server consolidation and increased utilization to the vast majority of servers in the enterprise.
- Superb resource partitioning for CPU, memory, and block and network I/O – this resource protection model leads to improved security because guests and drivers immune to denial of service attacks. Xen is fully open its security is continuously tested by the community. Xen is also the foundation for a multi-level secure system architecture being developed by XenSource, IBM and Intel.
- Extraordinary community support – industry has endorsed Xen as the de-facto open source virtualization standard and is backed by the industry's leading enterprise solution vendors.

VMWare Server

VMware Server is the server component of for free virtualization of Windows and Linux servers with enterprise-class support and (paid for) VirtualCenter management. VMware Server is a robust yet easy to use server virtualization product and is based on proven virtualization technology, its major features are:

- Runs on any standard ×86 hardware
- Supports 64-bit guest operating systems, including Windows, Linux, and Solaris
- Can be managed by VMware VirtualCenter to efficiently provision, monitor and manage infrastructure from a central management console
- Supports two-processor Virtual SMP, enabling a single virtual machine to span multiple physical processors
- Runs on a wider variety of Linux and Windows host and guest operating systems than any server virtualization product on the market
- Captures entire state of a virtual machine and rolls back at any time with the click of a single button
- Installs like an application, with quick and easy, wizard-driven installation

- Quick and easy, wizard-driven virtual machine creation
- Opens VMware or Microsoft virtual machine format and Symantec LiveState Recovery images with VM Importer
- Supports Intel Virtualization Technology

Each of the above is a viable choice for your virtualization needs; the choice to use either primarily depends on your infrastructure. If you are a predominantly Linux based shop then Xen would be an excellent choice. We chose to use VMWare due to the fact that the "player" runs on both Linux and Windows physical hosts expanding the number of systems the ITSox2 Toolkit VM can be successfully run on. From a SOX perspective virtualization as a technology presents some interesting solutions to the problems of availability and configuration management. In a virtual system, any service related to your financial reporting chain has the potential to be quickly migrated away from failing hardware to another system, thus drastically reducing your potential downtime in the event of an unexpected system crash. As far as configuration management is concerned, a virtual machine can be backed up or a "snapshot" can be taken just prior to a major upgrade, providing a point-in-time rollback capability if things go horribly awry with the intended changes.

Lessons Learned

When Not to Virtualize

Prior to SOX 404 a system admin could save company resources and money by analyzing server usage patterns and consolidating servers into virtual machines on a single hardware platform. It was a simple matter of identifying the least-utilized boxes through performance monitor logs over the course of a quarter by collecting and aggregating performance monitor data. If two servers were found running consistently under fifty percent of usable maximum they could be virtualized on a single hardware platform. This would free an entire box or blade to be used elsewhere.

Today's process contains all the above but includes the identification of applications and the users of those apps to ensure that they are not both part of an overall finance process. This requires another full-time position dedicated to gathering soft-data from users, their managers, and process owners and aggregating the findings with the original performance data. It should come as no surprise that in real-world scenario's this eliminates the possibility of virtualization for over fifty percent of the candidates.

The problem comes in when the virtual server is constructed. Each host machine has what's known as a hypervisor. This is the software that allows the "guest" operating systems to be installed and to make function-calls to the hardware. This software runs under a "host" operating system which has its own set of human administrators. Anyone with administrative access to the host by extension has administrative access to the guests. This grants one or more single persons both vision into, and potential control (authorized or not) over processes hosted on the virtualized servers.

SOX 404 prohibits any one person from having end-to-end control over any given process. Therefore low-usage machines that are part of a multi-step business process cannot be virtualized on the same server or under any host operating system where the hypervisor administrator is the same person(s). As a result, these machines often need to be left to their dedicated, low-utilization lifecycle; wasting un-tolled CPU cycles, storage, network bandwidth, electricity, and by extension, fossil fuels and the overall climate in the name of segregation of duties.

–John Scott

High Availability and Load Balancing

In the SOX vein of the desire to provide maximum availability of your financial services and supporting systems, high availability offers the ability to restore system or application functionality after a hardware or software failure in a reasonable amount of time - usually measured in increments such as 30, 60 or 90 seconds. Because high availability uses off-the-shelf components and standard servers this is typically far less expensive than fault tolerance (see below) to achieve. Even though there are many high availability products on the market however the approach is basically the same, the main differentiators being the number of failover hosts, whether it is an active/active cluster or active/passive, or what platforms are supported. The High-Availability Linux Project http://www.linux-ha.org provides the latest open source solution for high availability with its Heartbeat project. Figure 8.1 shows a typical Heartbeat HA active/active cluster under normal operations and in failover mode.

Figure 8.1 High Availability Cluster

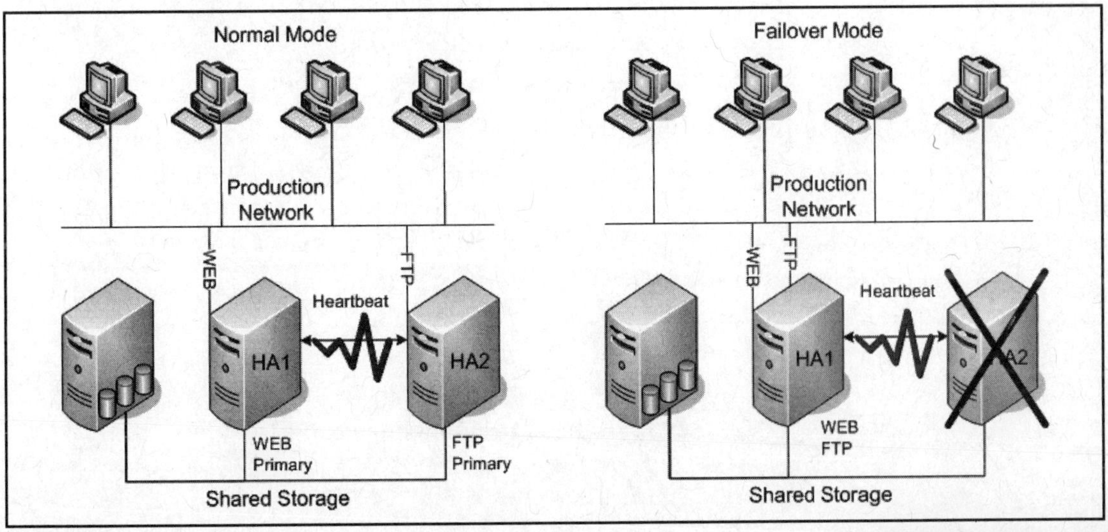

There are typically four elements needed to make an application or service highly available:

- Virtual IP Address – This is the IP address that clients query to utilize the service and which migrates with the service. In the event of a failure and the application is started on an alternate node, the new box must respond to queries on the same IP address transparently to the clients. This is usually achieved by aliasing the IP address on the network interface connected to the production network and sending the appropriate ARP broadcast to let clients know the physical MAC address has changed.
- Daemon/Executable and Configuration – All of the potential failover candidates must have the application installed with the identical configuration in order for a completely transparent failover to occur. In some instances you can have the application running on both boxes and just fail the IP, but if the application requires data it is usually better to not have the secondary service running since most applications will not run properly without the data being available, if you share the data between the nodes simultaneously you may have locking problems. This is dependent on whether the application has been explicitly designed to share data.
- Shared Storage – For applications that serve static data, such as html you might be able to replicate the data between the cluster nodes and avoid a shared storage system. The Linux Logical Volume Management (LVM) kernel module is an excellent solution for this need since it provides native snapshot capabilities that can easily be copied to other nodes. If the primary system goes down the secondary host can continue the service with its own copy of the data. In some cases the application has replication or shared data capabilities of its own eliminating the need for shared storage, in other cases where the data is dynamic you will need a shared storage mechanism such as a Storage Area Network (SAN) volume or an NFS or SMB share. In these cases the primary node reads and writes its data on the shared volume and in a failure situation the secondary takes over the data volume and resumes the application or service.
- Heartbeat – This is the mechanism in which cluster nodes talk to each other to determine the up/down status of the applications being managed in the cluster. These heartbeat paths can be proprietary however more common devices include network interfaces and serial cables. If you are using a network interface these should be dedicated heartbeat devices so that you do not introduce unnecessary traffic onto your production network, and you should have redundant devices to ensure a failover does not occur because the heartbeat interface fails. Finally you should also have a peer device in the heartbeat mechanism external to the cluster so that general network failures do not cause the application to failover.

NOTE

It is often the case that when a node fails in a cluster it does not completely die. It may become unresponsive for your HA application but still have files locked on your shared storage for example or the application is hung but the IP address cannot be released. This is where a STONITH device, short for "Shoot the Other Node in the Head" comes in handy. In this case as part of the failover process the node taking over the service sends a signal to a controller connected to the other system to completely power it off, thus ensuring the new node can completely take control of the IP and data. NetReach remote power control http://www.wti.com/power.htm provide HA compatible devices.

Load balancing, on the other hand, is similar to high availability in that multiple systems provide the same service, however the difference being these servers respond to queries either in a round robin fashion or as a single virtual service. For example, a cluster of web servers can appear as a single web server to end-users, or a set of geographically dispersed LDAP servers can respond to queries to clients in close proximity. The Linux Virtual Server project (http://www.linuxvirtualserver.org) provides mature load balancing software strategies on the Linux platform. Figure 8.2 and 8.3 illustrate these concepts.

Figure 8.2 Round Robin DNS or SRV Records

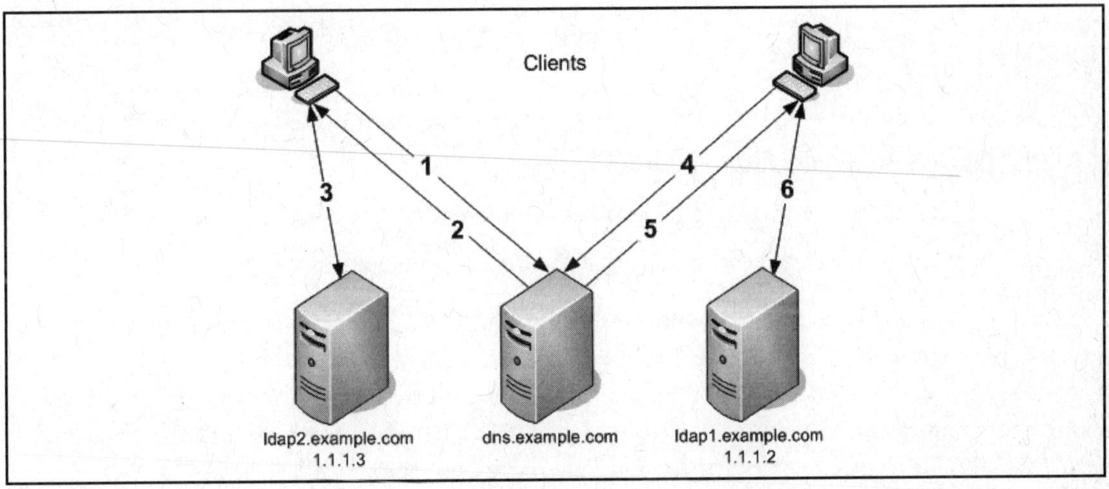

Client one requests DNS address for ldap.example.com

DNS returns 1.1.1.3

Client one performs LDAP query to 1.1.1.3, which is actually ldap2.example.com

Client two requests DNS address for ldap.example.com

DNS returns 1.1.1.2

Client two performs LDAP query to 1.1.1.2, which is actually ldap1.example.com

Figure 8.3 Linux Virtual Server

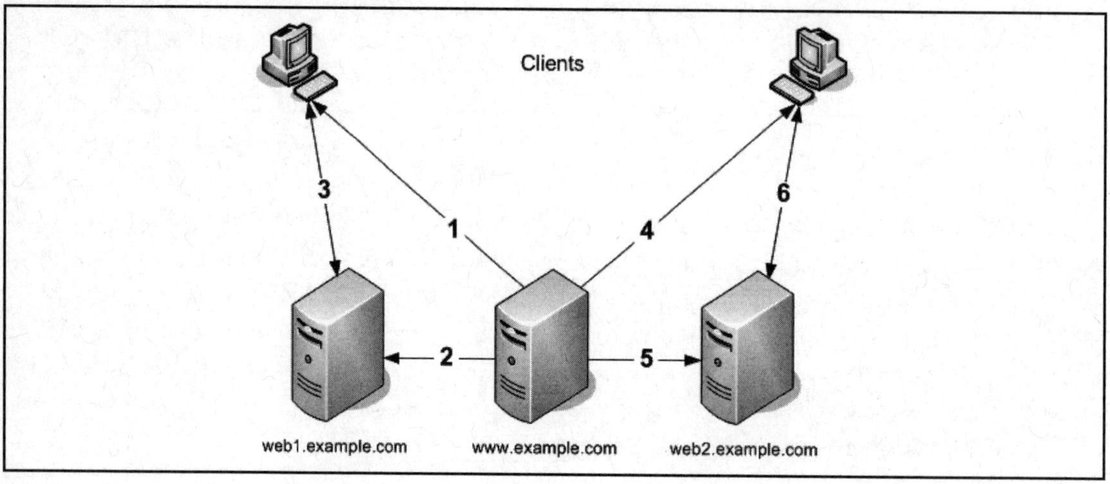

1. Client one requests www.example.com
2. www.example.com is acting as a redirector and sends client request to web1.example.com
3. web1.example.com responds with requested web page directly to client one
4. Client one requests www.example.com
5. www.example.com is acting as a redirector and sends client request to web1.example.com
6. web1.example.com responds with requested web page directly to client one

In some cases it is desirable to combine both load balancing and fault tolerance into one service. To accomplish this goal you will need to extend the LDAP example in Figure 8.3. Lets assume you are using LDAP to perform authentication services for your environment using the round robin DNS method to spread the load between multiple LDAP servers and you would want to ensure that 100% of client queries are successfully handled. If you just had load balancing and one out of your two end point LDAP servers died unexpectedly, 50% of your client LDAP requests would fail since the DNS server is unaware that the service is unavailable. SRV records add the ability to "weight" the services so you could send 75% of requests to server one and the other 25% to server two, however the client application has to have support for SRV records. The solution to the problem is to make sure all virtual IP's that can resolve to a mission critical service are protected with a failover strategy as illustrated in Figure 8.4. In this example LDAP.example.com resolves to either ldap1 or ldap2, however since ldap2 has failed ldap1 takes over the virtual IP address usually associate with ldap2 and responds to requests on both IP's. The Ultra Monkey Load Balancing and High Availability project http://www.ultramonkey.org aims to combine the Heartbeat and LVS projects into one managed framework.

Figure 8.4 High Availability with Load Balancing

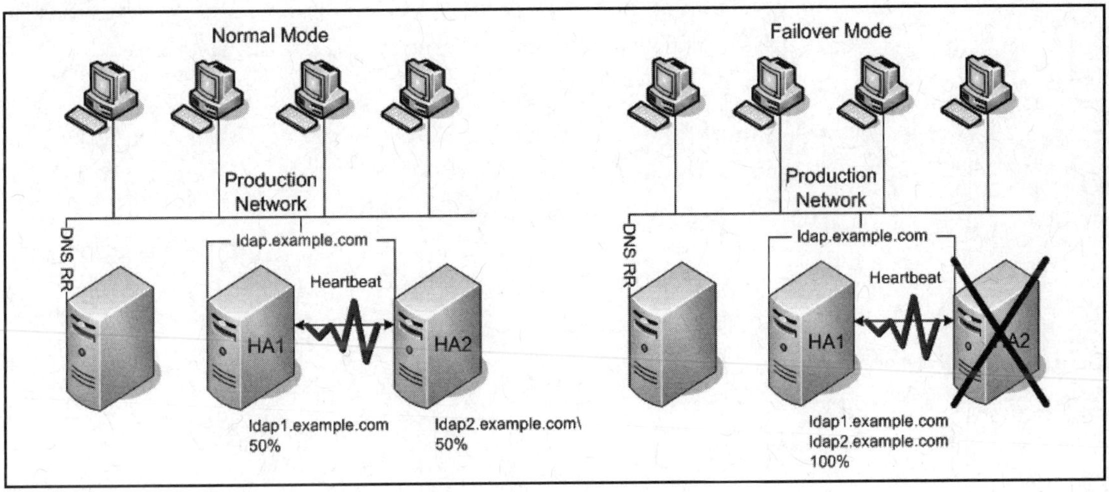

Fault Tolerance

Fault tolerance is the ability for a system or application to continue operating without interruption in the event of a hardware or software failure. A failure is defined as the service delivered to the users deviates from an agreed upon specification for an agreed upon period of time. To avoid any ambiguity this concept should be tied directly to your Service Level Agreements discussed in the previous section. A fault on the other hand is the condition or source that causes a failure. The distinction is that faults can exist without failures. Using a bridge as an example a design fault can exist when the designers of the bridge underestimate the amount of traffic they need to support, however the bridge ends up supporting more than the original specifications once deployed. Over time however the bridge would most likely need to be reinforced as a preventative measure since the extra traffic would most likely cause stress fractures to occur which would ultimately lead to a failure. The goal with fault tolerance is to prevent a fault from manifesting itself as a failure and there are several ways to approach this.

- Fault Containment – provide an automatic method for preventing failure by suppressing or limiting the original fault and preventing faults in one subsystem from affecting other subsystems. Consider the role of a network firewall. Its job is to contain certain TCP/IP traffic from traversing to or from the connected networks based on a set of rules. The rules in turn define what is considered faulty traffic. While you might not consider this as an example of fault tolerance it indeed is because it fulfills all the requirements to be classified as such. On one hand it prevents failures such as network denial of service, another examples is protection of business policies by preventing employees from visiting non-work related website sites.
- Fault Masking – prevent the consumer or end user of the system from knowing a fault has occurred. An example of this would be a the TCP/IP networking protocol, which has built in error correction at the packet level and the ability to request retransmission of corrupted packets, thus shielding the user from knowing a fault has occurred.
- Fault Compensation – provide an alternate method for the service to continue. One of the most common examples of this is a "redundant array of inexpensive disks" (RAID). Figure 8.5 presents an example of a RAID5 system which uses an n + 1 disk configuration where a parity checksum is "striped" across all disks so that if any one disk experiences a failure the system can continue to provide storage services by generating the missing data from the parity information on the remaining disks.

Figure 8.5 RAID5 Fault Compensation

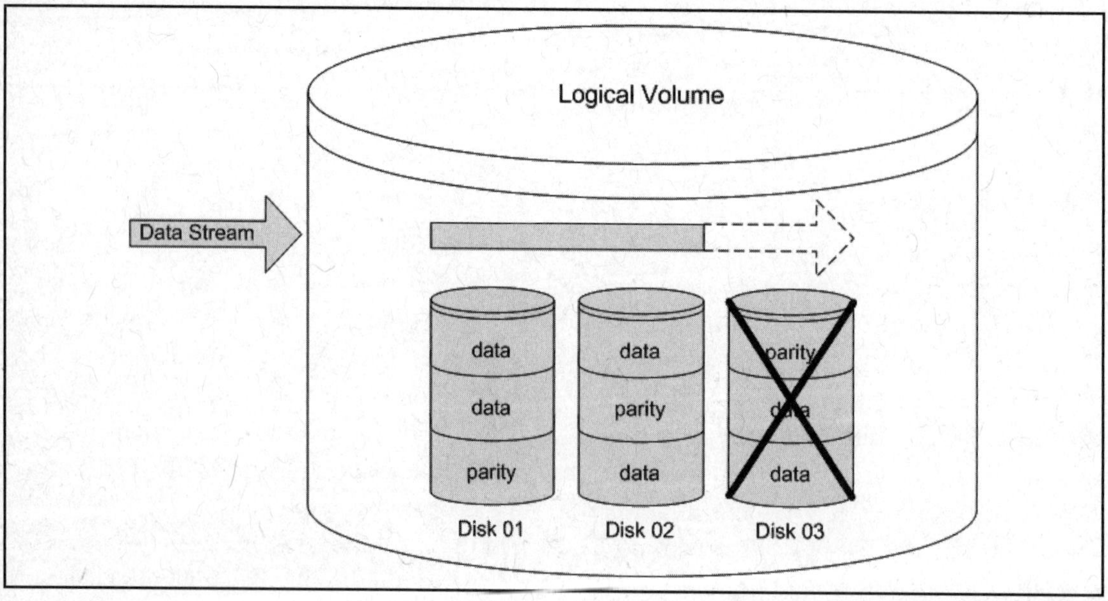

- Fault Repair – Automatically replace or repair the faulty item. Continuing our storage example most modern RAID systems provide for "hot spares" in which a failed disk in the array is automatically removed and another unused disk is configured into the array. As illustrated by Figure 8.6 the information on the original disk is rebuilt on the fly and the system returns to nominal operation without interruption.

Figure 8.6 RAID5 Fault Repair

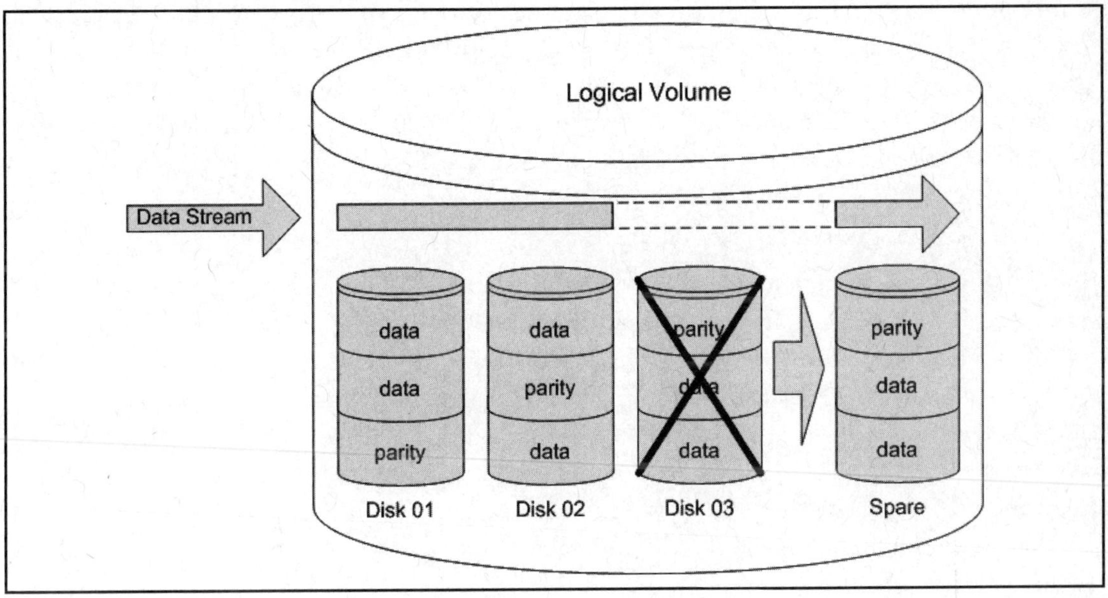

Typical computer systems may deploy one or more of these to increase overall system reliability, and while there is a penalty in terms of performance, fault tolerance is the most robust way to achieve agreed upon service levels. The main problem with fault tolerance however, particularly with hardware, is cost. While fault tolerance can be common in storage systems since it is relatively easy to provide because the unit cost for doing so usually boils down to the cost of the individual units, i.e. hard disks, other fully redundant hardware systems such as processors and memory are typically very difficult to deploy and expensive to acquire. With storage you can simulate all of the recovery options with software, although there are expensive hardware options for this as well, both Linux and Windows support software RAID systems natively. By contrast the design of redundant memory, processors, ASICs, and such can be much more complex so the next best thing to fault tolerance in providing that 99.99999% uptime Service Level Agreement is high availability.

Uninterruptible Power

In the vein of availability we all have experience power outages of some sort in our IT environments, so it almost goes without saying you need to protect your vital infrastructure from this type of threat. Not only is this a cause for potential downtime for the duration of the outage, if your systems are not properly shut down then the likelihood of a problem when these systems come back online is much higher due to incomplete disk writes, thus causing additional downtime as the file systems are checked and repaired. Those of you who are familiar with fsck, the Linux/UNIX file system check utility may know that it can take an incredibly long time to walk through a large file system, especially with today's explosion of storage requirements, often in terabytes. The main thing you need to ensure is you have either a centralized backup power protection system and/or local UPS units on your critical infrastructure, with power controllers that can send a shutdown signal to each of your systems so the risk of corrupted data is minimized - this is especially important for databases.

Security Considerations

The security considerations for performance, capacity and continuity are mainly physical access. You need to make sure an appropriate number of personnel have access to failed systems for example, while balancing this need with the SOX requirements that only authorized administrators have access to systems containing sensitive information. At the very least incidences of down time need to be recorded and documented as to who did what to facilitate the recovery of a system, and that the proper data owners are notified of the action taken. Yet another piece of documentation we know, but with automation systems such as workflow the added burden should be minimal.

Configuration Management and Control

Another important aspect of managing your operations is configuration management. The ability to be able to roll back to a known good configuration or a previous set of values for your systems, network devices and applications is essential to ensuring that you are able to recover from unexpected consequences as a result of intended or unintended changes to your environment.

Applying Changes

When applying changes to your configurations it is essential from a SOX perspective to work through your change management process on those systems that are material to your financial systems and reporting chain. The process will usually involves the use of a separate test system, which isn't necessarily required to be exactly the same as your production system but close enough such the results can be extrapolated to your production environment. Once the result have been verified in test and the necessary sign-off obtained on the Change Request for production the necessary changes can then be migrated to production.

Rollback to Previously Known Good Configuration

Prior to any changes, if you do not have one already you should create a copy of current configurations before any changes are applied so that a history is maintained to give you a measure of protection if a rollback is necessary due to unexpected results and to satisfy the SOX auditors. We have already seen some automatic version control in action with the Wiki module of eGroupware, and there are many methods in the Open Source world to maintain versions of configuration files that you can use to save copies of your configuration files and important data such as:

- KnowledgeTree (http://www.knowledgetree.com) – Although this is traditionally used for document management there is no reason not to consider the use of this to maintain versions of your system and application configuration files. We spotlight this application in chapter 6.
- Subversion (http://subversion.tigris.org) – Subversion is an open-source network-transparent version control system that uses a non-exclusive check-out model. Although this is typically used for software development, it can be used to version control any type of file and is suitable for storing configuration and system data. Additionally there are excellent windows and Linux client applications that make the process fairly easy. We spotlight this application later in the chapter.
- Kiwi CatTools (http://www.kiwisyslog.com) Kiwi CatTools is a freeware tool to capture and manage network device configurations. Kiwi CatTools allows you to perform automatic collection of data via SMNP protocol, which is typical of how these tools are designed and work. . Kiwi CatTools is not open source, but is free to use for up to 5 devices and is moderately priced if your needs exceed this number.

It is worth mentioning that the use of the above tools requires discipline on the part of the administrator to actually check out/in versions of the files you wish to manage. The typical use model being to start with a baseline version as the initial check in, then when a change is necessary the baseline configuration is checked out and modified for the new configuration. If the new configuration is successfully updated, tested and deployed the administrator would then check in the new configuration as the latest version. This way you retain full change history.

Managing Systems and Applications

One of the most important items you will need to demonstrate for your audit is the concept of "segregation of duties". For access control this means that the person approving access to sensitive information cannot be the same person who grants access, who cannot be the same person who reviews the access. In an ideal world you'd have plenty of staff to go around looking over each other's shoulders and making sure everyone is on the up and up. A more realistic compromise is to ensure you have a system to capture the approval of the owners of the data or application. Once again we turn to the workflow system to automate this. As you will see in on the example VM we have provided both system level and application level workflows to manage account requests, both activation and termination. The important thing an auditor wants to see is that the person creating the account or granting the access has someone (virtually) watching what he is doing and ensuring the proper people are aware of changes being made.

Identity Management

There are three A's to identity management. Authentication is the process of verifying identity, origin, or lack of modification of a subject or object. Authentication of a user is generally based on something the user knows (i.e. a password), is (i.e. eyes or fingerprints), or has (i.e. a key or ticket). Authorization on the other hand is the process of determining, by evaluating applicable access control information, whether a user is allowed to have the specified types of access to a particular resource. Usually, authorization is in the context of authentication. And finally Auditing is the capture of authentication and authorization attempts which can either be successes, failures or both. Since authentication is central to your security it should exhibit the following characteristics:

1. Security
   a. A potential attacker should not find authentication to be the weak link in your security footprint
   b. Network sniffing should not reveal plain text passwords on the network
2. Transparency
   a. The user should (ideally) not be aware that authentication is taking place
   b. The user should (ideally) prove his identity only once (per session or period)
3. Reliability
   a. Failure of authentication system should force a failure to gain access to requested resource(s)
   b. Should be designed to refuse authorization in the absence of authentication
4. Scalability
   a. System should be able to handle 100% of expected client requests
   b. System should not break due to a malformed authentication request

When considering an identity management system from an open source perspective, here is how the authentication mechanisms that Linux can use stack up.

NOTE

It is ideal to use more than one factor for authentication, such as biometrics combined with a password, or keyfabs such as SecureID combined with a fingerprint, however the implementation of these types of systems is typically very expensive, so we will limit our discussion to the "over the counter" solutions typically available in all distributions.

Password & Shadow Text File System

This can actually be fairly secure if used exclusively via a local terminal or remotely via SSH. The /etc/passwd and /etc/shadow system is time tested, however this quickly becomes an administrative nightmare once more than one server or application comes into play, so this is not an ideal choice for the enterprise. For systems falling under the SOX compliance where this type of access control is sufficient we see no problems in using this facility.

Network Information Systems (NIS)

NIS has built in scalability in that you can have multiple slave servers receive automatic updates from the master host. The main problem is that NIS is not secure since all passwords are transmitted in clear text. Sun Microsystems, the inventors of NIS, addressed this security problem with NIS+, but unfortunately this is unnecessarily complex to implement and never gained much traction. Sun in fact has announced end of life support for both NIS and NIS+, so the long-term prospects are also not good. From compliance standpoint NIS is an inappropriate mechanism for authentication due to these security concerns.

Lightweight Directory Access Protocol

As discussed in Chapter 5, LDAP is a protocol that was derived from the old cumbersome X500 DAP. Since it is optimized for read operations it is particularly well suited as an authentication mechanism since it supports Transport Layer Security on the wire, multiple password hashing algorithms, and robust access control lists and multiple client replication among other features. It is very well supported in cross platform applications and a great many open source and proprietary software application support LDAP authentication natively. We spotlight Fedora Directory Server in the ITSox2 VM Toolkit in chapter 5, additionally you might consider OpenLDAP http://openldap.org which is also a robust and mature open source LDAP server.

Kerberos

Kerberos represents the state of the art in enterprise authentication. Kerberos was developed by MIT and is freely available. The main design goal of Kerberos is to provide security over an insecure connection between two parties via strong cryptography. It accomplishes this via a trusted third party scheme for allowing users and services to prove their identity to each other while ensuring the integrity and confidentially of the data exchange, while only having to prove their identity to the Key Distribution Service once per session. An apt analogy could be an amusement park pass that you purchase at the front gate, providing proof of your identity only once and then using that same ticket to enjoy all of the rides the park has to offer. Each operator examines your ticket rather than ask you for proof of identity (trusted third party) and the ticket is only good for that single day. Kerberos also has built in support for replay detection so nobody can use a copy of your ticket to gain unauthorized access. Unfortunately there is not very much native support for Kerberos as yet from an application standpoint, however interestingly both Fedora Directory Server and OpenLDAP supports the use of Kerberos as the authentication layer, so you can deploy both and use either mechanism where it makes sense to do so. You can visit MIT's website http://web.mit.edu/kerberos/www for more information.

So what about Microsoft Active Directory? AD is an amalgamation of Kerberos and LDAP and makes a good choice for authentication for the same reasons you would deploy LDAP and Kerberos. You can authenticate UNIX clients that support Kerberos V5, however you cannot transparently authenticate windows clients to a standard Kerberos V5 KDC due to the use of some proprietary bits in Microsoft's implementation. You can get this to work but this requires quite a bit of setup and is not ideal from an administrative standpoint, particularly if you have a large number of Windows clients. For Linux and UNIX Microsoft provides AD4Unix which is a plug-in extension for Microsoft's Active Directory Server that enables UNIX-related authentication and user information to be stored in Active Directory. The main drawback of course is the requirement that sufficient client access licenses be available for your authentication needs, which can quickly become very costly. See the Microsoft website http://www.microsoft.com/windowsserver2003/technologies/directory/activedirectory/default.mspx for more information.

Another project to watch in this space is Samba version 4. The popular Samba suite is an implementation of Microsoft's SMB (Server Message Block)/CIFS (Common Internet File System) protocols that allow windows clients to access Linux/Unix file systems and printers. Samba 4 is the ambitious next version of the Samba suite that is being developed in parallel to the stable 3/x series with the main goal to support native Active Directory logon protocols used by Windows 2000 and above. Samba 4 has been in development for 4 years and is currently in "technology preview" version 5.

Samba 4 supports the server-side of the Active Directory logon environment used by Windows clients providing full domain join and domain logon operations. The domain controller implementation contains a built-in LDAP server and Kerberos key distribution center, and will support the existing Samba 3-like logon services provided over CIFS. Notably the Kerberos implementation of Samba 4 correctly deals with the infamous "Kerberos PAC (Privilege Access Certificate)" which is data field in the Kerberos authentication protocol which attracted controversy when critics claimed that Microsoft's "improper" use of this field was an attempt at vendor lock-in designed to tie users into its own version of Kerberos. Other improvements include the integration of Samba's Web-based administration tool (SWAT), a new scripting interface which allows JavaScript programs to interface with Samba's "internals", and new Virtual File System (VFS). The Samba 4 architecture is based around an LDAP-like database that can use a range of modular back ends. Samba 4 promises to provide a cross platform "Single Signon" environment without the limitations or cost of Active Directory.

Systems and Network Devices

Your application and file servers that hold your financial applications and their supporting systems will be the subject of scrutiny, both from a "who currently has access" standpoint and "how are you keeping unauthorized users out". If your systems are running Windows you can use Secure-It http://www.sniff-em.com/secureit.shtml which is a local Windows security hardening tool which secures Windows systems either by disabling the intrusion and propagation vectors proactively or reduce the attack surface by disabling underlying functionality malware uses to execute itself. For those systems running Linux the Bastille Hardening program http://www.bastille-linux.org is a script that helps you lock down your operating system, configuring the system for increased security and decreasing its susceptibility to compromise. Bastille can also assess a system's current state of hardening, granularly reporting on each of the security settings with which it works. In addition to Linux Bastille supports the HP-UX and Mac OS X operating systems.

Databases and File Shares

On the subject of file protection, this would be an appropriate time to mention anti-virus and anti-spyware measures. If you are using windows clients or servers your audit team will expect to see at least an anti-virus solution in place scanning both client and server files. They will also like to see automated virus definitions that do not require user interaction and a regular scanning schedule. There are a few open source and low cost tools to help you:

- ClamAV http://www.clamav.net – a command-line anti-virus scanner for Linux that supports on-access scanning, detection of over 36000 viruses, worms and trojans and built-in support for most standard compressed archive files
- ClamWin http://www.clamwin.com – a windows Anti-Virus scanner based on ClamAV

- Panda ActiveScan http://www.pandasoftware.com/activescan – free online anti-virus scanner for Windows
- AVG Anti-Virus http://www.grisoft.com – not free but very low cost anti-virus scanner for Windows
- Spybot-S&D http://www.safer-networking.org – malware and spyware remover for Widows, free for personal use
- Ad-Aware http://www.lavasoftusa.com/software/adaware – malware and spyware remover for Widows, free for personal use

Backup and Data Retention

Since there is no specific mention in the Sarbanes-Oxley Act as to methodology and/or architecture for backups this is yet another nebulous area with regards to the Sarbanes-Oxley Compliance. The only definitive backup requirement from a Sarbanes-Oxley perspective is the duration of retention, which is a period of five years. This being said, if your company has adopted industry standard retention periods for financial data this stipulation should not be an issue. So the medium that should be use will be strictly up to you to determine, as well as the off-site storage rotation. But what you needed to do is to ensure your business process owner is in agreement with your approach, frequency, off-site rotation, etc. Below are generally accepted guidelines for the retention of financial data:

- A/P and A/R 7 years
- Audit reports Permanent
- Bills of Lading 5 years
- Charts of accounts Permanent
- Fidelity and surety bonds 3 years
- Sales & tax returns 10 years
- Tax returns Permanent
- Payroll tax returns 7 years
- Expense records 7 years

As a side note, Lightweight Directory Access Protocol (LDAP) servers' logs can be used to support some of the user access control and auditing requirements of SOX. Therefore it might be beneficial to include the appropriate logs with your backup process.

Security Considerations

Linux traditionally follows the UNIX file permissions of read (r), write (w), and execute (x) permissions for the three user groups owner, group, and world or other. These Nine bits are used to determine the permission characteristics of all objects in the system. While this is usually adequate for most file access scenarios, some applications require a much more complex and robust permissions structure, such as the assignment of permissions to individual users or groups even if these do not correspond to the primary user or group owner. In a mixed platform environment it is particularly useful to be able to replicate the same access controls for both Linux and Windows clients. Access Control Lists are a feature of the Linux 2.6 kernel and are currently supported by ReiserFS, Ext2, Ext3, JFS, and XFS file systems. ACLs are maintained by the getfacl and setfacl utilities and provide the following access methods:

Permission Type	Text Form
User Owner	user::rwx
Named User	user:name:rwx
Group Owner	group::rwx
Named Group	group:name:rwx
Mask	mask::rwx
Other	other::rwx

For securing sensitive file systems and data related to the financial reporting chain you will need to fully document access controls by role. It is prudent to keep the permissions as simple as possible in order to keep the administration manageable over time since as we will see in the next chapter you will be required to periodically review these permissions and document any changes made.

VM Spotlight – Subversion

SUBVERSION

http://subversion.tigris.org/

Subversion is a free/open-source version control system that can manage files and directories, and more importantly the changes made to them, over time. This allows you to recover older versions of your data, or examine the history of how your data changed. Subversion can operate across networks, which allows it to be used by people on different computers. Although Subversion is traditionally used as a software revision control system the more interesting use from a SOX perspective is to maintain version control on system and application configuration files. It is very simple to setup a repository and use this for in place editing of configurations at the source, without the maintenance headache of using a document management system such as KnowledgeTree. With subversion to maintain your configuration files you basically:

- Change directory to where the files to be maintained are
- Create a "module" for the files you want to version control
- Use subversion commands to either commit changes you have made or
- Use subversion commands to rollback to a previous version.

We will have a closer look at the above procedure to maintain the apache configuration files as an example, but first let's explore subversions main features:

- Directory versioning – Some other version control systems such as CVS only tracks the history of individual files, but Subversion implements a "virtual" versioned filesystem that tracks changes to whole directory trees over time. Files and directories are versioned.
- True version history – With Subversion you can add, delete, copy, and rename both files and directories. And every newly added file begins with a fresh, clean history all its own.
- Atomic commits – A collection of modifications either goes into the repository completely, or not at all. This allows users to construct and commit changes as logical chunks, and prevents problems that can occur when only a portion of a set of changes is successfully sent to the repository.
- Versioned metadata – Each file and directory has a set of properties associated with it in the form of key/value pairs. You can create and store any arbitrary key/value properties you wish, and these are versioned over time just like file contents.
- Choice of network layers – Subversion has an abstracted notion of repository access, making it easy for people to implement new network mechanisms. Subversion can plug into the Apache HTTP Server as an extension module. This gives Subversion a big advantage in stability and interoperability, and instant access to existing features provided by that server—authentication, authorization, wire compression, and so on. A more lightweight, standalone Subversion server process is also available. This server speaks a custom protocol that can be easily tunneled over SSH.
- Consistent data handling – Subversion expresses file differences using a binary differencing algorithm, which works identically on both text (human-readable) and binary (human-unreadable) files. Both types of files are stored equally compressed in the repository, and only the differences are transmitted in both directions across the network.

Getting Data into your Repository

The svn import command is a quick way to copy an unversioned tree of files into a repository, creating intermediate directories as necessary. svn import doesn't require a working copy, and your files are immediately committed to the repository. This is handy when working with configuration files since you probably already have a "baseline" version to start from. Using our Apache configurations as an example we would first "create" the initial repository. This of course needs to be performed locally as root on the central server you will be using to hold the configuration files, however once the repository is setup you can then "expose" this to the network over http so that you can check in files from other systems.

```
[root@itsox2 ~]# mkdir -p /var/repos/configs
[root@itsox2 ~]# svnadmin create /var/repos/configs
```

Next we will add the current Apache configurations to the repository using the **svn** local command. Assuming Subversion has been setup for web access You could easily substitute file:// for http://<subversion_server_url>, but let's not get ahead of ourselves.

```
[root@itsox2 ~]# cd /etc
[root@itsox2 httpd]# svn import httpd \
      file:///var/repos/configs/itsox2/httpd/conf -m "initial import"
Adding          httpd/run
Adding          httpd/logs
Adding          httpd/conf
Adding          httpd/conf/magic
Adding          httpd/conf/httpd.conf
Adding          httpd/conf.d
Adding          httpd/conf.d/welcome.conf
Adding          httpd/conf.d/sslcert.pem
Adding          httpd/conf.d/proxy_ajp.conf
Adding          httpd/conf.d/zabbix.conf
Adding          httpd/conf.d/egroupware.conf
Adding          httpd/conf.d/ssl.conf
Adding          httpd/conf.d/python.conf
Adding          httpd/conf.d/perl.conf
Adding          httpd/conf.d/sslcert.key
Adding          httpd/conf.d/webalizer.conf
Adding          httpd/conf.d/php.conf
Adding          httpd/conf.d/README
Adding          httpd/modules
Committed revision 1.
```

The previous example copied the contents of the httpd directory into the repository, located at itsox2/httpd/*. You can verify and view these files with the svn command. To show the repository contents of the "conf" subdirectory for example type:

```
[root@itsox2 etc]# svn list file:///var/repos/configs/itsox2/httpd/conf
httpd.conf
magic
```

After the import is finished the original tree is *not* converted into a working copy. To start working with your configurations locally you still need to `svn checkout` a copy of the tree into your file system. Checking out a repository module creates a "working copy" of it on your local machine. This copy contains the HEAD (latest revision) of the Subversion repository that you specify on the command line, be sure to move the old directory out of the way first:

```
[root@itsox2 etc]# cd /etc/
[root@itsox2 etc]# mv httpd old.httpd
[root@itsox2 etc]# svn co file:///var/repos/configs/itsox2/httpd httpd
A       httpd/logs
A       httpd/run
A       httpd/conf
A       httpd/conf/magic
A       httpd/conf/httpd.conf
A       httpd/conf.d
A       httpd/conf.d/proxy_ajp.conf
A       httpd/conf.d/sslcert.pem
A       httpd/conf.d/welcome.conf
A       httpd/conf.d/zabbix.conf
A       httpd/conf.d/egroupware.conf
A       httpd/conf.d/python.conf
A       httpd/conf.d/ssl.conf
A       httpd/conf.d/perl.conf
A       httpd/conf.d/sslcert.key
A       httpd/conf.d/webalizer.conf
A       httpd/conf.d/php.conf
A       httpd/conf.d/README
A       httpd/modules
Checked out revision 1.
```

Congratulations, your httpd configuration files are now under revision control! You can now make changes to the files, and check the new versions in, or revert back to a previous version at any time. Usually the first thing you do prior to making a change is the "update" the directory with the current copy in the repository. This will also give you the version number you would need to revert to in case you need to roll back. In this next example we update, make a change, commit the change, then roll back to the previous version.

```
[root@itsox2 etc]# cd /etc/httpd
[root@itsox2 httpd]# svn update
At revision 1.
[root@itsox2 httpd]# cd conf
```

```
[root@itsox2 conf]# vi httpd.conf
[root@itsox2 conf]# svn commit -m "made example change to the httpd.conf file"
Sending     conf/httpd.conf
Transmitting file data.
Committed revision 2.
[root@itsox2 httpd]# cd /etc/
[root@itsox2 etc]# svn co --revision 1 \
           file:///var/repos/configs/itsox2/httpd httpd
U      httpd/conf/httpd.conf
Checked out revision 1.
```

As you can see it is fairly trivial to setup version control for configuration files against a subversion repository, make changes and even rollback back to an earlier version. The next step is to expose your repository via apache so that you can use your repository for other network systems.

Using Apache to Expose Your Repository

To expose your new repository via Apache the following configuration file needs to be placed under `/etc/httpd/conf.d/subversion.conf`:

```
LoadModule dav_svn_module     modules/mod_dav_svn.so
LoadModule authz_svn_module modules/mod_authz_svn.so
<Location /repos>
  DAV svn
  SVNPath /var/repos/configs
  SVNAutoVersioning On
  AuthBasicProvider ldap
  AuthType Basic
  AuthzLDAPAuthoritative off
  AuthName "IT Configuration Repository"
  AuthLDAPURL "ldap://itsox2.nustuff.com:\
389/DC=nustuff,DC=com?uid?sub?(objectClass=*)" NONE
  AuthLDAPBindDN "CN=admin"
  AuthLDAPBindPassword "letmein!1"
  Require valid-user
  <LimitExcept GET PROPFIND OPTIONS REPORT>
    require ldap-group CN=ITServices,dc=nustuff,dc=com
  </LimitExcept>
</Location>
```

NOTE

The above configuration contains the password for the directory administrator, so you will want to secure this file with the appropriate permissions (600) so that unauthorized users cannot access the file. Additionally you might want to setup an apache "service" account in LDAP that has read permission Access Control Information (ACI) in the tree to avoid putting the admin credentials here. See Chapter 5 for more information on ACI's.

As you can see the above configuration sets up Subversion to authenticate against the locally running Fedora Directory server. For security purposes we want to limit repository access to the members of the ITServices group. Once you have this file in place you just need to restart Apache with a `service httpd restart` command and you are pretty much done. Of course we would want to commit this new file to our configuration repository as well:

```
[root@itsox2 httpd]# svn add conf.d/subversion.conf
A        conf.d/subversion.conf
[root@itsox2 httpd]# svn commit -m "added subversion configuration"
Adding   conf.d/subversion.conf
Transmitting file data .
Committed revision 3.
```

Using the ViewVC Web Interface

https://itsox2.nustuff.com/viewvc/itsox2/httpd/?root=svn

ViewVC is a browser interface for CVS and Subversion version control repositories. It generates templatized HTML to present navigable directory, revision, and change log listings. It can display specific versions of files as well as diffs between those versions. Basically, ViewVC provides the bulk of the report-like functionality you expect out of your version control tool, but much more prettily than the average textual command-line program output. Figure 8.7 shows a sample view of our apache configuration repository.

Figure 8.7 View VC View of Apache Configuration Repository

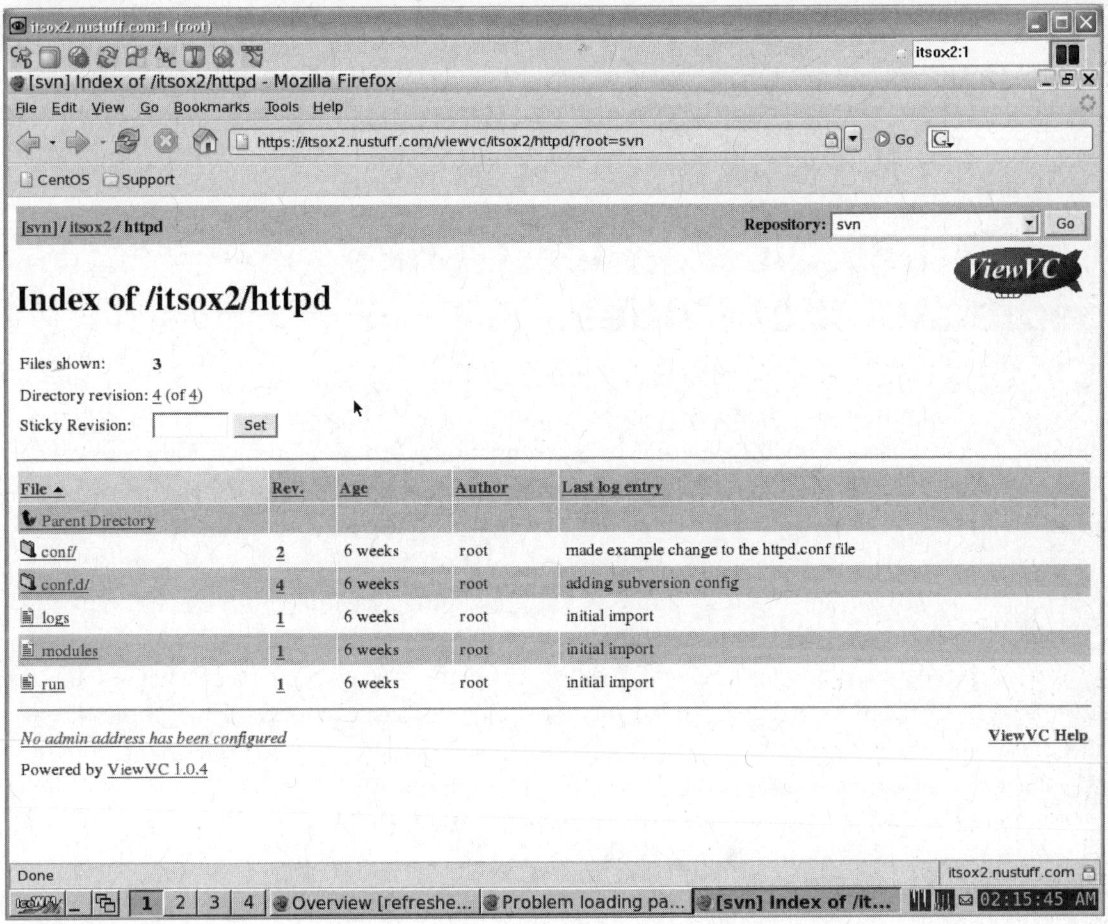

Some additional features of ViewVC include:

- Support for file system-accessible CVS and Subversion repositories.
- Individually configurable virtual host support.
- Line-based annotation/blame display.
- Syntax highlighting support (via integration with GNU enscript or Highlight).
- Bonsai-like repository query facilities.
- Template-driven output generation.
- Colorized, side-by-side differences.

- Tarball generation by revision for Subversion.
- I18N support based on the Accept-Language request header.
- Ability to run either as CGI script or as a standalone server.
- Regexp-based file searching.
- INI-like configuration files (as opposed to requiring actual code tweaks).

Case Study: NuStuff Electronics Segregation of Duties

Early on in NuStuff Electronics they realized that obtaining Sarbanes-Oxley compliance was an entirely different matter then sustaining compliance. As we have discussed previously, NuStuff Electronics as a small to medium size company had minimal staff and therefore not only had issues with workload but also with segregation of duties. We as it was clear how automation tools could effectively help them by reducing the amount of activities that the staff need to perform, it was not readily clear as to how they were going to overcome the segregation of duties issue. In order to try and address the segregation of duties NuStuff Electronics investigated various open source tools, but was unable to find one that met their requirements. After looking at the various different open source tools, without having any success NuStuff Electronics decided to re-look at the open source tools they had implemented and there capabilities. It was during this process that NuStuff Electronics realized that although Workflow did not have the capabilities that they need, with some code modification it could be tailored to work in a manner that would provide consistency, accountability, manageability and an audit trail. With this in mind NuStuff Electronics modified Workflow in such manner as to allow approval of activities by other not in their IT organization and to create an audit trail of these activities. In conjunction with our previous recommendation, we would like to make another one as it relates to implementing open source tools in your environment and it would be - If you have the skill set, prior to implementing another tool see if any of your existing tools can be modified to perform a required function. Below are some of the various roles and workflows that NuStuff Electronics customized Workflow to support.

Operations Workflows

https://itsox2.nustuff.com/egroupware/workflow/index.php

Account Activation Request

The Account Activation Request process is used to request the creation of a new account for an employee. The most basic use for this is when an employee is hired and they need a username and password set up so they can login to the network.

Workflow Roles

There are only two roles in the Account Activation Request process. The first role, HR Administrator, is the person who opens the ticket. When a person is hired or need an account created, the HR Administrator will initiate the start of the process by opening a new ticket. The other role, IT Administrator, is the person responsible for fulfilling the account request.

Workflow Activities

As far as activities are concerned, a four step process is used, which keeps the workflow very simple and straight forward. The first step of the process occurs when the HR Administrator opens a ticket and fills out the information needed to create an account. One of the variables that the HR Administrator specifies is the Department Manager, who is the person in charge of hiring the new employee. After the creation of the ticket, the Department Manager is asked to approve the request to approve the account request. The IT Administrator then creates the account in the "Fulfill Account Request" activity, and the HR Administrator then closes the ticket.

Account Termination Request

This process is almost identical to the Account Activation Request process, but, instead of creating an account, this process is used to request the deletion of one.

Workflow Roles

The roles of the Account Termination Request process are identical to the Account Activation Request process. Three people are needed to complete the process. The first two are the HR Administrator, who opens the ticket, and the IT Administrator, who actually removes the account. Department Manager approval is also needed to complete the account removal, but the HR Administrator determines who the Department Manager is when the ticket is opened, so there is no need to assign the role at design time.

Workflow Activities

All activities in this process are identical to the Account Activation Request process except that the information gathered within the activities is slightly different. There are four activities in the process again, and with the exception of the Department Manager approval activity, the process follows the traditional three step open, fulfill, and close layout.

Oracle Account Activation Request

When an employee needs an Oracle account, the HR Administrator will use this process to grant access to the user. This process does not require the HR Administrator to know exactly what type of account to grant because the "Request Oracle Access" activity only requires a detailed description of what the account will be used for.

Workflow Roles

Tickets for the Oracle Account Activation Request are opened by the HR Administrator or the IT Manager. The other users involved in the process are the IT Director, Finance Controller, and Business Analyst. The IT Director and Finance Controller roles are used only to get approval before the account is created. To fulfill the creation request, the Business Analyst is responsible for determining which rights the user needs to the database.

Workflow Activities

The process' flow is directed from the "Request Oracle Access" start activity to the "System Access Authorization Form", which is an important activity because it is the stage where the actual rights for the account are specified by the Business Analyst. After the Business Analyst completes the authorization form, the IT Director and Finance Controller approve or deny the request, and then the Business Analyst creates the account. Finally, the ticket opener completes the ticket by closing it.

Oracle Account Termination Request

The Oracle Account Termination Request workflow is very similar to the Oracle Account Creation Request, except that it is meant to remove an account rather than create one. All the same approval steps exist, and the people involved are all identical.

Workflow Roles

All the roles that were used for the Oracle Account Creation Request are again used for the termination request as well. The Finance Controller and IT Director are meant to approve the termination request, and the Business Analyst will remove the account. HR Administrators are to open the account termination request, and they also close the ticket when their request is complete.

Workflow Activities

Activities in the process remain unchanged from the creation workflow with one exception. The "System Access Authorization Form" activity is no longer required. The Business Analyst no longer needs to determine which rights to grant an account because the account is being removed instead. Therefore, after the termination request is made, the ticket is routed directly to the Finance Controller and IT Director for approval.

Data Access Request

This process is used to allow users to get access to data that is owned by someone else. For instance, if a user needs to read a network share to get specific documents off of a server, the user would need to request access to the data. Regardless of the data permissions model you choose to implement, from a SOX perspective once you have locked down your sensitive file systems or any data that is related to the financial reporting chain, you will need to quantify how people are granted access to this data, and who approves such access.

Workflow Roles

Only an IT Administrator role is needed for the Data Access Request process. Anyone within the company is allowed to request access to data, so the "All" role is used for the start activity, and the "Data Owner" role is determined dynamically when the ticket is already open, so there is no need to specify it at design time.

Workflow Activities

The layout of this process is pretty straight forward. After a user requests access to data, the IT Administrator determines who currently owns the data, and the ticket is then routed to the data owner. Once the data owner approves the access request, the IT Administrator then fulfills the access request and the ticket opener then closes the ticket to acknowledge that they have access to the data.

Data Restoration Request

When a user loses data, accidentally removes it, or somehow corrupts it then the Data Restoration Request process must be used to request that an IT Administrator restore the data.

Workflow Roles

The only role that this activity needs is the IT Administrator role. Anyone can open a data restoration ticket, so the "All" role is used for the start activity. The layout of the process follows a simple three step process. When the ticket is opened, the IT Administrator then restores the data, and the user then closes the ticket.

Workflow Activities

This workflow captures the requested path to the data. It is up to the administrator to determine how best to fulfill the request, either by restoring from a recent snapshot of the data or going back to the last full tape backup if necessary. This is a basic one step workflow process.

Report a Virus or Spyware

If a user's computer is infected by spyware or a virus then the user must open a "Report a Virus or Spyware" ticket to alert the IT department that their computer needs to be cleaned. This process is very simple and follows the General IT Request layout.

Workflow Roles

The only role in this process is the Anti-Virus administrator. The start activity uses the "All" role so that anyone can report a virus or spyware, and the closure activity is automatically assigned to the person who opened the ticket, so there is no need to specify a role for it at design time.

Workflow Activities

When a virus or spyware is reported, the ticket is opened and routed to the IT Administrator in charge of cleaning computers. Once the IT Administrator has attempted to clean the computer, they log the results and specify whether their efforts ended in success or failure. After the IT Administrator has completed the cleaning, the person who opened the ticket must then acknowledge the results of the ticket and close it.

VPN Access Request

Employees in remote locations may need to access the company's network. To do this they must fill out an VPN Access Request so that IT Administrators can verify that they have protected their computers with Firewall and Virus Scanner software.

Workflow Roles

The roles involved in this process are the IT Director role and the IT Administrator role. All users are allowed to request VPN Access so the start activity uses the "All" role and the closure activity is again automatically assigned to the person who opened the ticket.

Workflow Activities

When a user requests VPN Access, they specify their manager. The manager then must approve their request in the "Manager Approval" activity. Once the manager signs off on the request, the IT Manager approves the request. Finally, the IT Administrator fulfills the request in the "Fulfill VPN Access Request" activity. The requestor then closes the ticket and the process is complete.

Summary

In this chapter we have discuss COBIT and ITIL controls and why they are so important to not only Sarbanes-Oxley compliance activities but also why it is important from an IT Organization repositioning perspective. As part of this discussion we have identified what two of the biggest potential barriers to successfully executing the necessary Control Objective identified in this chapter. The concerns that:

- Given the number of IT resources can these activities be sustained?
- Is there really a need for all this bureaucracy?

In line with the discussion on the potential barriers we have also offer suggestions as to how to address these issues from and Executive Management and customer perspective. We have also provided guidelines by which the Control Objectives in the COBIT Delivery and Support Domain and the correlated ITIL components can be minimized and customized. Finally, we delved into what constitutes an SLA, what are the key elements of an SLA and the importance of SLAs as they relate to this chapter and the third domain of COBIT. The key elements of SLAs are:

- SLA metrics levels should be driven by business objectives and meet user requirements, be agreed upon by the parties involved, and be attainable.
- Executive Management should understand the correlation between IT funding and the ability to deliver agreed upon services and service levels.
- SLAs matrices should have performance cushions to allow for the recovery from breaches.
- To avoid user dissatisfaction it is essential the service levels defined are achievable and measurable.
- Committed to service levels should be monitored, managed and measured on a continual basis. Monitoring and alerting should be done in a proactive manner and should contain a performance cushion stated above.
- Document, document, document – any/all performance matrices should be included in the appropriate documentation, and if feasible contain sign-offs.
- Communicate, communicate, communicate – communication is essential. If a problem arise or an SLA can't be met proactive communication as to the problem and plan of action will go a long way in establish creditability for your organization.

We continue with a look at security and what this means to SOX compliance. We begin with a look at network security in the context of open source solutions such as firewalls, intrusion detection and prevention systems, and then take a brief look at network devices.

We continue with Enterprise identity management and how this affects your ability to prove provide the three aspects of identity, authentication, authorization, and auditing. We have a look at the various open source authentication systems such as LDAP and Kerberos and a brief look at Active Directory. Our security discussion continues with data and storage permissions and access controls, and we finish with a look at application security and a revisit of password policies, but this time we look more specifically on how to enforce the policy you devised in earlier chapters.

Configuration management is an important aspect of SOX compliance as it relates to both security availability concerns. We discuss how open source tools such as Webmin, Kiwi Cattools and CVS can help establish a baseline for configurations and manage changes in a controlled and structured environment while capturing the necessary audit trail via logs.

We then take a brief look at storage backups and data retention and give you some ideas on what you need to consider. Although SOX does not explicitly define what your data retention policy should be, it is important from an audit perspective that you have a policy in place.

Solutions Fast Track

All About Service

- ☑ IT Organizations can and should provide value to the company
- ☑ IT Organizations main goal should be to "To deliver IT product or service on time and in the condition which the client was led to expect."
- ☑ Most companies fail to see the value IT Organizations provide

Delivery & Support

- ☑ Systems should perform as expected upon implementation and continue to perform in accordance with expectations over time
- ☑ COBIT Domain III Delivery & Support and the correlated ITIL components will probably cause the most concern in a small to medium size company
- ☑ Open Source tools can provide gains in efficiencies, security, user satisfaction and environmental stability

Working The List

- ☑ Establishment of good SLAs will be critical to efforts to reposition an IT Organization
- ☑ eGroupware and Galaxia Workflow are key Open Source tools that can assist you in the COBIT Delivery & Support Domain III and the correlated ITIL components

Service Levels Agreements

☑ SLAs resources and services should be tied to IT funding and Executive Management should understand the correlation between IT funding and the ability to deliver agreed upon services and service levels

☑ Document, document, document – documentation of agreements should remove ambiguity and ease the reliance upon individual's recollections

☑ Communicate, communicate, communicate – communication is essential in the management of SLAs

Managing The Infrastructure

☑ The three aspects of identity management are authentication, authorization and auditing

☑ Authentication needs to be secure, for example a network sniffing program should not yield plain-text passwords on the wire

☑ Authentication should be transparent, in an ideal world the end user should not know that authentication is taking place and should only need to provide proof of identity once

☑ Authentication systems need to be reliable; failure of authentication should yield failure of authorization

☑ Authentication systems need proper scalability and should be able to handle 100% of expected client requests

☑ Kerberos and LDAP are two examples of robust and secure identity management mechanisms

☑ Data permissions need to be explicitly defined for both windows and Unix, these can be correlated as one overall policy with the use of Unix ACL's

☑ Password policies can be enforced on both Windows and Linux platforms

☑ Applications and System configurations can be managed with open source revision control systems such as Subversion

☑ Network Devices can be managed via SNMP with Kiwi CatTools

☑ Applications and Data should have a retention policy in place, although SOX does not explicitly define what that policy should be.

VM Spotlight – Subversion

- ☑ Subversion is a free/open-source version control system that can manage files and directories, and more importantly the changes made to them, over time.

Some of subversions' main features:

- ☑ Directory versioning – Some other version control systems such as CVS only tracks the history of individual files, but Subversion implements a "virtual" versioned filesystem that tracks changes to whole directory trees over time. Files and directories are versioned.
- ☑ True version history – With Subversion you can add, delete, copy, and rename both files and directories. And every newly added file begins with a fresh, clean history all its own.
- ☑ Atomic commits – A collection of modifications either goes into the repository completely, or not at all. This allows users to construct and commit changes as logical chunks, and prevents problems that can occur when only a portion of a set of changes is successfully sent to the repository.
- ☑ Versioned metadata – Each file and directory has a set of properties associated with it in the form of key/value pairs. You can create and store any arbitrary key/value properties you wish, and these are versioned over time just like file contents.
- ☑ Choice of network layers – Subversion has an abstracted notion of repository access, making it easy for people to implement new network mechanisms. Subversion can plug into the Apache HTTP Server as an extension module. This gives Subversion a big advantage in stability and interoperability, and instant access to existing features provided by that server—authentication, authorization, wire compression, and so on. A more lightweight, standalone Subversion server process is also available. This server speaks a custom protocol that can be easily tunneled over SSH.
- ☑ Consistent data handling – Subversion expresses file differences using a binary differencing algorithm, which works identically on both text (human-readable) and binary (human-unreadable) files. Both types of files are stored equally compressed in the repository, and only the differences are transmitted in both directions across the network.

Case Study: NuStuff Electronics

- ☑ Segregation of duties will likely be an issue at small to medium size companies.
- ☑ To address segregation of duties issue a process that promotes consistency, accountability, manageability and an audit trail will be needed.
- ☑ If the skill set is available, prior to implementing another tool check if your existing tools can be modified to perform a required function.

Frequently Asked Questions

Q: What if Executive Management or IT customer won't accept the changes need to comply with SOX?

A: Unfortunately the odds will be great that your company will not obtain SOX compliance.

Q: Why is the Delivery and Support Domain so important?

A: Not only is this Domain where you will need to walk the talk form a SOX perspective but it also affords the greatest opportunity to reposition an IT Organization.

Q: Will my auditor require workflow diagrams?

A: No, but they will require documentation and since workflows make great documentation and since you will need to develop them for your processes anyway you might as well use for documentation purposes.

Q: The budget process at my company has been completed for this year, what should I do?

A: Still ensure that funding for any additional resources or funding is still stipulated in your SLA and then negotiate with Executive Management or the customer for the additional funding.

Q: Are formal Service Level Agreements really necessary? Where can I find more information on writing effective SLAs?

A: The simple answer is yes. Keep in mind that the SLAs are the criteria by which your success will be assessed, and here are a few websites that can get you started:

- NextSLM IT Service Management Community http://www.nextslm.org/
- Darwin Guide to Service Level Agreements http://guide.darwinmag.com/technology/outsourcing/sla/
- Naomi Karten's Establishing Service Level Agreements http://www.nkarten.com/sla.html
- CIO Magazine's Put It In Writing http://www.cio.com/archive/111598_sla.html

Q: Can a small to medium size company really stay on top of all of the SOX requirements?

A: Yes, if the right Open Source tools are selected and the right processes put in place.

Chapter 9

Finally, We've Arrived

Solutions in this chapter:

- **Never Truly Over**
- **Monitoring In Theory**
- **Working The List**
- **Monitoring In Practice**
- **VM Spotlight – Zabbix Monitoring System**
- **Case Study: NuStuff – Oops, Still Not Right**

☑ **Summary**

☑ **Solutions Fast Track**

☑ **Frequently Asked Questions**

Never Truly Over

"Do it the hard way! Think ahead of your job. Then nothing in the world can keep the job ahead from reaching out for you. Do it better than it need be done. Next time doing it will be child's play. Let no one or anything stand between you and the difficult task, let nothing deny you this rich chance to gain strength by adversity, confidence by mastery, success by deserving it. Do it better each time. Do it better than anyone else can do it. We know this sounds old-fashioned. It is, but it has built the world"

–Harlow H. Curtice

So, have we finally reached the end of this process? Well, not quite. As we have discussed, a good quality process is a closed loop, which means the process itself never truly ends, but rather continually improves in each step based on input from the previous step. By continuing to improve each process, each successor process is improved through an iterative procedure. To illustrate this point in the chapter will we provide you with a summary of Deming's Quality Cycle – PDCA. In this chapter we will also look at the control objectives in the Monitoring Domain and based on our experience correlate them to ITIL guidelines and offer suggestions on how a small to medium size company might be able to reduce them to a manageable process.

The Transparency Test

The CFO Perspective

Information Technology plays a critical role in both the on-going operations of a company and in its Sarbanes-Oxley compliance. Given the investment companies make in IT, and the reliance on IT for critical business processes - the quality and efficiency of IT support should be monitored and managed. Strong control processes are absolutely required to protect and maintain a company's information. The use of utilization & efficiency metrics and Service Level Agreements are important tools to the IT management team to insure an effective organization.

–Steve Lanza

Monitoring In Theory

In chapter four, we established a high level definition for the COBIT "Monitoring Domain." "The monitoring phase uses the SLAs or baseline established in the subsequent phases to allow an IT organization not only gauge how they are performing against expectation but also provides them with an opportunity to be proactive". As with ITIL, although Monitoring was not explicitly addressed, monitoring was covered as part of the ITIL components in chapter four.

In previous chapters we have talked about good quality practices, Plan, Do, Check, Act and continuous improvement. We intend to accomplish, at the very least, three things 1) give you more information on Deming and his quality system, 2) illustrate more clearly PDCA and 3) demonstrate via Deming PDCA Cycle that "Monitoring" is not the end of the process but rather the beginning.

PDCA – Deming

In the 1950s W. Edwards Deming developed a quality system for continuous improvement of business processes. Deming's quality system contended that business processes should be analyzed and measured to identify sources of variations that cause products to deviate from customer requirements. He proposed that business processes be placed in a continuous feedback loop so that managers can identify and change the parts of the process that need improvements. To explain his continuous improvement system Deming developed a diagram using four arrows in a cyclical pattern; today this diagram is more commonly known as the PDCA cycle (Plan, Do, Check, Act). The cycle can be illustrated as in Figure 9.1.

Figure 9.1 Deming's PDCA

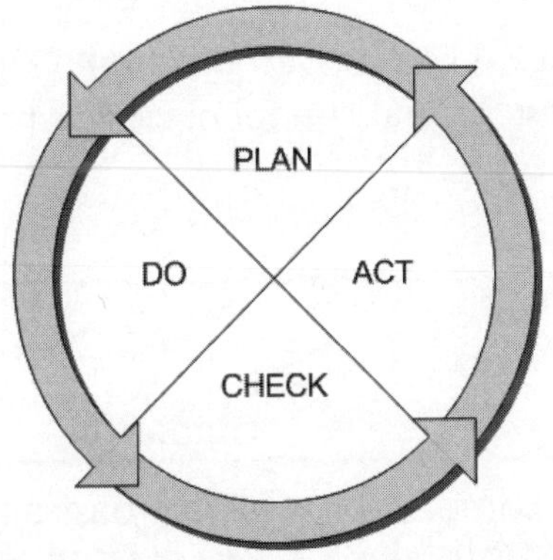

These sections are defined as:

- PLAN – Design or revise business process components to improve results
- DO – Implement the plan and measure its performance
- CHECK – Assess the measurements and report the results to decision makers
- ACT – Decide on changes needed to improve the process

Although Deming's focus was on industrial production processes, his methods and philosophies easily lend themselves toward modern business practices. Even though the COBIT and ITIL guidelines aren't explicitly express as a quality process, if we were to take the time and look carefully at the guidelines we would see a not so surprising resemblance to Deming PDCA model. Whether the COBIT and ITIL guidelines resemblances to Deming PDCA model was by intent or by accident, it illustrates the point that good quality practices endure the test of time, and drive to the point of monitoring. You might say this is all good and fine, but how does this apply to what needs to be accomplished for Sarbanes-Oxley compliance. The answer is relatively straightforward. most of the monitoring activities in COBIT IV Domain Monitoring and ITIL will come from SLAs. As much as possible, all monitoring activities should be automated via Open Source Tools – Nagios and eGroupware are just a few. Keep in mind that when you determine your thresholds, you may want to set them slightly below your SLA thresholds so that you have additional time to react and proactively correct prior to a service interruption.

In this chapter, we look at the specifics of each of the control objectives and attempt to summarize and condense the various control objectives to ones that lend themselves to small to medium-size companies. If a particular control objective of "Monitoring" or an individual item is not applicable to NuStuff Electronics and/or generally would not apply to a small to medium-size company it has not been listed as part of the guidelines below. For a complete list of the COBIT guidelines, please see Appendix A.

1. Monitor the Processes

This section as it pertains to COBIT deals with the monitoring processes and activities associated with ensuring that systems and control objectives previously defined perform and continue to perform as anticipated.

COBIT 1. Monitor the Processes	ITIL	Guidance	Objective
1.2. Assessing Performance	KPIs, CSFs, comparison with Operations, 4.6.1 The techniques The Business Perspective, Business/IS Alignment, 4.5.2 Benchmarking Planning to Implement, How Do We Check our Milestones Have Been Reached?, 6.1 Critical success factors and key performance indicators	COBIT basically states, key performance indicators and/or critical success factors should be developed and services delivered should be measured against these performance indicators. This process should be performed on a continuous basis. As discussed these performance indicators would be SLA based.	SOX & Repositioning

COBIT 1. Monitor the Processes	**ITIL**	**Guidance**	**Objective**
1.3. Assessing Customer Satisfaction	Customer satisfaction surveys, 4.4.8 Customer satisfaction analysis and surveys Service Delivery, Service Level Management, 4.5.2 Service review meetings	Customer satisfaction should be measured at regular intervals and any short fall should be addressed as part of a continuous improvement process. Again, the measurement criteria would be based on SLAs. Internal controls should be monitored for effectiveness as part of the normal course of operations through management and supervisory activities. As with Deming any deviations should require analysis and corrective action plan(s). Also, these deviations should be reported to the individual responsible for the function and also at least one level of management above. Any serious deviations should be immediately reported to Executive management. This particular Control Objective will be critical in the development of process and procedures for SOX compliance.	SOX & Repositioning

2. Assess Internal Control Adequacy

Once you have implemented the various policies, processes and procedures and have obtained your SOX compliance, you will need to sustain your new environment. This is where you will find that the various Open Source tools we've identified can pay additional dividends with regard to continuance of your SOX compliance. Since COBIT developed their guidelines in 1996, a lot of the Internal Control Adequacy assessing activities they recommended that an organization perform have been incorporated into SOX compliance. One could assume that since it will be part of the normal SOX compliance audit, there is no need for an organization to perform the assessments. Technically they would be correct. However, unless they like and live for surprises, it would not be advisable to take that approach. Later in this chapter we will introduce specific tools and configurations in the "Monitoring Domain".

3. Obtain Independent Assurance

Although the Control Objectives in this section have no bearing on Sarbanes-Oxley Compliance, they are noteworthy to review with regards to adding credence to the effectiveness of an IT organization after obtaining Sarbanes-Oxley Compliance and/or any repositioning efforts.

4. Provide for Independent Audit

The control objectives in this section aren't required to comply with Sarbanes-Oxley but since these Control Objectives are what Sarbanes-Oxley Compliance is all about, we felt compelled to list it and provide a few insights. As unfortunate as it is, most small to medium size companies can't afford the staffing on a full time basis to comply with this COBIT section or periodically perform self-audits. However, what might be more feasible and realistic is to designate an audit team made up of existing employees. The main caveat to keep in mind is that the employee performing the audit of a department cannot work within the audited department. If the luxury of budgetary funding does exist, we would advise the periodic use of an independent audit firm to ensure your controls are still effective, not one of the big four. The reason for using an independent audit firm is because the impartiality of the independent audit firm will lend more credence to the audit findings and your audit firm.

Working The List

By now we hope that it is clear that there is a difference between COBIT, ITIL frameworks and Sarbanes-Oxley compliance. That being said, the process for working the Control Objectives in COBIT Monitoring Domain and ITIL guidelines for Sarbanes-Oxley compliance is the same as the previous chapters.

The process for working the list of control objectives will be similar to the one discussed previously. We will be using our fictitious company to drive the process of customizing the control objectives in the COBIT Monitoring Domain. We have continued the practice of identifying the purpose of the various control objectives – Sarbanes-Oxley or Repositioning. Again, the example is just that; your particular environment and TDRA should drive your customization activities and you should work with your auditor prior to finalizing your efforts.

Though from this book and a Sarbanes-Oxley compliance perspective, the "Monitoring Domain" has the least amount of Control Objectives identified, we are by no means trivializing the importance of "Monitoring". Although you can look at the previous COBIT Domains and ITIL guidelines as forming the foundation, wall and door/windows of your compliance house, the "Monitoring Domain" is the roof. The "Monitoring Domain" serves as vital a role as the roof on a house in ensuring that the elements are kept out – in this instance those elements would be unexpected results. So, when you are customizing "Control Objectives for your environment use the following as over arching guidelines:

- The importance of the system, application and/or infrastructure component should drive the frequency or monitoring and auditing.
- The monitoring process should be automated as much as possible
- Monitoring capabilities should employ exception reporting
- Above all check

Lessons Learned

Monitoring Is Not The End

Generally speaking, when IT organizations are confronted with a problem or new requirement they usually view it as a point in time activity. By this I mean a process, solution or resolution is implemented and they are on to the next problem – this has been our typical experience as well. Very rarely–if at all–do we revisited (monitor) the activity to ensure that it continues to perform as expected or better yet, determine if we can derive additional benefits from it. This is a practice that has been happening in IT organization for decades without fail or change. Just as this age old practice is a fact, it is also a fact that IT organizations will have to change to continue to provide the type of support to their companies SOX dictates. So, the question isn't really whether your IT organization changes but rather if they embrace change and derive additional benefits from SOX to position themselves as being seen to provide value to the company.

Monitoring In Practice

When talking about monitoring, there are generally three distinct steps to consider. The first is the health monitoring of the actual servers, services and applications that comprise your infrastructure, or more specifically for Sarbanes-Oxley compliance, your financial reporting chain.

The second is the monitoring of configurations and data points to ensure change management and security parameters are being captured. Finally, compliance monitoring makes sure that the two previous types are properly observed and reviewed, and any anomalies are corrected accordingly. This three-step process is critical to sustain an ongoing healthy, managed Sarbanes-Oxley compliance strategy that can be quantified come audit time.

System Monitoring

There are several capable Open Source tools available for monitoring your environment. In the first edition, we highlighted Nagios, one of the most complete and mature solutions to provide network and host monitoring. Nagios runs on Linux and other UNIX operating systems, monitoring hosts and services on your network and provides you with an overview of the status of your systems, notifies you know when things are go wrong and ultimately allows you to resolve problems that occur in a predefined manner. Nagios can also provide historical reports and graphs of host or service down times, which can be used to support your Service Level Agreements. Over 150,000 copies of latest version have been downloaded, and many top installations monitor thousands of hosts and services.

Nagios is designed in a modular fashion. A central Daemon contains the monitoring logic and coordinates things like escalations and scheduling down time. Nagios executes external applications at regular intervals to actually handle the low-level monitoring of each individual item and other external commands can be triggered to manage alerts, state changes, and monitoring information. A CGI Web interface is included, which allows users to view status information via any browser, respond to alerts and manage schedules. With its plugin based architecture, Nagios allows you to monitor virtually any device that is accessible via TCP-IP, including Windows hosts and network devices. Figure 9.2 illustrates the conceptual architecture of Nagios.

Figure 9.2 Nagios Conceptual Design

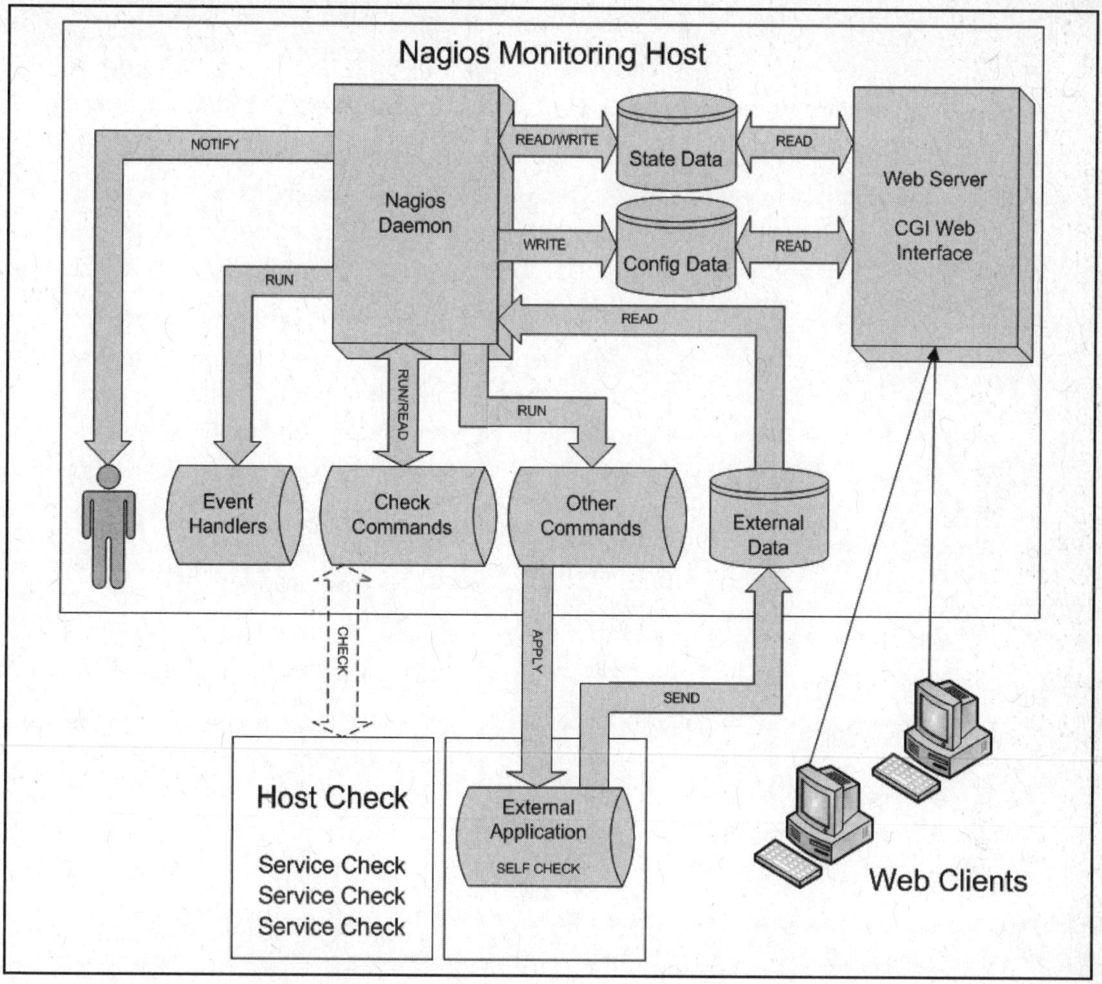

The only potential downside to Nagios is that it can be somewhat complex to set up, depending on your environment. However it is well worth the effort to deploy this excellent system. See http://www.nagios.org/ for more information. GroundWork Open Source has developed a supported infrastructure around Nagios as well; see http://www.groundworkopensource.com for more information.

There are a couple of recent additions in the Open Source monitoring space. Zenoss provides a complete suite of software and services, providing a single, integrated solution for monitoring your entire IT infrastructure such as network, servers and applications. Zenoss aims to manage your monitoring needs across the full lifecycle including inventory, configuration, availability, performance, events, logs and alerts. Zenoss is a commercial Open Source application, giving you the option to have a formally supported monitoring environment; their Website has more information http://www.zenoss.com. Here is a brief overview of the application:

- Configuration Management Database (CMDB). The CDMB houses a unified model of the IT environment and is basis of Zenoss' "model-driven" IT monitoring and approach.
- Inventory and Change Tracking - Zenoss provides inventory management and change tracking services that allow IT operators to maintain a current understanding of their IT environment with auto-discovery, automatic change history & detection and reporting
- Availability Monitoring - provides a complete suite of active availability monitors that test that resources, up and down the stack, are responding to client requests. These tests are performed without the need for additional software to be loaded on the test target. Through availability monitoring you gain a real-time of view of the availability of your IT environment. By combining end user perspective tests (e.g. URL monitoring) with element level monitoring (e.g. process and interface monitoring), you are able to quickly identify and address problems.
- Performance Monitoring - provides high-speed collection, historical graphing and real-time threshold analysis for any available metric in your IT environment.
- Event Management, Log Monitoring & Alerting - provides users the ability to aggregate log and event information from various sources including availability monitoring, performance monitoring, Syslog sources, SNMP trap sources, Windows Event log. Users can also custom create events from external applications through XML-RPC interface.

The final monitoring package in our list is Zabbix. We have chosen Zabbix as the subject of our ITSox2 VM Toolkit in this edition, as it provides an excellent platform for monitoring both high availability and performance. We will offer a more detailed look at Zabbix later in this chapter. Regardless of which package you choose, any one of these can provide the monitoring capability for your IT infrastructure. The specific SOX requirements are to ensure that critical systems and services remain available and have the ability to detect any crucial configurations and/or files are not changed without authorization, which leads us to our next subject.

Configuration Monitoring

There are other tools that you can use to assist you with your SOX compliance monitoring requirements. Here we discuss three separate examples, each designed to capture and monitor a different type of resource, system or application.

Syslog

Log files are the central place to find information about system events errors. With most Linux and Unix services, when anything of significant happens a message about it is reported to the syslogd facility. The problem lies in the fact that this creates thousands of events for each server, and is nearly impossible for an administrator to review by hand, so log checkers are needed as time savers and to make sure an important indication of trouble is not missed. There are analysis tools such as Swatch (http://swatch.sourceforge.net), which scan the logs periodically and take actions when it encounters certain log messages, such as send email or execute a predefined command or script. The key is to identify what is anomalous log traffic some of the items you might want to look for include root access attempts, account password failures, or hardware problems on critical financial systems for example.

A centralized Syslog server is also desirable so that from a security standpoint a potential attacker cannot clean up evidence of a penetration on the local system and it would take a further compromise of the syslog server to cover their tracks. Centralizing your Syslog and integrating with your monitoring system of choice so that you have a common capture and escalation footprint for your IT infrastructure largely solves your SOX compliance monitoring and data capture requirements. For windows systems the event log can provide some of this type of information as well, the same integration opportunities exist for all of the monitoring systems we have mentioned in this chapter.

Tripwire and AIDE

Once a baseline has been established for servers critical to reporting your financials, an important SOX deliverable is to demonstrate that the files do not change over time without a corresponding change management request. This is for security and availability purposes, but this would be a very tedious job indeed to attempt to validate by hand. This is where tools such as Tripwire and AID can be a great help.

Tripwire (http://sourceforge.net/projects/tripwire) is a system integrity checker, a utility that compares properties of designated files and directories against information stored in a previously generated database. Any changes to these files are flagged and logged, including those that were added or deleted, with optional email reporting. Additionally, support files (databases, reports, etc.) can be cryptographically signed. AIDE (http://sourceforge.net/projects/aide) is another tool similar to Tripwire in that it creates a database from a set of regular expression rules defined in a configuration file. Once this database is initialized it can be used to verify the integrity of the files using several message digest algorithms such as MD5 and SHA1 check the files for changes.

Both of these tools can be configured to specify what files and directories to keep track of and to what level of detail you want to watch for such as changes in size, access time, modification time, inode creation time, and content. Additionally you can check for changes to the permissions and file mode), inode, number of links, file owner and file group. These tools can be used in order to verify, monitor, and most importantly substantiate during an audit that the file systems of your security sensitive systems have not been tampered with.

Kiwi Cat Tools

Kiwi CatTools is a freeware application designed to run on a Windows platform that provides automated device configuration management on routers, switches and firewalls. It supports most major vendor network equipment such as Cisco, 3Com, Extreme, Foundry and HP. Kiwi CatTools can perform configuration backups and have any differences alert you via email. CatTools can also issue commands to devices via Telnet or SSH, change all of your device passwords in one session, and change configurations according to a user defined schedule. As we have seen in earlier chapters, change management is an important aspect of Sarbanes-Oxley compliance, and CatTools can help you achieve this, as well as the monitoring aspects that are applicable to this domain. The free version of Kiwi CatTools is limited to five devices, five scheduled activities, and two simultaneous TFTP sessions, so if you require more for your environment the pricing is reasonable. See http://www.kiwisyslog.com for more information and a complete list of supported devices.

Compliance Monitoring

The final class of monitoring is the ability to monitor compliance objectives over time. While this is not "monitoring" in the traditional sense, it is an extremely important aspect of your continuing efforts to achieve and maintain compliance. The key concept here is that SOX is not something you specifically "do" to get compliant, but rather a holistic and systematic approach to IT that naturally lends itself to compliance and gives you the ability to quantify that with your auditors. In this regard, it is important to generate evidence that your day-to-day activities and deliverables are being met consistent with your stated policies and procedures.

This is proof that you are demonstrably where you need to be. These activities are most likely being performed by your IT staff to some degree already as part of an IT quality process, but it is important to capture the information in a regular pre-defined way. It is equally important to make sure all of the people who have a stake in these processes—both inside and outside IT—are kept in the loop. For example, finance staff needs to maintain awareness of administrative activities on their financial systems. Relying on human resources and manual processes to keep this information flowing can be time consuming and prone to lapses, therefore we highly recommend the deployment of some automation to perpetuate a culture of compliance, while at the same time reducing the manual steps that might be prone to human error.

All that being said, we believe workflow is the perfect tool for the job. With workflow, you can predetermine each process that you need to regularly review and/or perform, and establish the roles necessary to complete the flow. We have included many workflow examples on the ITSox2VM Toolkit. However, this is not an exhaustive list; your needs will be determined by your environment and specific policies. You can, however, use the following examples as a point of reference in defining your own compliance workflows. In chapter 10, we provide a step-by-step guide to defining a workflow so you can begin to automate your own processes. One of the most important additions we made to the workflow engines was the ability to schedule workflows to automatically "fire" on a regular interval. These changes have also been made available as Open Source under the terms of the GPL.

Annual Oracle Admin Review

Our first example is the Annual Oracle Admin Review workflow as shown in Figure 9.3. This process is meant to remind IT management to review all Oracle administration accounts. In order to maintain a secure database the business analyst must periodically make sure that there are no expired administration accounts and that only all users who have access are valid. In defining your workflows you might consider adding additional approval activities, for example you might want to include the Finance controller in the "loop".

Figure 9.3 Annual Oracle Admin Review

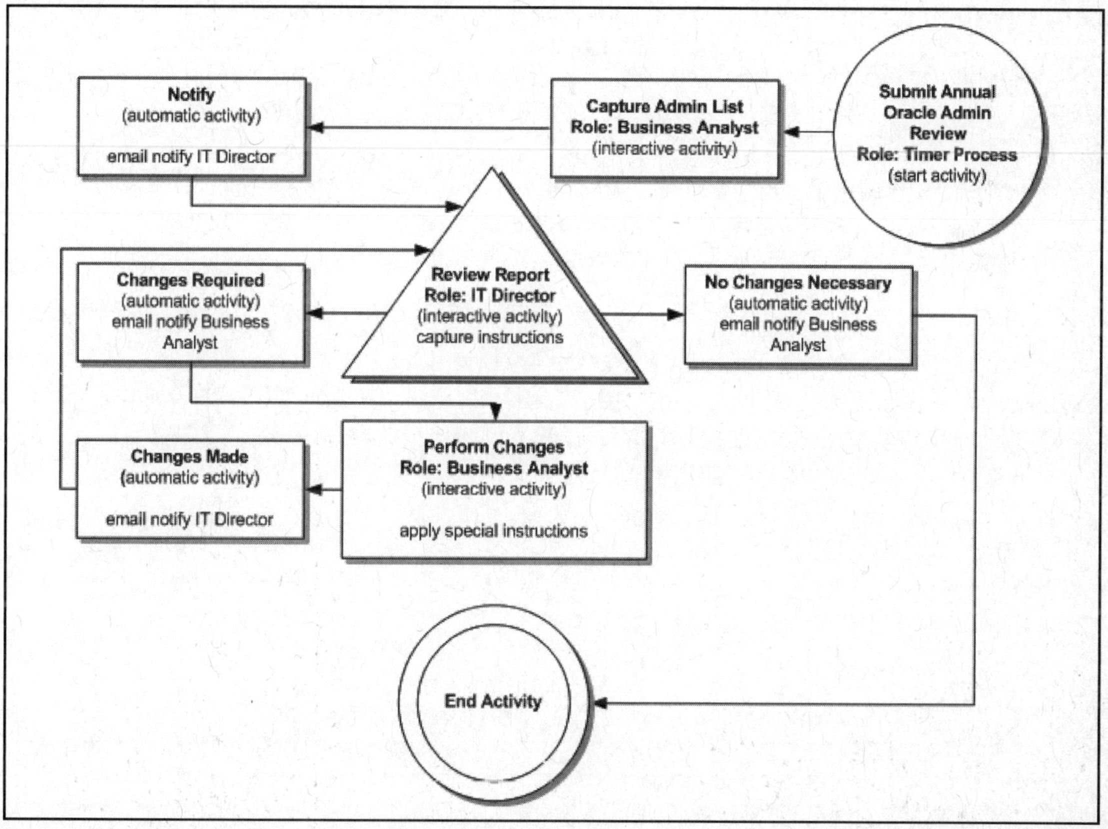

Role	Description	Member(s)
Business Analyst	The Business Analyst role is to be filled by the main administrator of the Oracle database system. In this process, Business Analysts are in charge of reviewing all the admin accounts that are currently active.	
IT Director	The IT Director is responsible for approving the account report provided by the Business Analyst, or requesting actions or changes as necessary to achieve approval.	

Activities	Description	Role
Start Activity	This workflow is automatically created on a timed interval, which notifies the Business Analyst to initiate an admin review.	Timer Process
Capture Admin List	The Business Analyst updates the workflow instance by attaching the current list of administrative access accounts. Once submitted this flows to the Review activity for approval.	Business Analyst
Review (Decision)	The IT Director is responsible for approving the account report provided by the Business Analyst, or requesting actions or changes as necessary to achieve approval.	IT Director
Approve	If the IT Director approves the account list the decision will be captured and the workflow will be directed to the End activity.	IT Director
Request Changes	If the IT Director wishes to request changes these are recorded in the workflow instance, which is then routed to the Perform Changes activity.	IT Director
Perform Changes	The Business Analyst performs any changes requested by the IT Director. Once changes have been made this is resubmitted to the Review activity.	Business Analyst
End Activity	The End activity marks the workflow as complete and records the final account list.	<none>

Bi-Annual IT Policy Review

The Bi-Annual IT Policy Review process is meant to remind the IT Director to review policies, as shown in Figure 9.4. If policies are stored in a Wiki, such as eGroupware's Wiki application, then any modifications made since the last review must be checked and validated.

Figure 9.4 Bi-Annual IT Policy Review

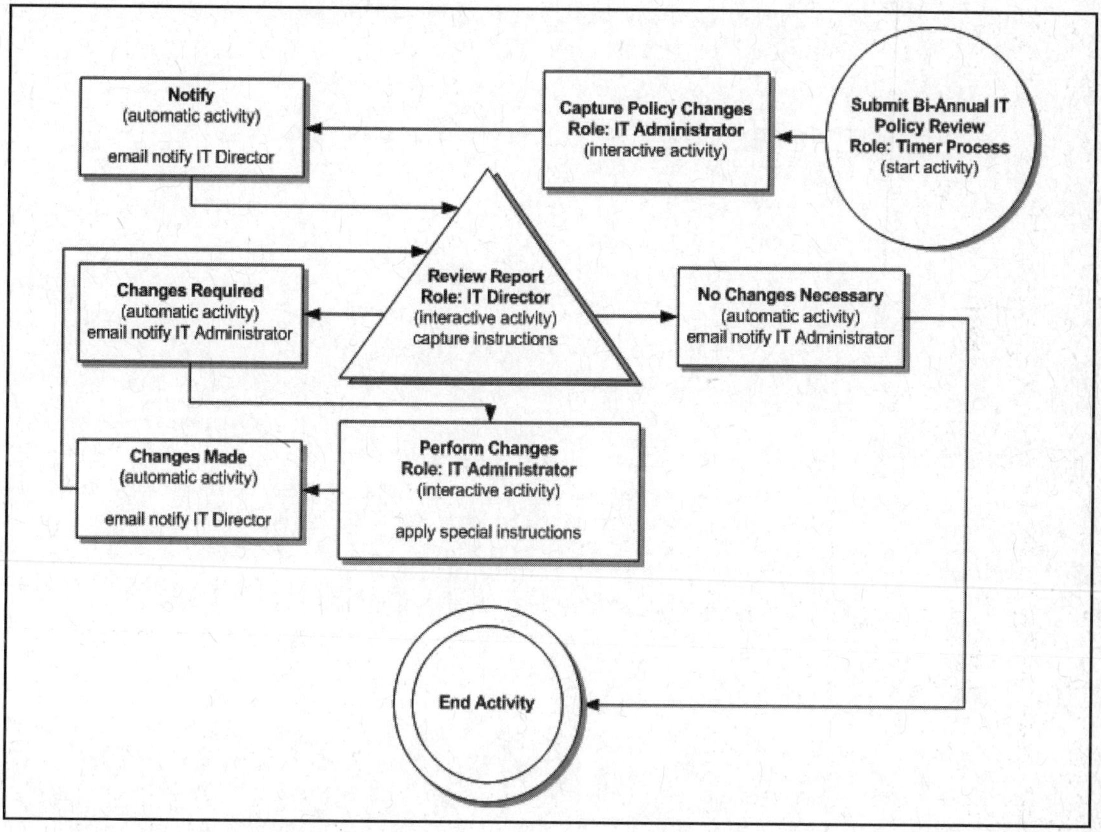

Role	Description	Member(s)
IT Administrator	The IT Administrator role is to be filled by the general IT staff. In this process the IT Administrator reviews the policies and records notes of any changes made since the last review	
IT Director	The IT Director is responsible for approving the Policy report(s) provided by the IT staff, or requesting actions or changes as necessary to achieve approval	

Activities	Description	Role
Start Activity	This workflow is automatically created on a timed interval, which notifies the IT staff to initiate a policy review.	Timer Process
	The IT staff initiates the workflow by attaching the current list of policies. Once submitted this flows to the Review activity for approval.	IT Administrator
Review (Decision)	The IT Director makes the decision to approve the admin list or specify actions or changes necessary to achieve approval	IT Director
Approve	If the IT Director approves the policies the decision will be captured and the workflow will be directed to the End activity	IT Director
Request Changes	If the IT Director wishes to request changes these are recorded in the workflow instance, which is then routed to the Perform Changes activity	IT Director
Perform Changes	The IT staff performs any changes requested by the IT Director. Once changes have been made this is resubmitted to the Review activity	IT Administrator
End Activity	The End activity marks the workflow as complete and records the final policy list	<none>

Monthly Data Restoration Test

In order to be prepared in the event of a system failure, it is necessary to run periodic data restoration tests from a backup mechanism. This process is meant to remind an IT Administrator to run such a test, and also to alert management that the backup devices in place are functioning properly. This workflow is initiated by a timer process that notifies the IT Administrator to perform the test as shown in Figure 9.5. Unlike the first two examples, there is no decision branch in this workflow, rather this provides an automated way to demonstrate to your auditors that the right "eyeballs" have seen this process and made notes.

Figure 9.5 Monthly Data Restoration Test

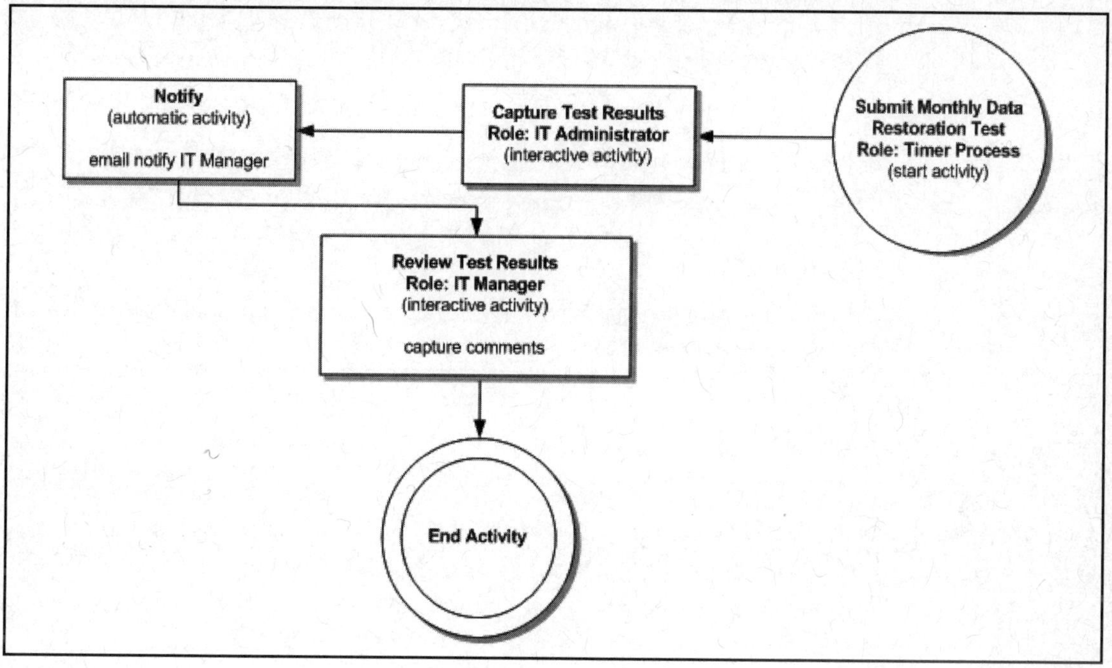

Role	Description	Member(s)
IT Administrator	The IT Administrator role is to be filled by the general IT staff. In this process the IT Administrator fulfills the backup test, and verifies that all data is restored and reading as expected	
IT Manager	The IT Director is responsible for reviewing the backup test results provided by the IT staff	

Activities	Description	Role
Start Activity	This workflow is automatically created on a timed interval, which notifies the IT staff to initiate a data restoration test.	Timer Process
Start Activity	The IT staff updates the workflow with the result of the test. Once submitted this flows to the Review activity for approval.	IT Administrator
Review	The IT Manager notes the result, making any desired comments before closing the workflow instance.	IT Manager
End Activity	The End activity marks the workflow as complete and records the final comments	<none>

Monthly Offsite Backup

In order to be sure that data and software are secure, offsite backups are required. This process is used to remind IT to back up any new software that has been purchased, as well as any data or serial numbers that need to be stored off-site. From a SOX perspective, this usually means your financial data, however this workflow can be used for all backup activities. This is very similar in workflow to the preceding data restoration test, shown in Figure 9.6, and is geared to be a "reminder" to perform this important function.

Figure 9.6 Monthly Offsite Backup

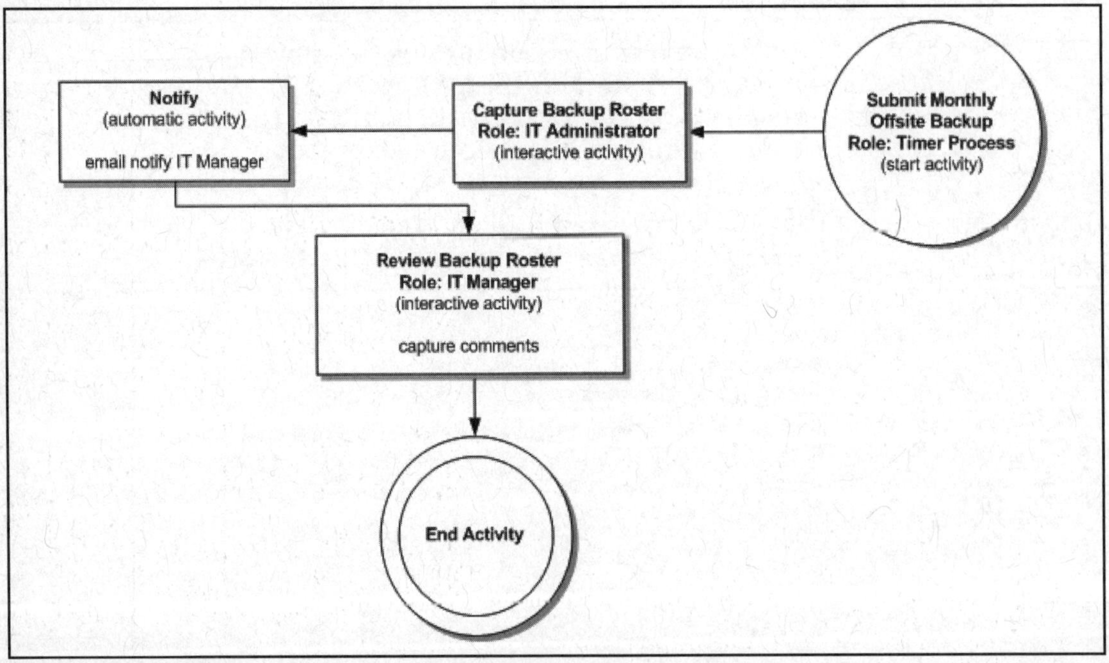

Role	Description	Member(s)
IT Administrator	The IT Administrator role is to be filled by the general IT staff. In this process the IT Administrator fulfills the backup, and completes the backup roster.	
IT Manager	The IT Director is responsible for approving the Policy report(s) provided by the IT staff, or requesting actions or changes as necessary to achieve approval	

Activities	Description	Role
Start Activity	This workflow is automatically created on a timed interval, which notifies the IT staff to initiate a data restoration test.	Timer Process
Start Activity	The IT staff updates the workflow with the result of the test. Once submitted this flows to the Review activity for approval.	IT Administrator
Review	The IT Manager notes the result, making any desired comments before closing the workflow instance.	IT Manager
End Activity	The End activity marks the workflow as complete and records the final comments	<none>

Monthly Oracle Active User Review

In principle, the Monthly Oracle Active User Review process is similar to the Annual Oracle Admin Review, as shown in Figure 9.7. The main difference is this workflow is used to review all accounts for the financials Oracle databases, and also as a way for management to specify any changes that they wish to implement in the account list. This workflow adds an approval step for Finance Controller, who should be aware of user access to the financial systems. This workflow occurs on a more frequent schedule.

Figure 9.7 Monthly Oracle Active User Review

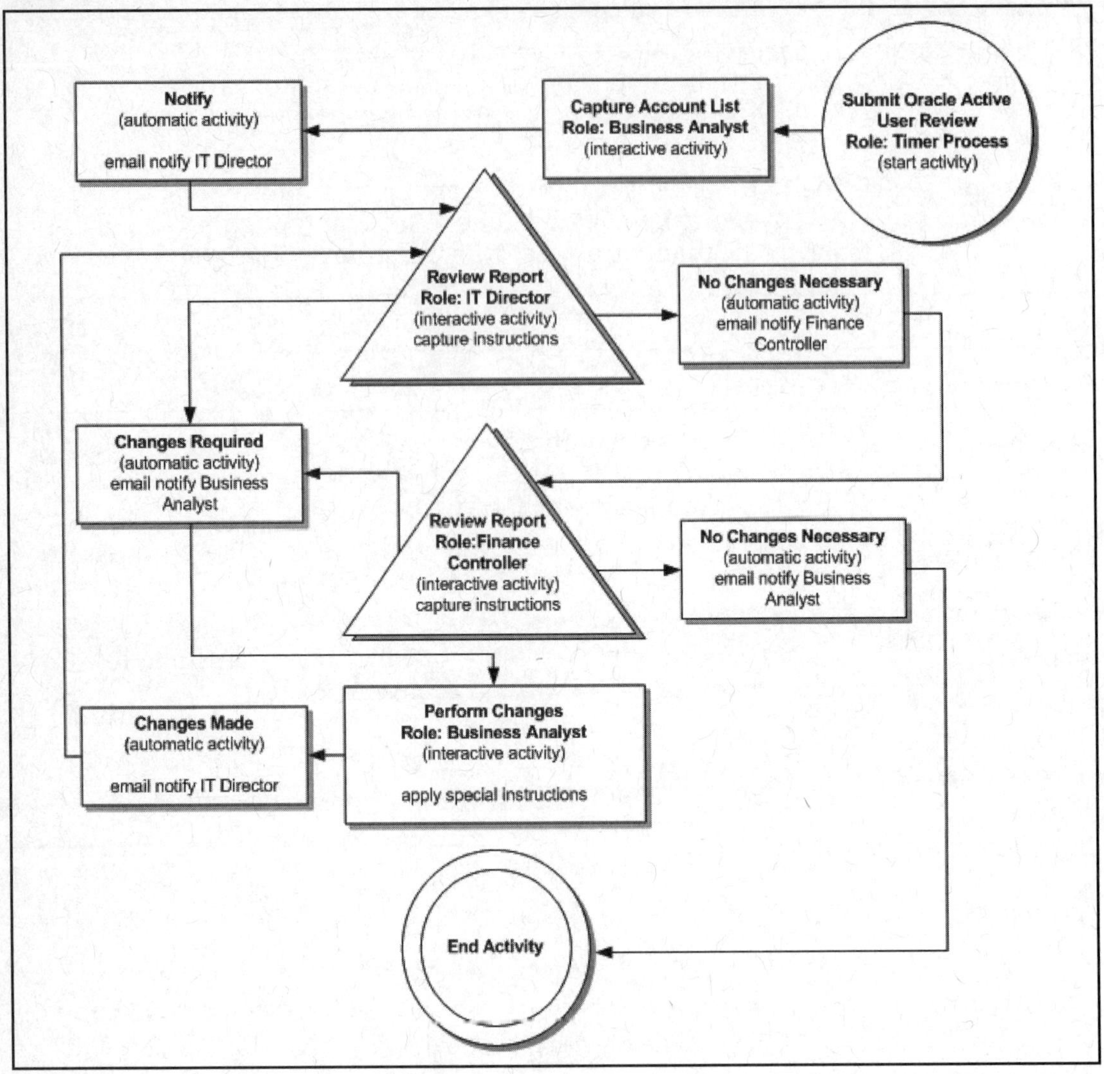

Role	Description	Member(s)
Business Analyst	The Business Analyst role is to be filled by the main administrator of the Oracle database system. In this process, Business Analysts are in charge of reviewing all the user accounts that are currently active.	
IT Director	The IT Director is responsible for approving the account report provided by the Business Analyst, or requesting actions or changes as necessary to achieve approval	
Finance Controller	The Finance Controller is also responsible for approving the account report provided by the Business Analyst, or requesting actions or changes as necessary to achieve approval	

Activities	Description	Role
Start Activity	This workflow is automatically created on a timed interval, which notifies the Business Analyst to initiate an account review.	Timer Process
Capture Active Account List	The Business Analyst updates the workflow instance by attaching the current list of administrative access accounts. Once submitted this flows to the Review activity for approval.	Business Analyst
IT Review (Decision)	The IT Director makes the decision to approve or request actions or changes as necessary to achieve approval	IT Director
IT Approve	If the IT Director approves the account list the decision will be captured and the workflow will be directed to the Finance Approve activity	IT Director
IT Request Changes	If the IT Director wishes to request changes these are recorded in the workflow instance, which is then routed to the Perform Changes activity	IT Director

Continued

Activities	Description	Role
Finance Review (Decision)	The Finance Controller makes the decision to approve or request actions or changes as necessary to achieve approval	Finance Controller
Finance Approve	If the Finance Controller approves the account list the decision will be captured and the workflow will be directed to the End activity	Finance Controller
Finance Request Changes	If the Finance Controller wishes to request changes these are recorded in the workflow instance, which is then routed to the Perform Changes activity	Finance Controller
Perform Changes	The Business Analyst performs any changes requested by the IT Director. Once changes have been made this is resubmitted to the Review activity	Business Analyst
End Activity	The End activity marks the workflow as complete and records the final account list	<none>

Quarterly AV Inventory Review

This workflow is designed to regularly review antivirus policies and events that occurred during the previous quarter. This is to review your systems to make sure they are protected, that every one has the right patch level, that virus definitions are up to date, and to review any incidents that occurred to determine the effectiveness of your current protection. The workflow is illustrated in Figure 9.8.

Figure 9.8 Quarterly AV Inventory Review

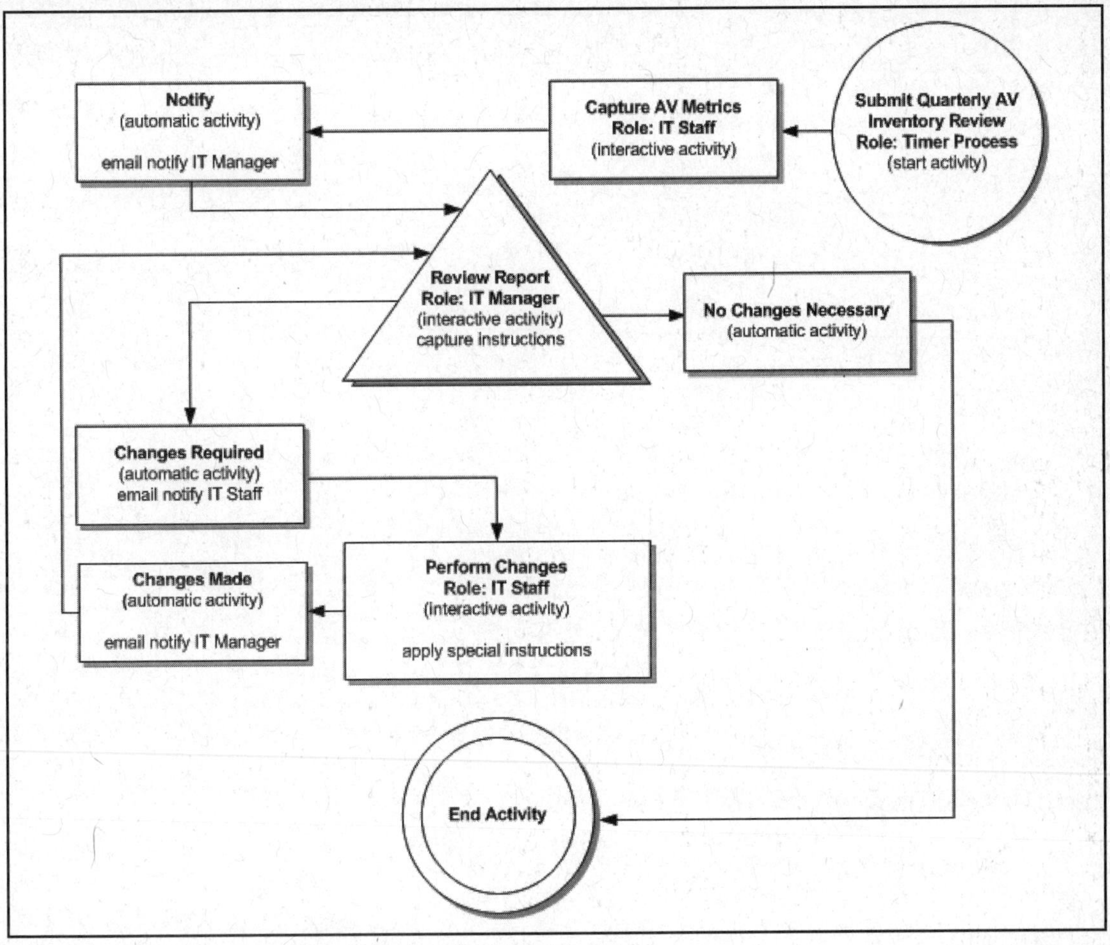

Role	Description	Member(s)
IT Staff	This role is to be filled by one of the regular IT staff members, who have access to run the reports provided by your antivirus management product.	
IT Manager	The IT Manger is responsible for approving the AV metrics provided by the IT Staff, or requesting actions or changes as necessary to achieve approval	

Activities	Description	Role
Start Activity	This workflow is automatically created on a timed interval, which notifies the IT Staff to initiate a metrics review.	Timer Process
Capture AV Metrics	The IT Staff updates the workflow instance by attaching the current metrics for the previous quarter. Once submitted this flows to the Review activity for approval.	IT Staff
IT Manager Review (Decision)	The IT Manager makes the decision to approve or request actions or changes as necessary to achieve approval	IT Manager
IT Approve	If the IT Manager approves the AV metrics the decision will be captured and the workflow will be directed to the end activity	IT Manager
IT Request Changes	If the IT Manager wishes to request changes these are recorded in the workflow instance, which is then routed to the Perform Changes activity	IT Manager
Perform Changes	The IT Staff performs any changes requested by the IT Manager. Once changes have been made this is resubmitted to the Review activity	IT Staff
End Activity	The End activity marks the workflow as complete	<none>

Quarterly Environmentals Review

This workflow is designed to regularly review environmental policies and to review any incidents that occurred during the previous quarter to determine the effectiveness of your current protection and take any necessary steps if improvement is needed, as illustrated in Figure 9.9.

Figure 9.9 Quarterly Environmentals Review

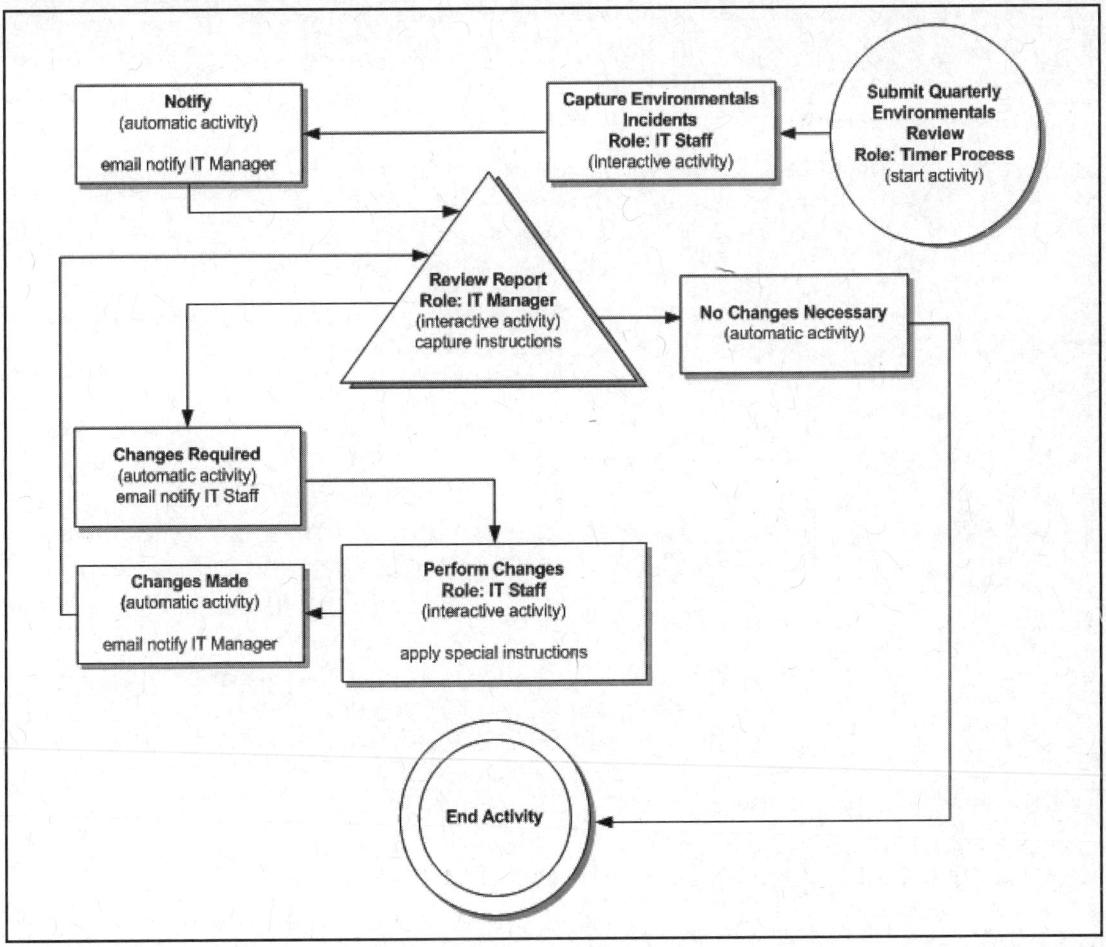

Role	Description	Member(s)
IT Staff	This role is to be filled by one of the regular IT staff members who have kept a log of environmentals incidents	
IT Manager	The IT Manger is responsible for approving the incident report provided by the IT Staff, or requesting actions or changes as necessary to achieve approval	

Activities	Description	Role
Start Activity	This workflow is automatically created on a timed interval, which notifies the IT Staff to initiate an incident review	Timer Process
Capture AV Metrics	The IT Staff updates the workflow instance by attaching the current metrics for the previous quarter. Once submitted this flows to the Review activity for approval	IT Staff
IT Manager Review (Decision)	The IT Manager makes the decision to approve or request actions or changes as necessary to achieve approval	IT Manager
IT Approve	If the IT Manager approves the incident report the decision will be captured and the workflow will be directed to the end activity	IT Manager
IT Request Changes	If the IT Manager wishes to request changes these are recorded in the workflow instance, which is then routed to the Perform Changes activity	IT Manager
Perform Changes	The IT Staff performs any changes requested by the IT Manager. Once changes have been made this is resubmitted to the Review activity	IT Staff
End Activity	The End activity marks the workflow as complete	<none>

Quarterly File Permissions Review

It is important to review the file permissions on your financial systems to make sure that there are no security holes in the current configuration. As shown in Figure 9.10, this workflow is designed to regularly make sure that only the persons who need to have access to these systems are defined in the access control lists, weeding out any staff changes as necessary. This is a somewhat generic workflow; it is up to you to determine what files need to be monitored and the methodology for reporting on the ACLs. This might be in the form of a script that checks for specific file permissions, or if you are storing data in a document management system such as KnowledgeTree, you would glean this information from the Web administration interface.

Figure 9.10 Quarterly File Permissions Review

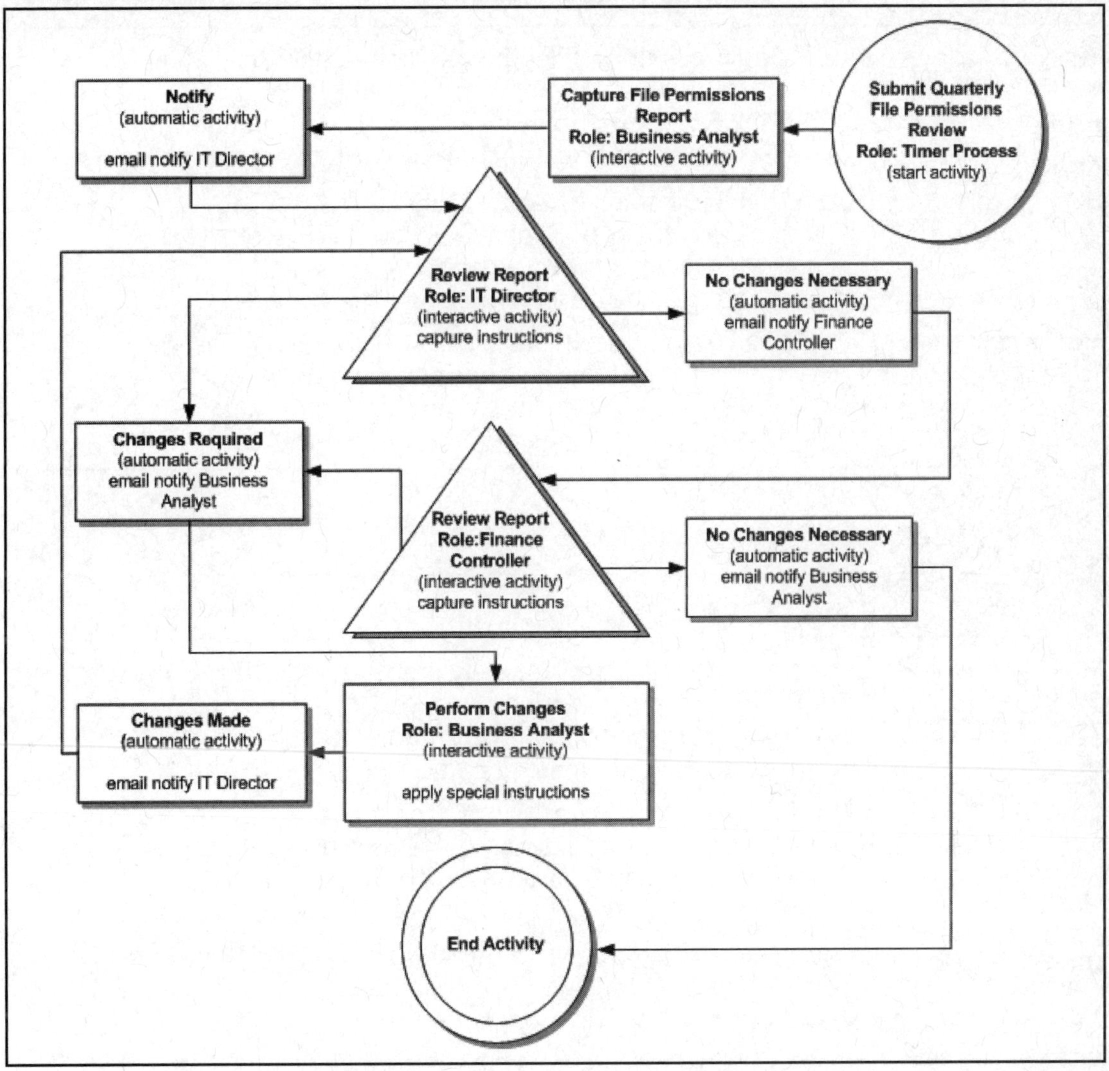

Role	Description	Member(s)
System Administrator	This role is to be filled by the main administrator of the financial systems, or this might be a finance person who is maintaining financial documents in your document control system.	
IT Director	The IT Director is responsible for approving the account report provided by the System Administrator, or requesting actions or changes as necessary to achieve approval	
Finance Controller	The Finance Controller is also responsible for approving the access report provided by the System Administrator, or requesting actions or changes as necessary to achieve approval	

Activities	Description	Role
Start Activity	This workflow is automatically created on a timed interval, which notifies the System Administrator to initiate an account review.	Timer Process
Capture Active Account List	The System Administrator updates the workflow instance by attaching the current report of ACLs. Once submitted this flows to the Review activity for approval.	System Administrator
IT Review (Decision)	The IT Director makes the decision to approve or request actions or changes as necessary to achieve approval	IT Director
IT Approve	If the IT Director approves the account list the decision will be captured and the workflow will be directed to the Finance Approve activity	IT Director
IT Request Changes	If the IT Director wishes to request changes these are recorded in the workflow instance which is then routed to the Perform Changes activity	IT Director
Finance Review (Decision)	The Finance Controller makes the decision to approve or request actions or changes as necessary to achieve approval	Finance Controller

Activities	Description	Role
Finance Approve	If the Finance Controller approves the account list the decision will be captured and the workflow will be directed to the End activity	Finance Controller
Finance Request Changes	If the Finance Controller wishes to request changes these are recorded in the workflow instance, which is then routed to the Perform Changes activity	Finance Controller
Perform Changes	The Business Analyst performs any changes requested by the IT Director. Once changes have been made this is resubmitted to the Review activity	System Administrator
End Activity	The End activity marks the workflow as complete and records the final account list	<none>

Quarterly Infrastructure Change Review

Any Infrastructure Change Requests that have been submitted over the past quarter must be reviewed. All changes that have been made must be double checked to be sure that no policy is being violated, and that no change has expired. This workflow is a collaborative review between the IT manager and IT Director, as shown in Figure 9.11. The source of the materials for review is completed Infrastructure Change Request workflows that have been submitted during the past quarter.

Figure 9.11 Quarterly Infrastructure Change Review

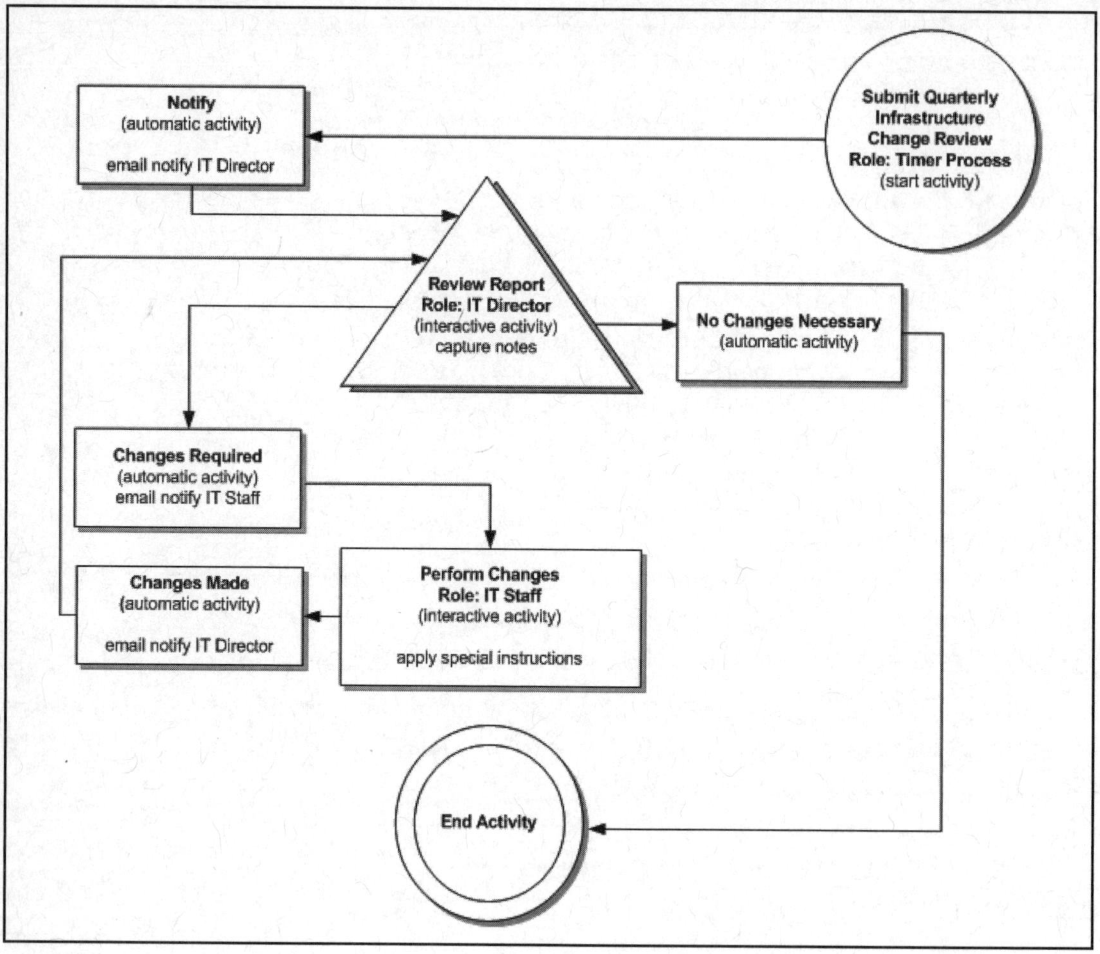

Role	Description	Member(s)
IT Staff	This role is to be filled by one of the regular IT staff members who might need to perform any changes as specified by the IT Director	
IT Director	The IT Director is responsible for performing the review to his or her satisfaction, or requesting actions or changes as necessary from the IT Staff to achieve approval	

Activities	Description	Role
Start Activity	This workflow is automatically created on a timed interval, which notifies the IT Staff to initiate an incident review.	Timer Process
IT Directory Review (Decision)	The IT Director makes the decision to approve or request actions or changes as necessary to achieve approval	IT Director
IT Approve	If the IT Director approves the incident report the decision will be captured and the workflow will be directed to the end activity	IT Director
IT Request Changes	If the IT Manager wishes to request changes these are recorded in the workflow instance, which is then routed to the Perform Changes activity	IT Director
Perform Changes	The IT Staff performs any changes requested by the IT Director. Once changes have been made this is resubmitted to the Review activity	IT Staff
End Activity	The End activity marks the workflow as complete	<none>

Additional Workflows

The remaining workflows are similar to the above in the sense that they perform some periodic check of a system or activity, the roles and steps necessary for the review process have been covered for the most part. Here is a list of additional workflows that you might consider:

- Quarterly Oracle DBA Review – this is identical to the Monthly Oracle Active User Review. The only obvious difference between the two processes is that this workflow is meant to review the Oracle DBA rather than the active users for the database.
- Quarterly Oracle System Defaults Review – All Oracle System Defaults must be reviewed periodically to maintain the database and make sure that it is secure. This process will be created once per quarter, and serves as a reminder for any Oracle Business Analysts who are in charge of reviewing the system defaults.
- Quarterly Security & Monitoring Review – In order to maintain a secure network, all monitoring devices and security implementations must be reviewed to make sure they are running properly. On a quarterly basis, the IT department is reminded

by this process to review the network and make sure that all security measures are functioning properly.

- Quarterly VPN Access Review – In order to keep a company's intranet secure, the IT department should periodically review all VPN access to make sure that no terminated employees have access, and that nothing has expired.

VM Spotlight – Zabbix Monitoring System

ZABBIX

http://www.zabbix.com

Zabbix is an enterprise-class monitoring solution for your entire IT environment. Not only does it have the ability to monitor exceptions and problems, Zabbix continually collects metrics on virtually any aspect of your systems and services to provide in-depth visualization of usage and problems over time. Although Zabbix is a feature rich and flexible monitoring solution, we like the fact that it is fairly simple and straightforward to set up, compared to other monitoring solutions such as Nagios due to its template design for items, triggers and graphs. We will drill down into the details of Zabbix, but here is an overview of Zabbix features:

- Web interface – The key to Zabbix ease of use is an excellent Web interface that offers minimal learning curve to deploy Zabbix in your environment. With the Web front-end you not only defined the items you wish to monitor, but this also provides the interface for visualizing the results of those checks over time.
- Flexible monitoring methods – Zabbix can perform monitoring via both "agent" and "agentless" methods. You choose how you wish to collect the information. Additionally, server-side external checks may be performed, and devices such as routers and switches may be monitored via SNMP (v1,v2,v3).
- Data visualization – Graphs, Screens and Slide Shows - Graphs for every data point collected are automatically available providing instant drill down from real-time status of IT Services. Historical trend and statistics may be visualized by adding graphs, which can be created using virtually any formula you desire. Screens may be defined to show many graphs, and several screens can be grouped into a slide show for better presentation.
- Flexible Actions – Multiple operations (notifications, script execution) per action are supported, with your choice of action calculation algorithm for each element.
- Auto-discovery – ZABBIX supports the discovery of systems on your network via IP ranges, service checks, agent and SNMP.

- Service Level Agreements – Zabbix provides a facility to demonstrate predefined SLA's for IT Services.
- Distributed Monitoring – ZABBIX server can spread load across several servers to better handle large enterprise environments. Visualization may then be aggregated into one holistic view. Groups of server side processes (discoverer, poller, HTTP poller, trapper, etc.) can be located on different physical servers for better performance and availability.
- Application monitoring – Enterprise applications such as Oracle, WebSphere, WebLogic, Exchange, Apache, etc. can be monitored using SNMP. In many cases, agentless technology can be used, cutting down on the deployment and management costs. ZABBIX agents can be easily extended to perform monitoring of any aspect of your applications. It is matter of writing a shell script, which would return required data back to the agent.
- WEB application monitoring – Zabbix has a sophisticated Web monitoring module for availability and performance of sites and applications, allowing the use of both GET and POST variables to accurately define and monitor the health of your Web processes for virtually any level of granular detail.
- Vendor agnostic database support – Zabbix may use many different SQL databases for storing configuration, collected data and trends such as MySQL, Postgrsql and Oracle.
- Many-to-many template linkage – flexible host-template linkage allows configuration of hosts to be more flexible and straight forward.
- XML data import/export – New XML data import and export functionality is an excellent way of sharing templates, hosts configuration and items/triggers related information.
- User permission schema – Users and Groups may be defined in the Zabbix Web front-end for access to various aspects of configuration and visualization. This is handy so that your process and data owners can see their own information on demand in a controlled and predefined way.

Zabbix in its entirely could be the subject of another book, so we will focus on the aspects of this application that we feel is important to support the effort of SOX compliance.

Zabbix Architecture

Figure 9.12 shows the architectural overview of how Zabbix works.

Figure 9.12 Simplified Zabbix Architecture

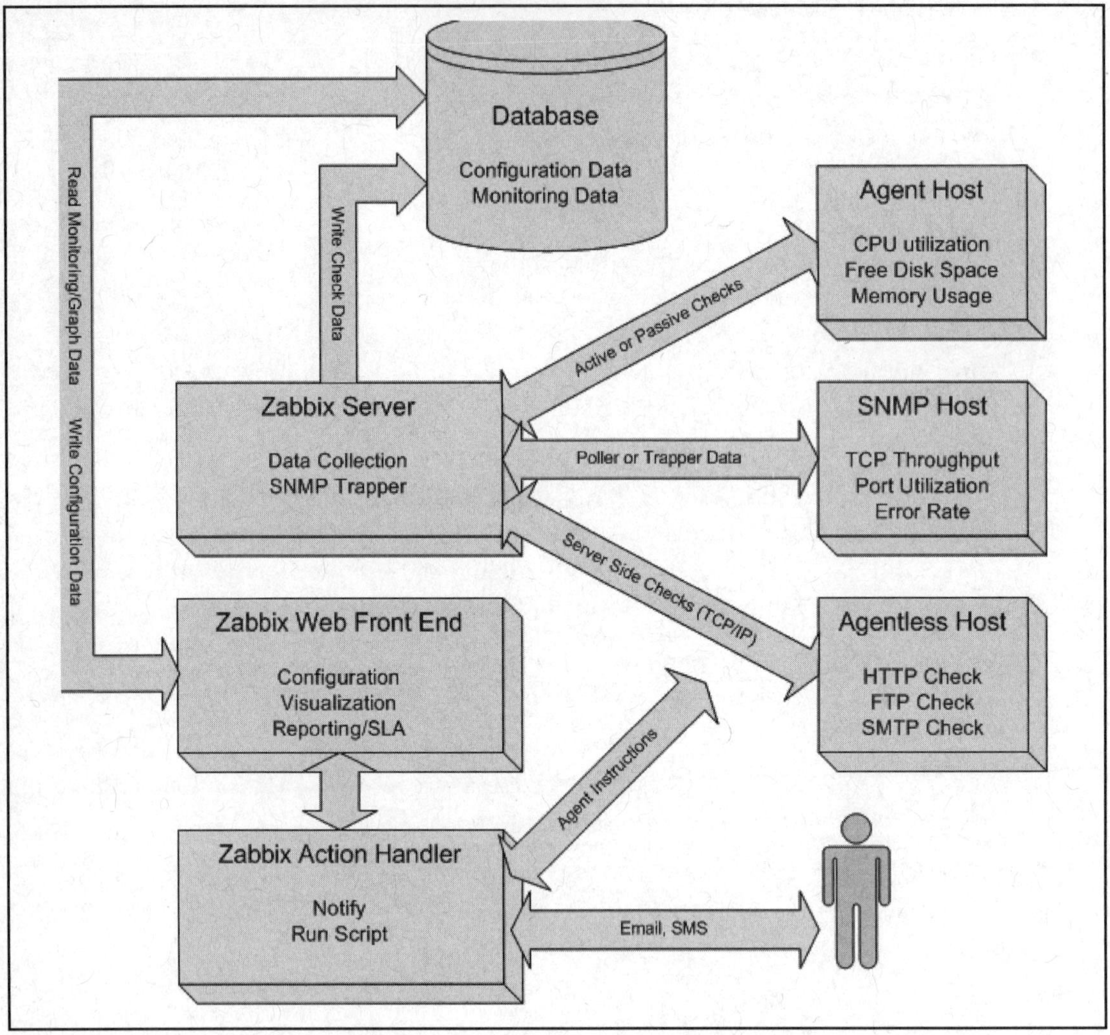

Typically the Zabbix server hosts the database, server, and Web front-end functions but this is not necessarily the case. Additionally, you may have multiple Zabbix servers reporting into a centralized facility to distribute the monitoring load, you might wish to have a Zabbix server in each physical location that reports back to a central HQ for metrics reporting for example. Zabbix basically performs three functions for any given service or data point you wish to collect:

- Item – This is the "thing" you wish to check, for example you could ask an agent for CPU utilization, perform a Web query from the Zabbix server or trap an SNMP parameter for port utilization from a configured switch

- Trigger – Once a data point is collected you can define triggers that go into on/off states based on sophisticated formulas calculated against the data point values over time. For example you could define a trigger to fire to the ON state when CPU utilization is 90% for the last five collected data points. It is important to note that a trigger does not "do" anything other than change state based on your defined criteria. One or more triggers can be defined for each data point, and aggregate triggers that involve more than one data point may also be defined
- Actions – Actions are defined against triggers. These are the things that actually do something when the trigger changes state. For example an email may be sent to one or more persons, instruct an agent to restart the service, or even remotely reboot a server.

The real power of Zabbix ease of use is in its template system for defining Items, Triggers, and Actions. Figure 9.13 shows the standard templates that ship with the base 1.4.2 version. In addition you can easily define your own templates and add standardized graphs to templates as well.

Figure 9.13 Zabbix Base Templates

TEMPLATES Group all

Name	Templates	Actions
Template_AIX		Select
Template_APC_Battery		Select
Template_App_MySQL		Select
Template_Cisco_PIX		Select
Template_Cisco_PIX_525		Select
Template_Dell_OpenManage		Select
Template_Dell_PowerConnect_6248		Select
Template_Dell_PowerEdge		Select
Template_FreeBSD		Select
Template_HPUX		Select
Template_HP_InsightManager		Select
Template_Linux		Select
Template_MacOS_X		Select
Template_Netware		Select
Template_OpenBSD		Select
Template_SNMPv1_Device		Select
Template_SNMPv2_Device		Select
Template_Solaris		Select
Template_Standalone		Select
Template_Tru64		Select
Template_Windows		Select

As you can see Zabbix comes with quite a bit of out of the box functionality, supporting many methods of checking a host or service:

- Agent – This is the most common type of check, where Zabbix asks and installed agent running on a host for a specific item to check. The default agent functionality comes with a very robust feature set to monitor virtually any aspect of the host and services. Additionally, the agent may be extended to run local scripts that return a value, making the agent able to monitor even items that do not come with the base functionality.
- Agent (Active) – In this mode, the role is reversed and the installed agent asks the server for a list of items that it should check for. In this manner the load on the Zabbix server is greatly reduced since a list of results is sent from the agent periodically rather than the server making an individual request on a per-item basis. The resulting functionality is identical, including any extended agent items defined.
- Simple Check – This type of check has predefined items that the server queries on an agentless host. For example you can define an item to check the availability of HTTP, FTP, SMTP, etc. You may also define a TCP check against an arbitrary host/port combination to see if a service is listening. For more robust checks see External below.
- SNMP (v1, v2, v3) – Zabbix can directly query any configured SNMP device for a data point based on its OID. SNMP is a large subject so we will refer you to the Zabbix manual for implementation details, but Zabbix comes with many predefined OID's for popular devices.
- Zabbix Trapper – Zabbix may also be the recipient of SNMP devices sending their data to a centralized SNMP collection host. This essentially is like the agent in "active" mode since the Zabbix server is the passive collector (trapper) and the clients are responsible for actually sending the data.
- Zabbix Internal – These checks are designed to collect metrics on the health of Zabbix itself, such as the number of values stored in history table or number of items in the current queue.
- Zabbix Aggregate – Zabbix has the ability to define data points against multiple values stored in the database. This does not involve checking a client but rather performing a query against data already collected, so you can have sophisticated conditions defined to fire a trigger.
- External Check – Where the simple checks leave off, the external checks pick up for whatever you can dream up. The Zabbix server performs an external check by running a script you define and storing the result as a normal data point. For example, you might want to query the response time for an LDAP server, but the agent and/or simple

checks do not have this defined. You can write a simple perl script using perl-ldap to perform and LDAP query and time the response, which would then be passed to Zabbix. With this methodology, your options are virtually limitless. You can even write wrapper scripts around the excellent Nagios plugins, so you do not need to reinvent the wheel for many services not natively supported by the Simple Check facility.

Zabbix Example Linux Template

Zabbix comes with many predefined Items and Triggers in the base templates. For example the following table lists the items from the default Linux Host template, you can add or subtract items as desired to tailor this to your environment.

Items	Zabbix Item Definition
Buffers memory	vm.memory.size[buffers]
Cached memory	vm.memory.size[cached]
Checksum of /etc/inetd.conf	vfs.file.cksum[/etc/inetd.conf]
Checksum of /etc/passwd	vfs.file.cksum[/etc/passwd]
Checksum of /etc/services	vfs.file.cksum[/etc/services]
Checksum of /usr/bin/ssh	vfs.file.cksum[/usr/bin/ssh]
Checksum of /usr/sbin/sshd	vfs.file.cksum[/usr/sbin/sshd]
Checksum of /vmlinuz	vfs.file.cksum[/vmlinuz]
CPU idle time (avg1)	system.cpu.util[,idle,avg1]
CPU nice time (avg1)	system.cpu.util[,nice,avg1]
CPU system time (avg1)	system.cpu.util[,system,avg1]
CPU wait time (avg1)	system.cpu.util[,wait,avg1]
CPU user time (avg1)	system.cpu.util[,user,avg1]
Email (SMTP) server is running	net.tcp.service[smtp]
Free disk space on /	vfs.fs.size[/,free]
Free disk space on /home	vfs.fs.size[/home,free]
Free disk space on /opt	vfs.fs.size[/opt,free]
Free disk space on /tmp	vfs.fs.size[/tmp,free]
Free disk space on /usr	vfs.fs.size[/usr,free]
Free disk space on /var	vfs.fs.size[/var,free]
Free disk space on / in %	vfs.fs.size[/,pfree]
Free disk space on /home in %	vfs.fs.size[/home,pfree]

Continued

Items	Zabbix Item Definition
Free disk space on /opt in %	vfs.fs.size[/opt,pfree]
Free disk space on /tmp in %	vfs.fs.size[/tmp,pfree]
Free disk space on /usr in %	vfs.fs.size[/usr,pfree]
Free disk space on /var in %	vfs.fs.size[/var,pfree]
Free memory	vm.memory.size[free]
Free number of inodes on /	vfs.fs.inode[/,free]
Free number of inodes on /home	vfs.fs.inode[/home,free]
Free number of inodes on /opt	vfs.fs.inode[/opt,free]
Free number of inodes on /tmp	vfs.fs.inode[/tmp,free]
Free number of inodes on /usr	vfs.fs.inode[/usr,free]
Free number of inodes on / in %	vfs.fs.inode[/,pfree]
Free number of inodes on /home in %	vfs.fs.inode[/home,pfree]
Free number of inodes on /opt in %	vfs.fs.inode[/opt,pfree]
Free number of inodes on /tmp in %	vfs.fs.inode[/tmp,pfree]
Free number of inodes on /usr in %	vfs.fs.inode[/usr,pfree]
Free swap space	system.swap.size[,free]
Free swap space in %	system.swap.size[,pfree]
FTP server is running	net.tcp.service[ftp]
Host boot time	system.boottime
Host information	system.uname
Host local time	system.localtime
Host name	system.hostname
Host status	status
Host uptime (in sec)	system.uptime
IMAP server is running	net.tcp.service[imap]
Incoming traffic on interface eth0	net.if.in[eth0,bytes]
Incoming traffic on interface eth1	net.if.in[eth1,bytes]
Free disk space on / in %	net.if.in[lo,bytes]
Incoming traffic on interface lo	kernel.maxfiles
Maximum number of opened files	kernel.maxproc
Maximum number of processes	net.tcp.service[nntp]
News (NNTP) server is running	proc.num[]

Items	Zabbix Item Definition
Number of processes	proc.num[,,run]
Number of running processes	proc.num[zabbix_agentd]
Number of running processes zabbix_agentd	proc.num[zabbix_server]
Number of running processes zabbix_server	proc.num[httpd]
Number of running processes apache	proc.num[inetd]
Number of running processes inetd	proc.num[mysqld]
Number of running processes mysqld	proc.num[sshd]
Number of running processes sshd	proc.num[syslogd]
Number of running processes syslogd	system.users.num
Number of users connected	net.if.out[eth0,bytes]
Outgoing traffic on interface eth0	net.if.out[eth1,bytes]
Outgoing traffic on interface eth1	net.if.out[lo,bytes]
Outgoing traffic on interface lo	agent.ping
Ping to the server (TCP)	net.tcp.service[pop]
POP3 server is running	system.cpu.load[,avg1]
Processor load	system.cpu.load[,avg15]
Processor load15	system.cpu.load[,avg5]
Processor load5	vm.memory.size[shared]
Shared memory	vfs.file.size[/var/log/syslog]
Size of /var/log/syslog	net.tcp.service[ssh]
SSH server is running	sensor[temp2]
Temperature of CPU 1of2	sensor[temp3]
Temperature of CPU 2of2	sensor[temp1]
Temperature of mainboard	vfs.fs.size[/,total]
Total disk space on /	vfs.fs.size[/home,total]
Total disk space on /home	vfs.fs.size[/opt,total]
Total disk space on /opt	vfs.fs.size[/tmp,total]
Total disk space on /tmp	vfs.fs.size[/usr,total]
Total disk space on /usr	vm.memory.size[total]

Continued

Items	Zabbix Item Definition
Total memory	vfs.fs.inode[/,total]
Total number of inodes on /	vfs.fs.inode[/home,total]
Total number of inodes on /home	vfs.fs.inode[/opt,total]
Total number of inodes on /opt	vfs.fs.inode[/tmp,total]
Total number of inodes on /tmp	vfs.fs.inode[/usr,total]
Total number of inodes on /usr	system.swap.size[,total]
Total swap space	vfs.fs.size[/,used]
Used disk space on /	vfs.fs.size[/home,used]
Used disk space on /home	vfs.fs.size[/opt,used]
Used disk space on /opt	vfs.fs.size[/tmp,used]
Used disk space on /tmp	vfs.fs.size[/usr,used]
Used disk space on /usr	vfs.fs.size[/var,used]
Used disk space on /var	vfs.fs.size[/,pused]
Used disk space on / in %	vfs.fs.size[/opt,pused]
Used disk space on /opt in %	vfs.fs.size[/tmp,pused]
Used disk space on /tmp in %	vfs.fs.size[/usr,pused]
Used disk space on /usr in %	vfs.fs.size[/var,pused]
Used disk space on /var in %	agent.version
Version of zabbix_agent(d) running	net.tcp.service[http]
WEB (HTTP) server is running	vm.memory.size[buffers]

We will not cover all of the above in great detail; rather we will focus on certain items that have direct pertinence to SOX compliance and that you might consider for your systems that are material to your financial reporting. In addition to the usual availability, disk usage and uptime checks you might consider the following, we have provided example triggers however you can define more than one for any item being checked. We usually define new templates that use the above items as well as our own in multiple smaller templates that we can pick and choose on a per-host basis. We also usually define two triggers for every item, one for the last value and one for the last five values so we can escalate a problem.

Item/Trigger	Discussion
File Checksum Item: vfs.file.cksum[<file name>] Trigger: {vfs.file.cksum[<file name>].diff(0)}>0	The file checksum ability of Zabbix is interesting from a SOX perspective since you can define triggers and alerts to warn of any changes to key files on your financials systems. With this check you can be alerted to an intrusion or an attempt to "root-kit" your server or detect any unauthorized changes to your systems.
Number of Running Processes Item: proc.num[<process name>] Trigger: {proc.num[<process name>].last(0)}<1	On your financial systems you can detect problems by either defining a trigger against a minimum and/or maximum number processes for a particular application. A common sign of problems is when an application dies, thus zero processes, or spawns multiple processes that "hang" causing system instability.
Host Information Item: system.uname Trigger: {system.uname.diff(0)}>0	The uname function may detect if a different kernel version gets installed, alerting you to unwanted system upgrades or other potentially nefarious changes to the underlying host.
Processor Load Item: system.cpu.load[,avg1] Trigger: {system.cpu.load[,avg1].last(0)}>5	Once you have established baseline usage of your financial systems, this type of item can alert you to activities that abnormally increase the CPU usage of the system. A typical scenario would be spyware or a virus on the box that is attempting to replicate itself or be used as a zombie-bot to mass email spam.
System Logs Item: log[<log file,pattern>] Trigger: {log[<log file,pattern>.last(0)>0	Zabbix provides flexible log file analysis, you can trap various events such as errors, security, and others. The syntax is basically to define an item to look for a specific value or class of values, and trigger an event if that value is found.

Continued

Item/Trigger	Discussion
System User Count Item: system.users.num Trigger: {system.users.num.last(0)}>50	On your financial systems you usually know in advance how many users are allowed to connect to the system. With this type of check you can ensure that this number does not exceed a predetermined number, alerting you if an unexpected number of users are connected to the system.

We recommend you use the existing templates when getting started with Zabbix. Over time you will develop your own items and triggers using the many predefined examples. One word of caution however. Since Zabbix is very powerful, you might begin to "over-use" the collection of data. A balance of good triggers and metrics should be thought out and planned so that you receive meaningful information, but do not overload your Zabbix server with unnecessary data that does not need to be retained. Speaking of retention, you can tune Zabbix to keep the data for a specified length of time for every individual data point, so you might consider keeping "triggerable" data retention to a minimum and meaningful visualization data points for metrics significantly longer. Zabbix performs regular housekeeping on the database to prune old information based on retention thresholds you define.

Zabbix Web Front End

The Web front end is where you will spend most of your time with Zabbix. Through this interface you define and configure virtually all aspects of Zabbix as well as use the monitoring and graphing capabilities for visualization and event handling. We will briefly explore each section of the Web interface to familiarize you with the system.

Administration

The administration section is used to configure the base Zabbix system and is made up of the following pieces:

- Users – These are the Zabbix system users. You can define individuals and groups with various levels of access to the application. For example, you could define Administrators in charge of a group of hosts and grant them full configuration access to a set list, or you could define users who may receive notifications (actions) for certain triggers.
- Media Types – These are the methods that Zabbix can use for actions. Zabbix comes with three predefined types: email, SMS, and Jabber however you are free to define your own additional types.

- Audit – This is an important facility that shows a running history of all things that occur in Zabbix, such as login attempts and changes/additions to any configurable aspect of Zabbix.
- Notifications – This is a list of the notification schedule for users defined in the system. Zabbix can be configured to only send certain actions based on severity, time the trigger occurred, and various other criteria. With this view you can see at a glance how many notifications went to whom, and can be used to fine tune your installation.
- Installation – This is the installation wizard for Zabbix, which runs when you first install the application. You can re-run the wizard if your Zabbix configuration substantially changes, but generally you only need to do this once.

Configuration

The configuration view is where you define all of the things that Zabbix will do in your environment. It is divided into the following sections:

- General – These items are for defining global items, such as how frequently the housekeeper process runs, working times, and the images available for the mapping feature (see below). An additional interesting feature you can define here are value maps, you can specify a trigger value other than ON/OFF states, mapping these to DOWN/RECOVERED for example.
- Web – The Web section is a new feature to Zabbix version 1.4. Essentially, you can define a series of POST and/or GET actions to take for a Web application, retrieving performance data and defining triggers on return data. Zabbix refers to these series of actions as scenarios and gives you the ability to emulate different browser agents to accomplish your desired result.
- Hosts – This is where you define the hosts and host groups for your environment. A host may have many item, triggers, actions, and graphs, and hosts may be grouped together for further visualization and administration tasks. There is a special host type for templates so you can group logical items together and apply them all at once to a real host or group.
- Items – These are the defined data points to be collected by the Zabbix server. They include all of the methods discussed earlier for capturing to the database. It is here that you define the datapoint, how often it is checked, how long to store the data, and what values to throw. Items can be globally enabled or disabled at this level.
- Triggers – Once items are defined, these are the various conditions based on the data that trigger to either ON or OFF. Zabbix provides a sophisticated formula syntax to define triggers. Triggers may also be dependent on other triggers, so if a parent trigger is fired then all children behind that trigger are not checked. This is

to reduce the actions taken, such as email, to only act on the root cause. Triggers are classified in levels of severity ranging from Information to Disaster.

- Actions – Actions are taken once a trigger changes state. Zabbix is very flexible in the way actions are evaluated such as severity level of the trigger, the time the trigger changed states, the host or group the trigger is owned by, and many more. See the Zabbix manual for a full list of options.
- Maps – This is an interesting feature of Zabbix where you can define maps of hosts like a network diagram that give visual clues to something gone wrong. The Zabbix Website as some examples of what you can do with maps, see http://www.zabbix.com/screenshots.php for some ideas.
- Graphs – Although Zabbix gives you instant graphs for every collected data point, sometimes it is desirable to have multiple items on one graph, or have the ability to aggregate graph items across multiple hosts. This is where the graphing facility comes in, where you can define graphs for virtually any criteria. If you apply these graphs to an appropriate template then all hosts that have this template defined automatically inherit the graph as well.
- Screens – Screens are a collection of items in a single view. You define a grid of blocks or sections that items will be embedded into, and these items may be comprised of graphs, plain text, other screens, maps, remote URLs, and many more. Screens provide a great visualization tool for a host or group of hosts for general IT monitoring and administration. We have defined a simple screen for the ITSox2 VM Toolkit server to give you an idea of what you can do with this facility.
- IT services – This is where you define your Service Level Agreements in Zabbix. You can specify a set of criteria to report on and Zabbix will provide availability reporting for this so you can demonstrate SLA compliance or identify potential problems over time.
- Discovery – This is where you define auto-discovery in Zabbix. Discovery can be used for a range of IP addresses, searching for new Zabbix agents, SNMP devices, specific TCP/IP services, etc.
- Export/Import – This facility gives you the ability to import and export both host and template information in XML format. This gives you the ability to reuse templates developed by the active Zabbix user community and share your Item, Trigger, and Graph definitions between multiple Zabbix servers and the community as well.

Monitoring

The monitoring section is where you see all of the visualization aspects of Zabbix, and is divided into the following sections:

- Overview – The Zabbix Overview gives you a dashboard view of Triggers for all hosts or host groups and gives visual clues to their current states. As depicted in Figure 9.14 the screen will blink red on server triggers where the state has changed to ON, and similarly will show green if the trigger is in its normal OFF state. The screen refreshes every minute automatically to give you up to date information.

Figure 9.14 Zabbix Overview Dashboard

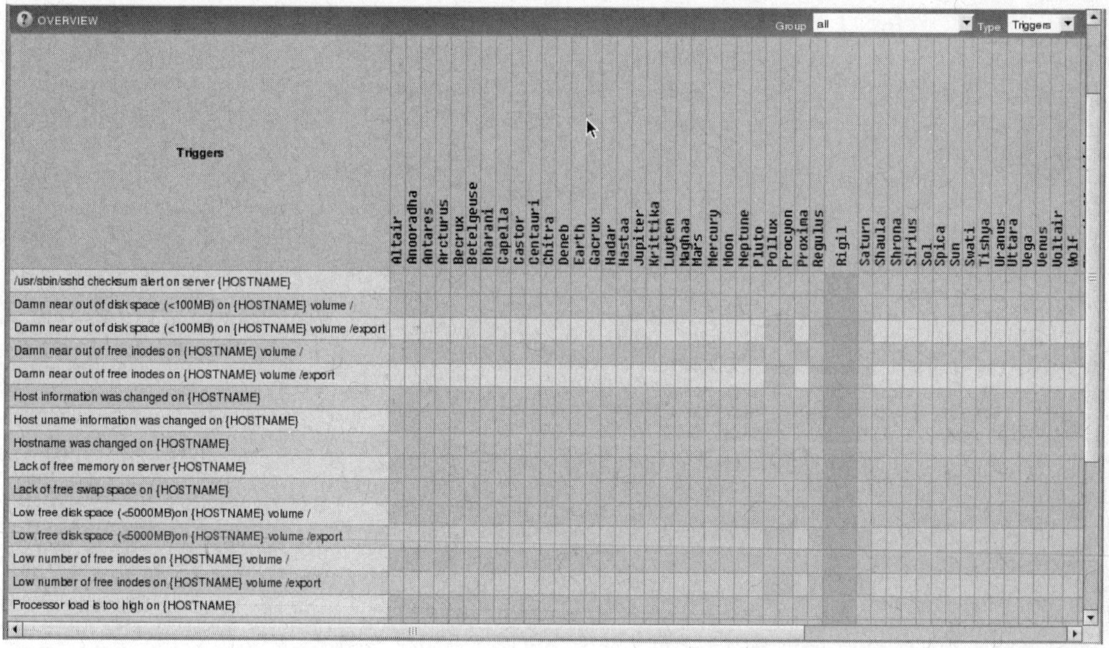

- Latest data – With this view you can see at a glance both the raw data of the items being collected as well as an instant graph of that data over time. Simply by applying a template to a host and collecting the data you get instant visualization of your data, and example is depicted in Figure 9.15.

Figure 9.15 Zabbix Example Graph

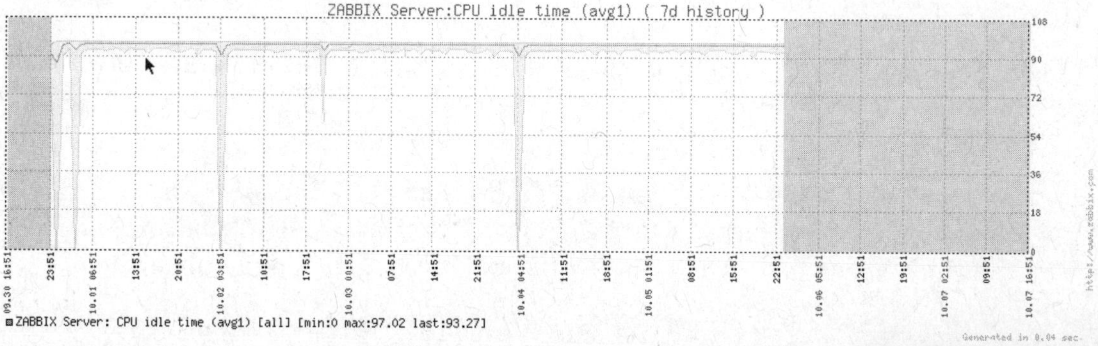

- Web – This provides the visual piece of the Web applications you may have defined in the configuration section, showing the states of the various steps taken in the application check.
- Triggers – This gives you a dashboard view of the current trigger states. By default this view shows you the triggers that are currently ON, together with the severity and gives you the ability to acknowledge the event with comments. You can also show actions that have taken place due to trigger state changes.
- Queue – This is an overview of the current queue of actions pending processing, separated by check type and age of queue item. A periodic check of this screen will alert you to any problems your Zabbix server might be having in processing the current queue load, ideally you want none or a very low number of items in the processing queue.
- Events – This is a running history of the triggers that have changed state. With this you can see a list of trigger changes in chronological order.
- Actions – This is a running history of actions that have taken place as a result of trigger state changes. With this you can see a list of actions performed in chronological order, and the disposition of that action such as success or fail.
- Maps – This provides the visual piece of the maps you may have defined in the configuration section, showing a graphical network diagram of hosts and various item states.
- Graphs – This provides the visual piece of any graphs you may have defined in the configuration section. You can specify the time period for the graph such as 1 hour, 1 day or one week, shift the view left or right, or specify a specific date and time to begin the graph from. We have defined a few simple graphs for CPU and Network utilization of the ITSox2 VM Toolkit for as examples.
- Screens – This provides the visual piece of any screens you may have defined in the configuration section. As discussed earlier screens may be comprised of many related elements to give you a holistic view of your environment. We have a simple screen defined for the ITSox2 VM Toolkit that includes a map and several graphs as an example.
- Discovery – This is a running history of discovery actions that have been performed by the Zabbix server. With this you can see a list of hosts and their current disposition from a discovery standpoint.
- IT services – This provides the visual piece of the SLA's you may have defined in the configuration section, showing a seven day percentage of uptime for those items being monitored. You can also retrieve instant graphs of this information.

In Conclusion

In addition to the above, Zabbix provides a simple mechanism for tracking inventory, storing information about your hosts such as Name, asset tag, MAC address, hardware and configuration, contact and location information, etc. Zabbix also provides several reports such as Zabbix status, availability reports for items, as well as the top 100 busy triggers. As you can see, Zabbix is a very capable monitoring tool and can be used not only for your general IT requirements but can be leveraged to provide metrics and alerts for compliance monitoring needs as well. Zabbix has an active user community and is very responsive to questions posted in the forums, and formal support is available from Zabbix SIA, see http://www.zabbix.com for more information.

Case Study: NuStuff – Oops, Still Not Right

When Sarbanes-Oxley legislation and the PCAOB standard (AS2) were initially introduced, it was very general and vague, and the interpretation varied not only from auditor to auditor but from audit firm to audit firm. Now, after almost five years and a new PCAOB standard (AS5), one might think that the compliance requirements are much better understood and universally applied. However, nothing could be further form the truth. Although we are encouraged by the PCAOB's new AS5 standard, we believe that more work needs to be done with regards to establishing straightforward standards and setting up guidelines that are clear such that they can be applied in a consistent manner. Since it is more than likely that the Sarbanes-Oxley Act will be around in one form or another for a very long time and the consequences quite grave for any company that does not pass, it would seem to us that some sort of certification of auditors and audit firms would not be out of the question. Before we go any further, we would like to re-emphasize that we believe that the new AS5 standard is definitely a step in the right direction, but there is still more work to done. To illustrate our point, the following is an excerpt from an article published in the February 2007 edition of *Compliance Week*.

"Consulting firm Accretive Solutions told the PCAOB in a comment letter that the new standard is clear enough in calling for a top-down, risk-based audit, but doesn't adequately describe how a risk-based assessment of internal controls should be performed."

"Implementation of the standard as currently proposed will continue to require a high level of subjectivity due to a lack of definitive guidance as to the practical application of the top-down, risk-based approach," wrote Dirk Hobgood, senior vice president for Accretive. "This lack of definitive guidance will lead to inconsistent application by public filers and their external auditors."

As it has been our object to try and cut through the regulatory jargon to provide you with what you need to know, we would like to attempt to extract the key points of AS5:

- Audit should be done based TDRA and therefore focused mostly on materiality and risk.
- Recognizes that smaller companies have simpler processes and therefore should not be to the same rigorous audit procedures.

As you can see there was, and still is, a lack of clarity for Sarbanes-Oxley compliance. Although NuStuff Electronics did an excellent job at identifying and implementing the right set of Control Objectives, in following the doctrine of the Quality Cycle and reassessing their environment and implemented key controls, they themselves discovered there was more work that could be done in the second year audit to further reduce their number of Control Objectives. The point is, with the exception of the aforementioned changes as a result of AS5, Sarbanes-Oxley compliance is still in the same state it was in almost five years ago – requirements are unclear and auditors are still learning. Therefore, as with NuStuff Electronics, you should not be afraid to further modify your key controls, hopefully a decrease. Some tips we would like to offer in this regard are:

- If at all possible do not change key controls during the middle of an audit period. NuStuff's experience was to change their key controls for the coming year during the previous year audit.
- If it is not possible to postpone a potential change for the audit period, we hope that you have documented periodic reviews defined in your policies, thereby allowing for changes as a result of reviews.
- Document, document, document. Make sure the who, what and why of the key control change. We might even suggest that the change be treated no differently then any other changes in your environment and follow your Change Management Process.
- Communicate with your auditor.

Summary

Deming's continuous quality improvement process is predicated on a closed loop process. In their totality, the COBIT and ITIL frameworks have a not so surprising resemblance to Deming Quality cycle. This resemblance illustrates the point that if an IT organization has been following and implementing sound quality practices all along, they are already halfway to Sarbanes-Oxley compliance. Although it may not be easy, with the utilization of the Open Source tools listed in this book, they should be able to obtain the remaining half required for Sarbanes-Oxley compliance.

A look at the various control objectives of the "Monitoring Domain" and the items that relate specifically to Sarbanes-Oxley compliance are illustrated in the context of our sample companies. We looked at monitoring in practice and some of the Open Source tools that can be leveraged to meet your compliance goals. One important tool is Nagios, which is an enterprise ready monitoring and escalation tool. Others that can be useful are centralized Syslog that can be scanned for security anomalies and system problems, Tripwire and AIDE, which can help establish a baseline of files on your critical systems and report any unauthorized changes, and Kiwi CatTools, which can monitor and manage changes made to your networking devices.

Monitoring is not limited to just hosts and services, as there are quite a few ongoing compliance objectives to meet and maintain over time. This is where compliance monitoring workflows can be of great assistance, as they remind you of important recurring compliance activities you may need to perform, combined with review routing, to make sure the proper management team keeps up their requirements. This makes a valuable addition to your SOX processes. We looked at some of the samples workflows provided on ITSox2 VM Toolkit.

Finally, we close with the VM Spotlight discussing Zabbix monitoring capabilities and our Case Study discussing the possible need to modify existing Control Objectives.

Solutions Fast Track

Never Truly Over

- ☑ Monitoring process should be close looped
- ☑ Deming developed PDCA Quality cycle
- ☑ Sarbanes-Oxley compliance drives a closed loop process
- ☑ COBIT and ITIL frameworks resemblance to Deming PDCA model

Monitoring In Theory

- ☑ SLA thresholds so that you may have additional time to react and proactively correct prior to a service interruption
- ☑ In the 1950s W. Edwards Deming developed a quality system for continuous improvement of business processes
- ☑ Deming's quality system contended that business processes should be analyzed and measured to identify sources of variations that cause products to deviate from customer requirements
- ☑ Deming stated that business processes should be placed in a continuous feedback loop so that managers can identify and change the parts of the process that need improvements

Working The List

- ☑ SOX is different from compliance COBIT and ITIL
- ☑ Do not trivializes the importance of "Monitoring"
- ☑ Keep an eye on the ability to sustain compliance

Monitoring In Practice

- ☑ The three types of monitoring are services, configuration and compliance
- ☑ Service and host monitoring can be accomplished by Nagios, which is an enterprise worthy monitoring and escalation tool
- ☑ Configuration monitoring can be achieved with the use of centralized syslog, Tripwire and AIDE
- ☑ Compliance monitoring can be assisted with the use of the eGroupware workflow application

Frequently Asked Questions

Q: If my company does not have a top down Quality methodology in place can I still use Deming?

A: Yes, although it would be more beneficial if it were top down Deming can be used within an organization.

Q: My company has implemented another Quality methodology - can I use the one already implemented at my company?

A: Yes, as long as it drives continuous improvement.

Q: My organization already performs Customer surveys – can I use these for SOX compliance?

A: Yes, as long as you have history that captures the results and any actions.

Q: Are there any other Quality systems?

A: Yes, Taguchi, Six Sigma and even ISO.

Q: What monitoring thresholds should I set?

A: Monitoring thresholds should be based on criticality of the systems or applications that are being monitored.

Q: Is it really necessary to strive for continuous improvement?

A: Yes, not only for SOX but also as a customer satisfaction tool.

Chapter 10

Putting It All Together

Solutions in this chapter:

- Analysis Paralysis
- Organization – Repositioning
- Policies, Processes and SLAs
- Control Matrices, Test Plan & Components
- Return On Investment (ROI)

☑ Summary

☑ Solutions Fast Track

☑ Frequently Asked Questions

Analysis Paralysis

"While an open mind is priceless, it is priceless only when its owner has the courage to make a final decision which closes the mind for action after the process of viewing all sides of the question has been completed. Failure to make a decision after due consideration of all the facts will quickly brand a man as unfit for a position of responsibility. Not all of your decisions will be correct. None of us is perfect. But if you get into the habit of making decisions, experience will develop your judgment to a point where more and more of your decisions will be right. After all, it is better to be right 51% of the time and get something done, than it is to get nothing done because you fear to reach a decision."

– H. W. Andrews

We chose this particular quote by H.W. Andrews for the last chapter of this book for three reasons:

- Inordinate amounts of decisions will need to be made as part of the process of obtaining Sarbanes-Oxley compliance.
- As the quote states, "None of us is perfect." So, it is inevitable that you will make incorrect decisions.
- Even with the PCAOB's adoption of AS5, the auditors are still unclear of the impact on small to medium-size companies and subsequently what constitutes Sarbanes-Oxley compliance for these companies.

Based on the above, most people would begin to ask, if the odds are skewed in such a manner, can Sarbanes-Oxley compliance really be achieved? Worse yet, why should I even bother? Well, the answer to the first question is a definite yes! As for second question, we have already addressed it in Chapter 3 but I will summarize the major reason below:

- Exposure to the possibility of lawsuits and negative publicity.
- If a corporate officer, even if unintentional, files an inaccurate certification they are subject to a fine up to $1 million and ten years in prison.
- If a corporate officer intentionally files an inaccurate certification, the fine can be as much as $5 million and possible twenty years in prison.

In this final chapter we attempt to "Put It All Together" by delving more in-depth in to some areas such as Repositioning and Policies, Processes and SLAs. We also look at in detail an example "Test Plan & Components" and finally we will discuss the concept of ROI and whether it is applicable to SOX.

The Transparency Test

The CEO Perspective

Theodore Roosevelt once said "The best executive is one who has sense enough to pick good people to do what he wants done, and self-restraint enough to keep from meddling with them while they do it." This is an excellent comment on the challenges, needed skill sets and why, utilizing marketing and strategy, the good CIO/I.T. executive will succeed.

A typical IT executive tenure at any company, in my opinion, is at best five years. The first six months to gain insight into the company and its political nature. Another six months to structure their team and determine those systems requiring change. Then, they have a three-year period to implement and educate the company, six months to deal with the fallout and dissatisfaction of the participants, and finally six months to prepare their resume and look for their next opportunity. IT executive themselves are the most adverse to risk within the organization, they bring with them personnel that have performed in the past (trusting your direct reports early in the game is paramount to success). They use the same brand hardware and software that has worked for them in the past. However, in doing so they place themselves and their organization in jeopardy. Sarbanes-Oxley alone will force the CIO/IT executive to go outside their comfort zone to not only champion change within their organization but to also build support for their objectives among their peers and management.

The problems is, of course, if CIO/IT executives play the political game with their peers, they take their eye off the ball within their department and rapidly lose support of their own personnel. Once this occurs, rapid turnover will soon follow and then of course the CIO becomes the fall guy. If they spend their time working with their department personnel for their change agenda objectives, they will inevitably lose ground with the corporate executive rank and file membership and find themselves "begging" for funding and marketing their agenda in a one on one environment, rather than a solid sell to the executive team up front. If a CIO/IT executive is to be successful they will need to go beyond merely being a technologist and bring to the business environment skill sets and knowledge that encompass political science, finance, human resources and marketing. A CIO/IT executive's demise need not be inevitable. What is inevitable, however, is that the compliance phenomena will test them in ways in which they've not been tested before.

–Bill Haag

Organization – Repositioning

In general, the majority of CFOs, CIOs or IT Directors have been confronted with the questions of, What do those IT people do? What value does the IT organization provide? And the granddaddy of them all – why shouldn't we outsource the IT organization? As we alluded to previously, as part of your Sarbanes-Oxley Compliance, if you choose, you can elect to reposition your IT organization via COBIT, ITIL and Sarbanes-Oxley Compliance. Now if I were reading this book, I would certainly be questioning the intelligence of its authors - they must be joking. They have spent the majority of the book regaling me with innumerable hurdles, the difficulties and just how much work needs to be done to obtain Sarbanes-Oxley Compliance. Clearly, you will need to assess your own individual needs and priorities to reposition your IT organization. But if you do elect to take on this challenge, take comfort in the knowledge that the majority of the activities you need to do are ones you to need to do to obtain Sarbanes-Oxley Compliance. OK, you've identified the Control Objectives and you at least calmed my fear a little, so now what? Well, in order not to be redundant, we will assume that you are committed to the processes we've discussed in prior chapters in this book, particularly the Quality Process. That being said, the remaining task is really just old-fashioned marketing. The mantra should be leverage, leverage, leverage. For example, once you have determined what your performance matrices are, post them, distribute them, and do whatever is needed to make your customers and management aware of how you are doing. This is one example that you can leverage, use the example for availability, response time, services requests, etc. Your awareness effort might require that you produce separate reports for your Executive Management versus your customers, depending upon the level at which perceptions need to be changed.

As we stated, the majority of the tasked needed to effectively reposition IT are the same ones involved with Sarbanes-Oxley Compliance and good old fashioned marketing. Below are traditional marketing items – and without exception they either match or can be easily matched to SOX compliance:

- Executive Summary
- Situation Assessment
- Competition Assessment – Benchmark services
- Statement of Objectives
- Develop Action Plan
- Develop Budget

Along with your awareness activities, you will also want to ensure that you continually update:

Policies, Processes and SLAs

We have correlated COBIT and ITIL to "A Good Quality Process" and even to Deming's Plan, Do, Check, Act (PDCA) quality cycle. Although we have discussed the correlation of Policies, Processes and SLAs, this concept still may not be clear. So, in this section it is not only our intent to illustrate this concept but also to provide you with some practice guideline and recommendations.

Although we have listed repositioning IT as one of the objectives of this book, it is imperative for you to remember that your first priority is to pass your Sarbanes-Oxley compliance audit. With that, let us restate the intent and goal of Sarbanes-Oxley Act and Compliance. The intent of the Sarbanes-Oxley Act is to ensure that public companies have auditable business processes and effective internal controls over their financial systems that are material in reporting their financials. It also requires that public companies have an ongoing process to identify, evaluate and correct control deficiencies. At this point, you may be asking how many more times do you need to emphasis ongoing processes as an integral part of Sarbanes-Oxley Compliance. Well, the answer is just one more time. So, here it is for the last time, "if you cannot sustain an activity by no means state it as a process or one of your Control Objectives – this is one of the surest ways to fail your audit."

Now you might be thinking in an ideal world, I might be able to review and "Soxerize" all of my processes and build in ongoing sustainability but given my particular variables such as time, complexity and resources I can't do it. The best approach as you look at your environment and start to develop and implement processes for Sarbanes-Oxley Compliance is to take a holistic view. The value of taking a holistic view is not only the ability to build sustainability into your processes but also to minimize the amount of work that will be required for your next audit. Conversely, you can also approach Sarbanes-Oxley Compliance as a point in time event and merely perform spot correction or develop short-term processes. Although we strongly advise against the latter, it will not preclude you from using the methods described in this book.

As with Deming's PDCA Quality Model, each phase of our SOX Process Flow model not only feeds into a ensuing phase but it is also feed from the preceding phase. The SOX Process Flow model is straightforward and fairly simple to follow. But for the sake of clarity with the process, if after following the flow of the model you find that the process did not yield the desired results (i.e. didn't produce evidence or produced the wrong evidence) you will need to look at the previous phase to ensure that the correct input was received by the ensuing phase. This may be an iterative process but a necessary one if you are to determine why the process did not yield the expected results.

SOX Process Flow

Figure 10.1 SOX Process Flow represents our interpretation of the flow for Sarbanes-Oxley compliance as it relates to Deming's Plan, Do, Check and Act quality model. By merely substituting the various categories of PDCA with terminology more suited toward SOX and frameworks you will hopefully begin to see as we stated previously, that COBIT, ITIL and SOX are based on a Quality Process. It is also our hope you will also see why SLAs play such an important role in our Sarbanes-Oxley Compliance efforts and this book as well.

Figure 10.1 SOX Process Flow

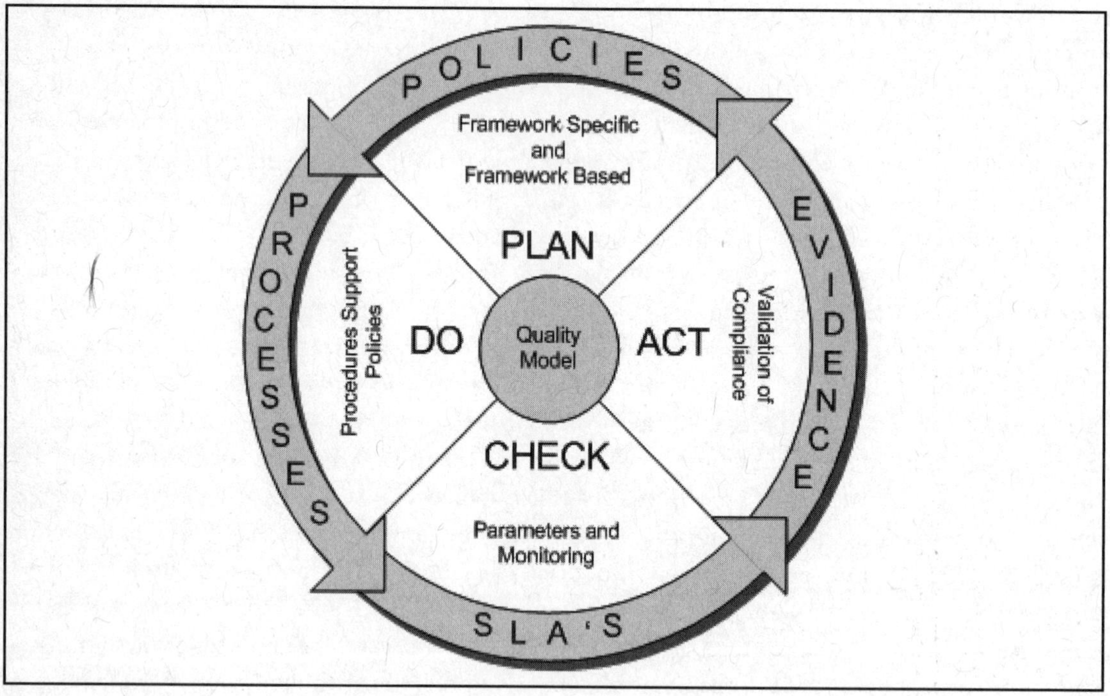

As you proceed with your process of identifying, documenting and ultimately implementing your processes for Sarbanes-Oxley Compliance, you will want to pay particular attention to certain areas. Below are the areas our experience has yielded as being important; yours may differ. If you want to jump-start your assessment process and at the same time identifying areas that may be hot buttons for your auditor we would suggest that you review previous audits and capture any unresolved items. From our experience, these are items that will surely come up in your audit, so you will need to either address them or be prepared to mitigate them.

- Change management process
- Data integrity
- Disaster recovery practices
- Electronic records retention policy

- Patch management
- Process/work flows – internal & partners
- Security policies and practices
    1. User Access
    2. Data Security

It is important to note that disaster recovery mentioned above is not critical for initial SOX compliance *per se;* however, BCP will more then likely be critical in subsequent audits.

Control Matrices, Test Plan & Components

As part of this section, we will be examining three of the fundamental forms you will utilize to capture your Control Objectives in order to prepare for your Sarbanes-Oxley audit. The forms we decided to use were created in an Excel format but there is no requirement for you to do so. As long as you have the required fields, you can use whatever application you prefer. Please keep in mind that since the PCAOB has adopted COBIT as the de facto audit standard, the sample forms were developed with this in mind. However, with the COBIT to ITIL correlations we provided in previous chapters, you should not have a problem replacing the COBIT control with the ITIL guideline. The following are the forms:

- Control Matrix – The intent of the Control Matrix form is to summarize your processes and control for your auditors. Beyond documenting or developing your control processes, it is absolutely critical that what is stated in your documentation matches what is described in this form.
- Gap and Remediation – The intent of the Gap and Remediation form is to capture and track any identified deficiencies with your documented control objectives. You are likely to use this form as part of your internal discovery process, as well as a result of your auditors' activities.
- Test Plan – Here is where the rubber meets the road. The intent of the Test Plan form is to identify what Controls are in place to mitigate the known risk and the methodology for testing the Control.

Control Matrix

Figure 10.2 is an example Control Matrix which represents a form you might use to capture and represent the information contain in your documented processes. Although this form may appear to be daunting, it really is not. The category that will require the most amount of thought and may provide the most difficulty is determining the correct Control Activity.

As you have probably gathered, the Control Activity should be derived from your documentation and/or narratives. What may not be so intuitive is that the Control Activity should address the Control Objective, which in turn mitigates the Business Risk.

Figure 10.2 Sample Control Matrix

Windows & Unix Authentication Password Controls										
CobiT Ref	Key Ctrl	Ctrl Pt.	Business Risk	Control Objective	Control Activity	Control Performer	Control Evidence	Application/ IT Depend/ Manual	Prevent/ Detect	Frequency
SDS 5.4	X	01	Unauthorized user access may lead to unauthorized access to financial data.	Procedures exist and are followed to maintain the effectiveness of authentication and access mechanisms (e.g., regular password changes).	Authentication to the NuStuff Windows domain is required to obtain access to any NuStuff files stored on any of the file servers.	System Administrator	Server Configuration	Application	Prevent	Real-time
SDS 5.4	X	02			Corporate Domain (Active Directory), Password Policy Windows passwords must comply with the following standards, which are automatically enforced: (a) Minimum length of 8 characters; (b) Multiple character sets; (c) History verification of three;	System Administrator	Server Configuration	Application	Prevent	Real-time
SDS 5.4	X	03			When a new user's system is prepared by Client Services, they expire the initial password, which forces the user to change their password on first login.	Help Desk Manager	User password history log	Manual	Prevent	Upon Demand
SDS 5.4	X	04			Unix Server, Local Password Policy Unix must comply with the following standards, which are automatically enforced: (a) Minimum length of six characters; (b) Multiple character sets; (c) Expiration of 13 weeks;	System Administrator	Server Configuration	Application	Prevent	Real-time
SDS 5.4	X	05			Oracle DB Password Policy Oracle DB passwords are automatically enforced: (a) Minimum length of six characters; (b) Multiple character sets; (c) Expiration of 13 weeks;	Oracle DBA	Database Configuration	Application	Prevent	Real-time

- COBIT Ref. – relates to the COBIT Domain that the Control Objective is contained
- Key Ctrl – is this considered to be a key Control Objective
- Ctrl Pt. – the Ctrl Pt. merely enumerates the items
- Business Risk – represents the risk to the business in the absence of effective controls as defined by COBIT
- Control Objective – represents the Control Objective as defined by COBIT
- Control Activity – the control activity should be verbiage gleaned from your narratives as to what processes are in place in your organization to mitigate the defined risk
- Control Performer – denotes who or what performs the stated Control
- Control Evidence – denote what evidence is produced to demonstrate that the control works. It is important to keep in mind that auditors' preference is to have system-generated evidence.

- Application/IT Depend/Manual – denotes what performs the control. As with Control Evidence, the auditors' preference would be to have as many application driven processes
- Prevent/Detect – denotes the type of the process. Your Control Objectives will utilize one of the two categories. Either it will fall under the category of Prevent or Detect. Based on our experience we can't necessarily say one is preferred over the other, but what we can say is auditor generally like it when there are multiple Prevent and Detect Controls in place for a Control Objective
- Frequency – denotes when the control is executed

Gap and Remediation

Figure 10.3 Gap and Remediation form represents an example of a form you might use to capture any/all deficiencies that would be yielded from either your initial testing or your auditor testing. This form is relatively straightforward as to what information should be contained in the respective fields. The main thing that you will want to keep in mind is that your auditor will be monitoring how well you manage the actions and due dates of the items captured. Be sure that you adhere to these actions and due dates.

Figure 10.3 Gap and Remediation Form

NuStuff
21 Windows & Unix Authentication Password Controls

Ref	Control Pt.	Gap	Action	Resp	Due Date
1		None			
2					
3					
4					

- Ref. – merely enumerates the number of items
- Control Pt. – the Ctrl Pt. merely enumerates the items. Should match your Control Matrix and Test Plan
- Gap – denotes any identified deficiencies in Control Objectives
- Action – denotes actions to be taken to remediate identified deficiencies in Control Objectives
- Resp – denotes the responsible person for resolution of the identified deficiencies in Control Objectives
- Due Date – denotes the date which the identified deficiencies in Control Objectives are to be resolved

Test Plan

Figure 10.4 Test Plan represents an example of a form you might use to capture and represent the information contain in your documented processes.

Figure 10.4 Test Plan

NuStuff

21 - Windows & Unix Authentication Password Controls

Key Ctrl	Ctrl Pt.	Control Activity	Control Performer	Control Evidence	Application/ IT Depend/ Manual	Frequency	Test Period Start	Pop Size	Sample Size	Test Plans	Tester	Test Date
X	01	Authentication to the NuStuff Windows domain is required to obtain access to any NuStuff files stored on any of the file servers.	System Administrator	Server Configuration	Application	Real-time	1/1/2004	1	1	Browse to Windows file server without connecting to the Company Corporate Domain. Verify it is not possible to connect to Windows file server.		
X	02	Corporate Domain (Active Directory), Password Policy Windows passwords must comply with the following standards, which are automatically enforced: (a) Minimum length of 8 characters; (b) Multiple character sets; (c) History verification of three;	System Administrator	Server Configuration	Application	Real-time	1/1/2004	1	1	1. Verify the Windows password configuration point that automatically enforces these standards. 2. Attempt to change an existing password to a password which for each of the rules (a)-(c) complies with the other standards but not that particular standard. Verify this is not possible.		
X	03	When a new user's system is prepared by Client Services, they expire the initial password, which forces the user to change their password on first login.	Help Desk Manager	User password history log	Manual	Upon Demand (add freq)	1/1/2004	1	1	Select a sample of users who have recently been assigned new accounts. BA to view password file and verify that all pass words are not the initial password granted at initial login. Note: Only the BA may conduct this test and sight this data. No record of user passwords is to be printed.		
X	04	Unix Server, Local Password Policy Unix must comply with the following standards, which are automatically enforced: (a) Minimum length of six characters; (b) Multiple character sets; (c) Expiration of 13 weeks;	System Administrator	Server Configuration	Application	Real-time	1/1/2004	1	1	1. Verify the Unix password configuration point that automatically enforces these standards. 2. Attempt to change an existing password to a password which for each of the rules (a)-(c) complies with the other standards but not that particular standard.		
X	05	Oracle DB Password Policy Oracle DB passwords are automatically enforced: (a) Minimum length of six characters; (b) Multiple character sets; (c) Expiration of 13 weeks;	Oracle DBA	Database Configuration	Application	Real-time	1/1/2004	1	1	1. Verify the Oracle DB password configuration point that automatically enforces these standards. 2. Attempt to change an existing password to a password which for each of the rules (a)-(c) complies with the other standards but not that particular standard.		

- Key Ctrl – is this considered to be a key Control Objective
- Ctrl Pt. – the Ctrl Pt. merely enumerates the items
- Control Activity – the control activity should be verbiage gleaned from your narratives as to what processes are in place in your organization to mitigate the defined risk
- Control Performer – denotes who or what performs the stated Control
- Control Evidence – denote what evidence exists to demonstrate that the control works. It is important to keep in mind that auditors' preference is to have system-generated evidence
- Application/IT Depend/Manual – denotes what performs the control. As with Control Evidence, the auditors' preference would be to have as many application driven processes
- Frequency – denotes when the control is executed

- Test Period Start – denotes the date which the control was implemented
- Pop Size – denotes the total number of auditable evidence
- Sample Size – denotes the number of Pop Size selected to audit
- Test Plan – denotes the actions and/or steps that are to be taken to demonstrate the effectiveness of the control
- Tester – denotes the person or persons responsible for performing and documenting the test results
- Test Date – denotes the date which the test was performed

What Makes a Good Test Plan

Since there is no specific format for performing Sarbanes-Oxley test and capturing the results if you elect to you can develop a format that may suit your particular company better then the example we have provided in this book. If for whatever reason you do elect to develop your own form and format for Control Objective testing the criteria below are key elements that you should keep in mind:

- Documented test plan
- Appropriate Controls identify
- Appropriate test for Control
- Selection of correct sample size
- Rational of sample size
- Method to capture test results.

Return On Investment (ROI)

There are countless periodicals and Internet-based articles concerning Sarbanes-Oxley Compliance and ROI. In order to frame the discussion and illustrate our opinion, we have selected a few excerpts from various articles in order to form a basis. The following is an excerpt from an article published in the August 2007 edition of Intelligent Enterprise.

"Sarbanes-Oxley Act compliance costs decreased overall in 2006, the out-of-pocket costs associated with compliance rose from 2005 to 2006, according to a study by Foley & Lardner."

"Foley & Lardner's fifth annual study measuring the financial impact of Sarbanes-Oxley on public companies finds that the cost of audit fees, board compensation, and legal fees continue to rise, despite an overall plateau in compliance costs that companies tend to see following initial implementation of Section 404 financial controls."

"The burden of compliance is prompting an increasing number of respondents at the 93 public companies surveyed to consider going private or selling the company. This year, 23% of those answering the survey said they're considering transactions to take their company private, 16% said selling their company was a possibility, and 14% said they were considering a merger."

Here are more Sarbanes-Oxley cost related points from an Article in SearchSMB.com "Sarbanes-Oxley compliance costs drop, better processes credited":

- The cost of compliance with Section 404 of the Sarbanes-Oxley Act (SOX) declined by 23% in fiscal 2006, according to a survey by Financial Executives International. The Florham Park, N.J.-based organization found the average company spent $2.9 million on SOX compliance in 2006, versus $3.8 million in 2005 and $4.5 million in 2004.
- "Technology has a lot to do with the cost reduction," said Sanjay Anand, chairperson of the Sarbanes-Oxley Institute. Public companies "are actually automating their controls. A good 20 to 30%, even as much 40%, of the cost reduction is actually coming from automated controls rather than manual controls."
- The cost reductions have come despite the fact that auditors' fees have remained relatively steady, the research revealed. External auditor fees dropped by just 11% in 2006, from $1.35 million to $1.2 million.

Additionally a survey by the American Institute of Certified Public Accounts found that 28% of Total Sarbanes-Oxley compliance costs were spent on technology. If we were to summarize the information above into concise statements we would conclude several things:

- Sarbanes-Oxley Compliance costs for businesses are still significant
- Sarbanes-Oxley Compliance costs for small businesses will be significantly more
- Most companies required to comply with Sarbanes-Oxley believe they will need to acquire software to sustain their compliance

As you can see there is still need for small to medium-sized companies to have concern over Sarbanes-Oxley compliance costs perspective.

We will define ROI in its simplest terms – a change that requires funding that will by cost savings over time pay for itself. If we look at the data above, we can only conclude that there is truly no quantitative way to do an ROI for Sarbanes-Oxley Compliance. Based on a recent ComputerWorld article "Sarb-Ox Adds To Cost, Length Of IT Projects" one would even conclude that not only is there no real way to calculate ROI for Sarbanes-Oxley Compliance, and if there were, there would be no ROI for SOX. We are by no means suggesting that Sarbanes-Oxley Compliance has no value. What we are saying is that the value of Sarbanes-Oxley Compliance is qualitative and not quantitative. This opinion seems to be supported by many periodicals and Internet based articles as well.

So, if there is no way to justify Sarbanes-Oxley Compliance, how do I handle questions about my company's compliance activities affect the bottom line? The answer is by shifting the ROI from SOX and cost savings to Open Source and cost avoidance. As we have stated previously, a decision point of whether to comply with Sarbanes-Oxley or not does not exist. But what you can more or less decide is how much software acquisition costs are and whether you will need additional headcount to sustain your environment after compliance. So, the next logical question would be how much savings could a small company reasonably expect to save in cost avoidance by utilizing Open Source? The answer to this question would depend on various factors at the particular company, but based on what we have seen at our fictitious company, the savings could start at $100,000.00 for the first year. This number represents $55,000.00 in salary and $45,000.00 in application acquisition costs.

There is however life after SOX. The journey to Sarbanes-Oxley Compliance will probably be one of the most difficult challenges you will have to face in your career. Not necessarily due to the technical aspect of the activity but rather as a result of the amorphous nature of the requirements and the professional challenges of dealing with auditors, customers and management in a way in which you have probably never had to do before. Although you might not see it at the time, there is light at the end of the tunnel.

Lessons Learned

SOX and Threat Mitigation

Microsoft alone issued 78 security updates last year – up from 55 in 2005. Add to that vulnerabilities found in Java, Firefox, Yahoo, AOL, iTunes, Quicktime, RealPlayer, and Adobe and you have yourself one heck of a job safeguarding your environment against these threats. All the while your IT staff is scrambling to keep up with the constant barrage of security updates; the bad guys are exploring new doors as fast as your staff can close the old ones. So far, with the help of software vendors, they have may have managed to keep ahead of the curve. Now add Sarbanes-Oxley compliance to the mix. The staff that used to simply read bulletins, test and apply patches across your infrastructure can now suddenly be a threat to your organization.

"How so?" you may ask. The problem is that this department is in control of a process from end to end, and therein lies the conflict in the matter of segregation of duties. How does one truly confirm that employee X has been diligent in his/her duties? Would it not be easier for said employee to choose a compliance marker that would always make his metrics look good, regardless of the actual state of any

Continued

given machine? Sarbanes-Oxley compliance requires that duties be divided so that the person who dictates the success marker for security compliance is not the same person who performs and/or updates the audit/compliance tool. Yet another person from a wholly disinterested department is required to test the patch before it can be handed off to the person/people who finally deploy it. This eliminates the segregation of duties issue and is part of walking the talk. Furthermore, one of these people also now must produce an internal, publicly available document that declares which updates are required for internal systems. Caution: this document should not contain the markers by which your audit mechanism determines compliance. This would provide an opening for misunderstanding, which may cause compliance markers to be faked by a well-intentioned or frustrated user.

If this sounds expensive, that's because it is, and it's getting more expensive every year as security threats and exploits rise. Over the course of the last 5 years I have watched staff numbers in this area grow form one or two people in a large enterprise (50K+) to more than 10 full time positions. It is because of 404 requirements such as segregation of duties and auditable paper trails that many companies are forced to add staff at a time when they can scarcely afford to do so. You might ask then, how do companies who have successfully obtained and maintain Sarbanes-Oxley compliance handle this? In order for a company, small or large, to become compliant without adding cost that would exceed the penalties of non-compliance, they must implement automation wherever and whenever possible and a process for continual process improvement and simplification.

We can all agree that the well-publicized corporate scandals were awful and hurt a lot of people. It seems strange then, that the legislation put in place to protect those very same people from that sort of thing ever happening again is the same legislation that–by virtue over-burdening companies, especially small to medium sized ones with excessive compliance cost–is now putting those people out of their jobs.

–John T. Scott

Summary

In "Putting It All Together," we have discussed the hows and whats of repositioning an IT Organization utilizing COBIT, ITIL and SOX. We have also illustrated quality processes, the tie between Policies, Processes and SLAs. As a subset of quality, we have reviewed in detail an example test plan and its components. Objective testing the criteria below are key elements that you should keep in mind:

- Documented test plan
- Appropriate Controls identify
- Appropriate test for Control
- Selection of correct sample size
- Rational of sample size
- Method to capture test results.

Our final ITSox2 VM Toolkit spotlight leads you through a step-by-step process in defining your own workflows, which we feel is an important component in automating many of the activities surrounding compliance, reducing labor overhead and providing a consistent mechanism to capture and retain evidence of compliance. Finally we discussed Sarbanes-Oxley and Open Source ROI and which is more feasible to measure in a traditional ROI format.

Solutions Fast Track

Analysis Paralysis

☑ Inordinate amount of decisions will need to be made as part of the process of obtaining Sarbanes-Oxley compliance.

☑ As the quote states "None of us is perfect." So, it is inevitable that you will make incorrect decisions.

☑ The auditors are still unclear and learning what truly constitutes Sarbanes-Oxley compliance.

Organization - Repositioning

☑ Repositioning will require only a few additional activities beyond SOX.

☑ Repositioning is just old fashion marketing.

☑ The mantra for repositioning should be leverage, leverage, leverage.

Policies, Processes and SLAs

☑ Sarbanes-Oxley Compliance can be approached as a point in time activity

☑ The best approach for Sarbanes-Oxley Compliance is holistic

☑ SLA can provide evidence of compliance

Control Matrices, Test Plan & Components

☑ Control Matrix – The intent of the Control Matrix form is to summarize your processes and control for your auditors. Beyond documenting or developing your control processes, it is absolutely critical that what is stated in your documentation matches what is described in this form.

☑ Gap and Remediation – The intent of the Gap and Remediation form is to capture and track any identified deficiencies with your documented control objectives. You will more then like use this form as part of your internal discovery process as well as a result of your auditor's activities.

☑ Test Plan – Here is where the rubber meets the road. The intent of the Test Plan form is to identify what Controls are in place to mitigate the known risk and the methodology for testing the Control.

Return On Investment (ROI)

☑ Sarbanes-Oxley Compliance costs for businesses will be significant

☑ Sarbanes-Oxley Compliance costs for small businesses will be significantly more

☑ Almost 70% of companies required to comply with Sarbanes-Oxley believe they will need to acquire software to sustain their compliance

Frequently Asked Questions

Q: Can Sarbanes-Oxley compliance be obtained even if deficiencies are found in my environment?

A: Yes, you can almost rest assure that deficiencies will be found. What your objective should be is to not have any Material Weaknesses, Significant Deficiencies, or too many Deficiencies.

Q: Considering the drain of cost of compliance to business is there anything being done?

A: Yes, but not from the SEC. If any relief from SOX is given it will more then likely come from Capitol Hill.

Q: Can I use the test plan examples?

A: Yes, however you should develop your own using what we have provided as guidance, just keep in mind the component we covered.

Q: Can SOX Compliance really be accomplished as a point in time project?

A: Yes, but that not the question. The question is do you want to do it all over again?

Q: Quality seems to be a major theme in the majority of the chapters, is it really that important?

A: Yes, Quality and Processes will either make or break your compliance activities.

Q: Can I really achieve Cost Avoidance utilizing Open Source?

A: Yes, recently a lot of major companies have started to realize the ROI on Linux and Open Source. To name a few: Google, E-bay, IBM, even Sun Microsystems.

Q: Where can I find more information on Sarbanes-Oxley Compliance?

A: There are multitudes of site on the Internet where you can find more general Sarbanes-Oxley compliance information, however here are a few websites that can get you started:

- http://www.sarbanes-oxley-act-compliance.com/sarbanesoxleydefinition/
- http://techrepublic com/5208-6230-0.html?forumID=101&threadID=227253&start=0
- http://www.controlit.org/modules.php?name=Forums&file=viewtopic&p=266
- http://www.sarbanes-oxley-forum.com/

Appendix A

COBIT Control Objectives

PLANNING & ORGANIZATION

1. Define a Strategic IT Plan
    - 1.1. IT as Part of the Organization's Long- and Short-Range Plan
    - 1.2. IT Long-Range Plan
    - 1.3. IT Long-Range Planning—Approach and Structure
    - 1.4. IT Long-Range Plan Changes
    - 1.5. Short-Range Planning for the IT Function
    - 1.6. Communication of IT Plans
    - 1.7. Monitoring and Evaluating of IT Plans
    - 1.8. Assessment of Existing Systems
2. Define the Information Architecture
    - 2.1. Information Architecture Model
    - 2.2. Corporate Data Dictionary and Data Syntax Rules
    - 2.3. Data Classification Scheme
    - 2.4. Security Levels
3. Determine Technological Direction
    - 3.1. Technological Infrastructure Planning
    - 3.2. Monitor Future Trends and Regulations
    - 3.3. Technological Infrastructure Contingency
    - 3.4. Hardware and Software Acquisition Plans
    - 3.5. Technology Standards
4. Define the IT Organization and Relationships
    - 4.1. IT Planning or Steering Committee
    - 4.2. Organizational Placement of the IT Function
    - 4.3. Review of Organizational Achievements
    - 4.4. Roles and Responsibilities
    - 4.5. Responsibility for Quality Assurance
    - 4.6. Responsibility for Logical and Physical Security
    - 4.7. Ownership and Custodianship
    - 4.8. Data and System Ownership
    - 4.9. Supervision

- 4.10. Segregation of Duties
- 4.11. IT Staffing
- 4.12. Job or Position Descriptions for IT Staff
- 4.13. Key IT Personnel
- 4.14. Contracted Staff Policies and Procedures
- 4.15. Relationships

5. Manage the IT Investment
   - 5.1. Annual IT Operating Budget
   - 5.2. Cost and Benefit Monitoring
   - 5.3. Cost and Benefit Justification
6. Communicate Management Aims and Direction
   - 6.1. Positive Information Control Environment
   - 6.2. Management's Responsibility for Policies
   - 6.3. Communication of Organization Policies
   - 6.4. Policy Implementation Resources
   - 6.5. Maintenance of Policies
   - 6.6. Compliance with Policies, Procedures and Standards
   - 6.7. Quality Commitment
   - 6.8. Security and Internal Control Framework Policy
   - 6.9. Intellectual Property Rights
   - 6.10. Issue-Specific Policies
   - 6.11. Communication of IT Security Awareness
7. Manage Human Resources
   - 7.1. Personnel Recruitment and Promotion
   - 7.2. Personnel Qualifications
   - 7.3. Roles and Responsibilities
   - 7.4. Personnel Training
   - 7.5. Cross-Training or Staff Back-up
   - 7.6. Personnel Clearance Procedures
   - 7.7. Employee Job Performance Evaluation
   - 7.8. Job Change and Termination

8. Ensure Compliance with External Requirements
    - 8.1. External Requirements Review
    - 8.2. Practices and Procedures for Complying with External Requirements
    - 8.3. Safety and Ergonomic Compliance
    - 8.4. Privacy, Intellectual Property and Data Flow
    - 8.5. Electronic Commerce
    - 8.6. Compliance with Insurance Contracts
9. Assess Risks
    - 9.1. Business Risk Assessment
    - 9.2. Risk Assessment Approach
    - 9.3. Risk Identification
    - 9.4. Risk Measurement
    - 9.5. Risk Action Plan
    - 9.6. Risk Acceptance
    - 9.7. Safeguard Selection
    - 9.8. Risk Assessment Commitment
10. Manage Projects
    - 10.1. Project Management Framework
    - 10.2. User Department Participation in Project Initiation
    - 10.3. Project Team Membership and Responsibilities
    - 10.4. Project Definition
    - 10.5. Project Approval
    - 10.6. Project Phase Approval
    - 10.7. Project Master Plan
    - 10.8. System Quality Assurance Plan
    - 10.9. Planning of Assurance Methods
    - 10.10. Formal Project Risk Management
    - 10.11. Test Plan
    - 10.12. Training Plan
    - 10.13. Post-Implementation Review Plan

11. Manage Quality
    - 11.1. General Quality Plan
    - 11.2. Quality Assurance Approach
    - 11.3. Quality Assurance Planning
    - 11.4. Quality Assurance Review of Adherence to IT Standards and Procedures
    - 11.5. System Development Life Cycle Methodology
    - 11.6. System Development Life Cycle Methodology for Major Changes to Existing Technology
    - 11.7. Updating of the System Development Life Cycle Methodology
    - 11.8. Coordination and Communication
    - 11.9. Acquisition and Maintenance Framework for the Technology Infrastructure
    - 11.10. Third-Party Implementor Relationships
    - 11.11. Program Documentation Standards
    - 11.12. Program Testing Standards
    - 11.13. System Testing Standards
    - 11.14. Parallel/Pilot Testing
    - 11.15. System Testing Documentation
    - 11.16. Quality Assurance Evaluation of Adherence to Development Standards
    - 11.17. Quality Assurance Review of the Achievement of IT Objectives
    - 11.18. Quality Metrics
    - 11.19. Reports of Quality Assurance Reviews

ACQUISITION & IMPLEMENTATION

1. Identify Automated Solutions
    - 1.1. Definition of Information Requirements
    - 1.2. Formulation of Alternative Courses of Action
    - 1.3. Formulation of Acquisition Strategy
    - 1.4. Third-Party Service Requirements
    - 1.5. Technological Feasibility Study
    - 1.6. Economic Feasibility Study
    - 1.7. Information Architecture
    - 1.8. Risk Analysis Report

- 1.9. Cost-Effective Security Controls
- 1.10. Audit Trails Design
- 1.11. Ergonomics
- 1.12. Selection of System Software
- 1.13. Procurement Control
- 1.14. Software Product Acquisition
- 1.15. Third-Party Software Maintenance
- 1.16. Contract Application Programming
- 1.17. Acceptance of Facilities
- 1.18. Acceptance of Technology

2. Acquire and Maintain Application Software
   - 2.1. Design Methods
   - 2.2. Major Changes to Existing Systems
   - 2.3. Design Approval
   - 2.4. File Requirements Definition and Documentation
   - 2.5. Program Specifications
   - 2.6. Source Data Collection Design
   - 2.7. Input Requirements Definition and Documentation
   - 2.8. Definition of Interfaces
   - 2.9. User-Machine Interface
   - 2.10. Processing Requirements Definition and Documentation
   - 2.11. Output Requirements Definition and Documentation
   - 2.12. Controllability
   - 2.13. Availability as a Key Design Factor
   - 2.14. IT Integrity Provisions in Application Program Software
   - 2.15. Application Software Testing
   - 2.16. User Reference and Support Materials
   - 2.17. Reassessment of System Design
3. Acquire and Maintain Technology Infrastructure
   - 3.1. Assessment of New Hardware and Software
   - 3.2. Preventative Maintenance for Hardware

- 3.3. System Software Security
- 3.4. System Software Installation
- 3.5. System Software Maintenance
- 3.6. System Software Change Controls
- 3.7. Use and Monitoring of System Utilities

4. Develop and Maintain Procedures
   - 4.1. Operational Requirements and Service Levels
   - 4.2. User Procedures Manual
   - 4.3. Operations Manual
   - 4.4. Training Materials
5. Install and Accredit Systems
   - 5.1. Training
   - 5.2. Application Software Performance Sizing
   - 5.3. Implementation Plan
   - 5.4. System Conversion
   - 5.5. Data Conversion
   - 5.6. Testing Strategies and Plans
   - 5.7. Testing of Changes
   - 5.8. Parallel/Pilot Testing Criteria and Performance
   - 5.9. Final Acceptance Test
   - 5.10. Security Testing and Accreditation
   - 5.11. Operational Test
   - 5.12. Promotion to Production
   - 5.13. Evaluation of Meeting User Requirements
   - 5.14. Management's Post-Implementation Review
6. Manage Changes
   - 6.1. Change Request Initiation and Control
   - 6.2. Impact Assessment
   - 6.3. Control of Changes
   - 6.4. Emergency Changes
   - 6.5. Documentation and Procedures

- 6.6. Authorized Maintenance
- 6.7. Software Release Policy
- 6.8. Distribution of Software

DELIVERY & SUPPORT

1. Define and Manage Service Levels
   - 1.1. Service Level Agreement Framework
   - 1.2. Aspects of Service Level Agreements
   - 1.3. Performance Procedures
   - 1.4. Monitoring and Reporting
   - 1.5. Review of Service Level Agreements and Contracts
   - 1.6. Chargeable Items
   - 1.7. Service Improvement Program
2. Manage Third-Party Services
   - 2.1. Supplier Interfaces
   - 2.2. Owner Relationships
   - 2.3. Third-Party Contracts
   - 2.4. Third-Party Qualifications
   - 2.5. Outsourcing Contracts
   - 2.6. Continuity of Services
   - 2.7. Security Relationships
   - 2.8. Monitoring
3. Manage Performance and Capacity
   - 3.1. Availability and Performance Requirements
   - 3.2. Availability Plan
   - 3.3. Monitoring and Reporting
   - 3.4. Modeling Tools
   - 3.5. Proactive Performance Management
   - 3.6. Workload Forecasting
   - 3.7. Capacity Management of Resources
   - 3.8. Resources Availability
   - 3.9. Resources Schedule

4. Ensure Continuous Service
   - 4.1. IT Continuity Framework
   - 4.2. IT Continuity Plan Strategy and Philosophy
   - 4.3. IT Continuity Plan Contents
   - 4.4. Minimizing IT Continuity Requirements
   - 4.5. Maintaining the IT Continuity Plan
   - 4.6. Testing the IT Continuity Plan
   - 4.7. IT Continuity Plan Training
   - 4.8. IT Continuity Plan Distribution
   - 4.9. User Department Alternative Processing Back-up Procedures
   - 4.10. Critical IT Resources
   - 4.11. Back-up Site and Hardware
   - 4.12. Off-site Back-up Storage
   - 4.13. Wrap-up Procedures
5. Ensure Systems Security
   - 5.1. Manage Security Measures
   - 5.2. Identification, Authentication and Access
   - 5.3. Security of Online Access to Data
   - 5.4. User Account Management
   - 5.5. Management Review of User Accounts
   - 5.6. User Control of User Accounts
   - 5.7. Security Surveillance
   - 5.8. Data Classification
   - 5.9. Central Identification and Access Rights Management
   - 5.10. Violation and Security Activity Reports
   - 5.11. Incident Handling
   - 5.12. Reaccredidation
   - 5.13. Counterparty Trust
   - 5.14. Transaction Authorization
   - 5.15. Non-Repudiation
   - 5.16. Trusted Path

- 5.17. Protection of Security Functions
- 5.18. Cryptographic Key Management
- 5.19. Malicious Software Prevention, Detection and Correction
- 5.20. Firewall Architectures and Connections with Public Networks
- 5.21. Protection of Electronic Value

6. Identify and Allocate Costs
   - 6.1. Chargeable Items
   - 6.2. Costing Procedures
   - 6.3. User Billing and Chargeback Procedures
7. Educate and Train Users
   - 7.1. Identification of Training Needs
   - 7.2. Training Organization
   - 7.3. Security Principles and Awareness Training
8. Assist and Advise Customers
   - 8.1. Help Desk
   - 8.2. Registration of Customer Queries
   - 8.3. Customer Query Escalation
   - 8.4. Monitoring of Clearance
   - 8.5. Trend Analysis and Reporting
9. Manage the Configuration
   - 9.1. Configuration Recording
   - 9.2. Configuration Baseline
   - 9.3. Status Accounting
   - 9.4. Configuration Control
   - 9.5. Unauthorized Software
   - 9.6. Software Storage
   - 9.7. Configuration Management Procedures
   - 9.8. Software Accountability
10. Manage Problems and Incidents
    - 10.1. Problem Management System
    - 10.2. Problem Escalation

10.3. Problem Tracking and Audit Trail

10.4. Emergency and Temporary Access Authorizations

10.5. Emergency Processing Priorities

11. Manage Data

11.1. Data Preparation Procedures

11.2. Source Document Authorization Procedures

11.3. Source Document Data Collection

11.4. Source Document Error Handling

11.5. Source Document Retention

11.6. Data Input Authorization Procedures

11.7. Accuracy, Completeness and Authorization Checks

11.8. Data Input Error Handling

11.9. Data Processing Integrity

11.10. Data Processing Validation and Editing

11.11. Data Processing Error Handling

11.12. Output Handling and Retention

11.13. Output Distribution

11.14. Output Balancing and Reconciliation

11.15. Output Review and Error Handling

11.16. Security Provision for Output Reports

11.17. Protection of Sensitive Information During Transmission and Transport

11.18. Protection of Disposed Sensitive Information

11.19. Storage Management

11.20. Retention Periods and Storage Terms

11.21. Media Library Management System

11.22. Media Library Management Responsibilities

11.23. Back-up and Restoration

11.24. Back-up Jobs

11.25. Back-up Storage

11.26. Archiving

11.27. Protection of Sensitive Messages

- 11.28. Authentication and Integrity
- 11.29. Electronic Transaction Integrity
- 11.30. Continued Integrity of Stored Data

12. Manage Facilities
    - 12.1. Physical Security
    - 12.2. Low Profile of the IT Site
    - 12.3. Visitor Escort
    - 12.4. Personnel Health and Safety
    - 12.5. Protection Against Environmental Factors
    - 12.6. Uninterruptible Power Supply
13. Manage Operations
    - 13.1. Processing Operations Procedures and Instructions Manual
    - 13.2. Start-up Process and Other Operations Documentation
    - 13.3. Job Scheduling
    - 13.4. Departures from Standard Job Schedules
    - 13.5. Processing Continuity
    - 13.6. Operations Logs
    - 13.7. Safeguard Special Forms and Output Devices
    - 13.8. Remote Operations

MONITORING

1. Monitor the Processes
    - 1.1. Collecting Monitoring Data
    - 1.2. Assessing Performance
    - 1.3. Assessing Customer Satisfaction
    - 1.4. Management Reporting
2. Assess Internal Control Adequacy
    - 2.1. Internal Control Monitoring
    - 2.2. Timely Operation of Internal Controls
    - 2.3. Internal Control Level Reporting
    - 2.4. Operational Security and Internal Control Assurance

3. Obtain Independent Assurance
    - 3.1. Independent Security and Internal Control Certification/Accreditation of IT Services
    - 3.2. Independent Security and Internal Control Certification/Accreditation of Third-Party Service Providers
    - 3.3. Independent Effectiveness Evaluation of IT Services
    - 3.4. Independent Effectiveness Evaluation of Third-Party Service Providers
    - 3.5. Independent Assurance of Compliance with Laws and Regulatory Requirements and Contractual Commitments
    - 3.6. Independent Assurance of Compliance with Laws and Regulatory Requirements and Contractual Commitments by Third-Party Service Providers
    - 3.7. Competence of Independent Assurance Function
    - 3.8. Proactive Audit Involvement
4. Provide for Independent Audit
    - 4.1. Audit Charter
    - 4.2. Independence
    - 4.3. Professional Ethics and Standards
    - 4.4. Competence
    - 4.5. Planning
    - 4.6. Performance of Audit Work
    - 4.7. Reporting
    - 4.8. Follow-up Activities

Appendix B

ITIL Framework Summary

The Five ITIL Volumes

Service Strategy

The Service Strategy book provides a view of ITIL that aligns business and information technology. It specifies that each stage of the service lifecycle must stay focused upon the business case, with defined business goals, requirements and service management principles.

Service Design

The Service Design book provides guidance upon the production/maintenance of information technology policies, architectures, and documents.

Service Transition

The Service Transition book focuses upon change management role and release practices, providing guidance and process activities for the transition of services into the business environment.

Service Operation

This book focuses upon delivery and control process activities based on a selection of service support and service delivery control points.

Continual Service Improvement

This book focuses upon the process elements involved in identifying and introducing service management improvements, as well as issues surrounding service retirement.

Service Support

1. Change Management
   - 1.1. Scope
     - 1.1.1. Hardware
     - 1.1.2. Communications equipment and software
     - 1.1.3. System software
     - 1.1.4. All documentation and procedures associated with the running, support and maintenance of live systems.
   - 1.2. Activities
     - 1.2.1. Filtering changes
     - 1.2.2. Managing changes and the change process

1.2.3. Chairing the CAB and the CAB/Emergency committee

1.2.4. Reviewing and closing of Requests for Change

1.2.5. Management reporting and providing management information

2. Release Management

    2.1. Scope

        2.1.1. Major software releases and hardware upgrades

        2.1.2. Minor software releases and hardware upgrades

        2.1.3. Emergency software and hardware fixes

        2.1.4. Delta Release

        2.1.5. Full Release

        2.1.6. Packaged Release

    2.2. Scope

        2.2.1. Plan the rollout of software

        2.2.2. Design and implement procedures for the distribution and installation of changes to IT systems

        2.2.3. Effectively communicate and manage expectations of the customer during the planning and rollout of new releases

        2.2.4. Control the distribution and installation of changes to IT systems

3. Problem Management

    3.1. Scope

        3.1.1. Error Control Process: an iterative process to diagnose known errors until they are eliminated by the successful implementation of a change under the control of the Change Management process.

        3.1.2. Problem Control Process: identify the root cause of incidents and reports it to the service desk

            3.1.2.1. Problem identification and recording

            3.1.2.2. Problem classification

            3.1.2.3. Problem investigation and diagnosis

    3.2. Activities

        3.2.1. Trend analysis

        3.2.2. Targeting support action

        3.2.3. Providing information to the organization.

4. Incident Management
    4.1. Scope
        4.1.1. Incident detection and recording
        4.1.2. Classification and initial support
        4.1.3. Investigation and diagnosis
        4.1.4. Resolution and recovery
        4.1.5. Incident closure
        4.1.6. Incident ownership, monitoring, tracking and communication
    4.2. Classification Examples
        4.2.1. Application
            4.2.1.1. service not available
            4.2.1.2. application bug
            4.2.1.3. disk-usage threshold exceeded
        4.2.2. Hardware
            4.2.2.1. system-down
            4.2.2.2. automatic alert
            4.2.2.3. printer not printing
        4.2.3. Service requests
            4.2.3.1. request for information/advice/documentation
            4.2.3.2. forgotten password
5. Configuration Management
    5.1. Attributes
        5.1.1. Technical – Data that describes the CI's capabilities which include software version and model numbers, hardware and manufacturer specifications and other technical details like networking speeds and data storage size. Keyboards, mice and cables are considered consumables.
        5.1.2. Ownership – Part of financial asset management, ownership attributes record purchase date, warranty, location and responsible person for the CI. Identification numbers like bar codes and type, like software, hardware and documentation are also ownership attributes.
        5.1.3. Relationship – The relationships between hardware items (e.g. a printer), software (e.g. drivers), and users (i.e. Alice).

5.2. Activities

5.2.1. Planning: The CM plan covers the next three to six months in detail, and the following twelve months in outline. It is reviewed at least twice a year and will include a strategy, policy, scope, objectives, roles and responsibilities, the CM processes, activities and procedures, the CMDB, relationships with other processes and third parties, as well as tools and other resource requirements. The number of CI categories to track in the CMDB determines the scope. The detail of the CI information is the depth.

5.2.2. Identification: The selection, identification and labeling of all CIs which creates a parts list of every CI in the system. This covers the recording of information about CI's, including hardware and software versions, documentation, ownership and other unique identifiers. CIs should be recorded at a level of detail justified by the business need, typically to the level of "independent change". This includes defining the relationships of the CIs in the system.

5.2.3. Control: This gives the assurance that only authorized and identifiable CIs are accepted and recorded from receipt to disposal. It ensures that no CI is added, modified, replaced or removed without the appropriate controlling documentation e.g. approved Requests for Change of a CI, updated specification. All CIs will be under Change Management (ITSM) control.

5.2.4. Monitoring: The status accounting and reporting of all current and historical data concerned with each CI throughout its life-cycle. It enables changes to CIs and tracking of their records through various statuses, e.g. ordered, received, under test, live, under repair, withdrawn or for disposal.

5.2.5. Verification: The reviews and audits that verify the physical existence of CIs, and checks that they are correctly recorded in the CMDB and parts list. It includes the process of verifying Release Management (ITSM) and CM documentation before changes are made to the live environment.

6. Service Desk

6.1. Scope

6.1.1. Providing a single (informed) point of contact for customers

6.1.2. Facilitating the restoration of normal operational service with minimal business impact on the customer within agreed SLA levels and business priorities.

6.2. Activities

6.2.1. Receiving calls, first-line customer liaison

6.2.2. Recording and tracking incidents and complaints

6.2.3. Keeping customers informed on request status and progress

6.2.4. Making an initial assessment of requests, attempting to resolve them or refer them to someone who can

6.2.5. Monitoring and escalation procedures relative to the appropriate SLA

6.2.6. Identifying problems

6.2.7. Closing incidents and confirmation with the customers

6.2.8. Coordinating second and third line support

Service Delivery

1. Service Level Management

1.1. Scope

1.1.1. ensuring that the agreed IT services are delivered when and where they are supposed to be

1.1.2. liaising with Availability Management, Capacity Management, Incident Management and Problem Management to ensure that the required levels and quality of service are achieved within the resources agreed with Financial Management

1.1.3. producing and maintaining a Service Catalog (a list of standard IT service options and agreements made available to customers)

1.1.4. ensuring that appropriate IT Service Continuity plans have been made to support the business and its continuity requirements.

2. Capacity Management

2.1. Scope

2.1.1. business capacity management

2.1.2. service capacity management

2.1.3. and component capacity management

2.2. Activities

2.2.1. Monitoring the performance and throughput or load on a server, server farm, or property

2.2.2. Performance analysis of measurement data, including analysis of the impact of new releases on capacity

2.2.3. Performance tuning activities to ensure the most efficient use of existing infrastructure

2.2.4. Understanding the demands on the Service and future plans for workload growth (or shrinkage)

2.2.5. Influences on demand for computing resources

2.2.6. Capacity planning – developing a plan for the Service

3. IT Service Continuity Management

3.1. Scope

3.1.1. Risk Analysis

3.1.2. Contingency Plan Management

3.1.3. Contingency Plan Testing

3.1.4. Risk Management.

4. Availability Management

4.1. Scope

4.1.1. Reliability: how reliable is the service? Ability of an IT component to perform at an agreed level at described conditions.

4.1.2. Maintainability: The ability of an IT Component to remain in, or be restored to an operational state.

4.1.3. Serviceability: The ability for an external supplier to maintain the availability of component or function under a third party contract

4.1.4. Resilience: A measure of freedom from operational failure and a method of keeping services reliable. One popular method of resilience is redundancy.

4.1.5. Security: A service may have associated data. Security refers to the confidentiality, integrity, and availability of that data

4.2. Activities

4.2.1. Realize Availability Requirements

4.2.2. Compile Availability Plan

4.2.3. Monitor Availability

4.2.4. Monitor Maintenance Obligations.

5. Financial Management
    5.1. Budgeting
    5.2. IT Accounting
        5.2.1. Capital Costs: Any type of purchases which would have a residual value as hardware and building infrastructure
        5.2.2. Operational Costs: Day to day recurring expenses cost like rental fees, monthly electrical invoices and salaries.
        5.2.3. Direct Costs: Any cost expenses which are directly attributed to one single or specific service or customer. A typical example would be the purchase of a dedicated server which cannot be shared and is needed to host a new application for a specific service or customer.
        5.2.4. Indirect Costs: One specific service provision which cost needs to be distributed in between several customers in a fair breakdown. A fair example is the cost associated to overall Local Area Network on which every customer are connected to. Breakdown could be done using total amount of users per customer or total amount of bandwidth usage per customer to accurately distribute the cost of providing this service.
        5.2.5. Fixed Costs: Any expenses established for long periods of time like annual maintenance contracts or a lease contracts.
        5.2.6. Variable Costs:
    5.3. Charging

Appendix C

GNU General Public Licenses

GPL Version III

GNU GENERAL PUBLIC LICENSE

Version 3, 29 June 2007

Copyright (C) 2007 Free Software Foundation, Inc. <http://fsf.org/>
Everyone is permitted to copy and distribute verbatim copies of this license document, but changing it is not allowed.

Preamble

The GNU General Public License is a free, copyleft license for software and other kinds of works.

The licenses for most software and other practical works are designed to take away your freedom to share and change the works. By contrast, the GNU General Public License is intended to guarantee your freedom to share and change all versions of a program–to make sure it remains free software for all its users. We, the Free Software Foundation, use the GNU General Public License for most of our software; it applies also to any other work released this way by its authors. You can apply it to your programs, too.

When we speak of free software, we are referring to freedom, not price. Our General Public Licenses are designed to make sure that you have the freedom to distribute copies of free software (and charge for them if you wish), that you receive source code or can get it if you want it, that you can change the software or use pieces of it in new free programs, and that you know you can do these things.

To protect your rights, we need to prevent others from denying you these rights or asking you to surrender the rights. Therefore, you have certain responsibilities if you distribute copies of the software, or if you modify it: responsibilities to respect the freedom of others.

For example, if you distribute copies of such a program, whether gratis or for a fee, you must pass on to the recipients the same freedoms that you received. You must make sure that they, too, receive or can get the source code. And you must show them these terms so they know their rights.

Developers that use the GNU GPL protect your rights with two steps: (1) assert copyright on the software, and (2) offer you this License giving you legal permission to copy, distribute and/or modify it.

For the developers' and authors' protection, the GPL clearly explains that there is no warranty for this free software. For both users' and authors' sake, the GPL requires that modified versions be marked as changed, so that their problems will not be attributed erroneously to authors of previous versions.

Some devices are designed to deny users access to install or run modified versions of the software inside them, although the manufacturer can do so. This is fundamentally incompatible with the aim of protecting users' freedom to change the software. The systematic pattern of such abuse occurs in the area of products for individuals to use, which is precisely where it is most unacceptable. Therefore, we have designed this version of the GPL to prohibit the practice for those products. If such problems arise substantially in other domains, we stand ready to extend this provision to those domains in future versions of the GPL, as needed to protect the freedom of users.

Finally, every program is threatened constantly by software patents. States should not allow patents to restrict development and use of software on general-purpose computers, but in those that do, we wish to avoid the special danger that patents applied to a free program could make it effectively proprietary. To prevent this, the GPL assures that patents cannot be used to render the program non-free.

The precise terms and conditions for copying, distribution and modification follow.

TERMS AND CONDITIONS

0. Definitions

- "This License" refers to version 3 of the GNU General Public License.
- "Copyright" also means copyright-like laws that apply to other kinds of works, such as semiconductor masks.
- "The Program" refers to any copyrightable work licensed under this License. Each licensee is addressed as "you". "Licensees" and "recipients" may be individuals or organizations.
- To "modify" a work means to copy from or adapt all or part of the work in a fashion requiring copyright permission, other than the making of an exact copy. The resulting work is called a "modified version" of the earlier work or a work "based on" the earlier work.
- A "covered work" means either the unmodified Program or a work based on the Program.
- To "propagate" a work means to do anything with it that, without permission, would make you directly or secondarily liable for infringement under applicable copyright law, except executing it on a computer or modifying a private copy. Propagation includes copying, distribution (with or without modification), making available to the public, and in some countries other activities as well.
- To "convey" a work means any kind of propagation that enables other parties to make or receive copies. Mere interaction with a user through a computer network, with no transfer of a copy, is not conveying.

- An interactive user interface displays "Appropriate Legal Notices" to the extent that it includes a convenient and prominently visible feature that (1) displays an appropriate copyright notice, and (2) tells the user that there is no warranty for the work (except to the extent that warranties are provided), that licensees may convey the work under this License, and how to view a copy of this License. If the interface presents a list of user commands or options, such as a menu, a prominent item in the list meets this criterion.

1. Source Code

The "source code" for a work means the preferred form of the work for making modifications to it. "Object code" means any non-source form of a work.

A "Standard Interface" means an interface that either is an official standard defined by a recognized standards body, or, in the case of interfaces specified for a particular programming language, one that is widely used among developers working in that language.

The "System Libraries" of an executable work include anything, other than the work as a whole, that (a) is included in the normal form of packaging a Major Component, but which is not part of that Major Component, and (b) serves only to enable use of the work with that Major Component, or to implement a Standard Interface for which an implementation is available to the public in source code form. A "Major Component", in this context, means a major essential component (kernel, window system, and so on) of the specific operating system (if any) on which the executable work runs, or a compiler used to produce the work, or an object code interpreter used to run it.

The "Corresponding Source" for a work in object code form means all the source code needed to generate, install, and (for an executable work) run the object code and to modify the work, including scripts to control those activities. However, it does not include the work's System Libraries, or general-purpose tools or generally available free programs which are used unmodified in performing those activities but which are not part of the work. For example, Corresponding Source includes interface definition files associated with source files for the work, and the source code for shared libraries and dynamically linked subprograms that the work is specifically designed to require, such as by intimate data communication or control flow between those subprograms and other parts of the work.

The Corresponding Source need not include anything that users can regenerate automatically from other parts of the Corresponding Source.

The Corresponding Source for a work in source code form is that same work.

2. Basic Permissions

All rights granted under this License are granted for the term of copyright on the Program, and are irrevocable provided the stated conditions are met. This License explicitly affirms your unlimited permission to run the unmodified Program. The output from running a covered work is covered by this License only if the output, given its content, constitutes a covered work. This License acknowledges your rights of fair use or other equivalent, as provided by copyright law.

You may make, run and propagate covered works that you do not convey, without conditions so long as your license otherwise remains in force. You may convey covered works to others for the sole purpose of having them make modifications exclusively for you, or provide you with facilities for running those works, provided that you comply with the terms of this License in conveying all material for which you do not control copyright. Those thus making or running the covered works for you must do so exclusively on your behalf, under your direction and control, on terms that prohibit them from making any copies of your copyrighted material outside their relationship with you.

Conveying under any other circumstances is permitted solely under the conditions stated below. Sublicensing is not allowed; section 10 makes it unnecessary.

3. Protecting Users' Legal Rights From Anti-Circumvention Law

No covered work shall be deemed part of an effective technological measure under any applicable law fulfilling obligations under article 11 of the WIPO copyright treaty adopted on 20 December 1996, or similar laws prohibiting or restricting circumvention of such measures.

When you convey a covered work, you waive any legal power to forbid circumvention of technological measures to the extent such circumvention is effected by exercising rights under this License with respect to the covered work, and you disclaim any intention to limit operation or modification of the work as a means of enforcing, against the work's users, your or third parties' legal rights to forbid circumvention of technological measures.

4. Conveying Verbatim Copies

You may convey verbatim copies of the Program's source code as you receive it, in any medium, provided that you conspicuously and appropriately publish on each copy an appropriate copyright notice; keep intact all notices stating that this License and any non-permissive terms added in accord with section 7 apply to the code; keep intact all notices of the absence of any warranty; and give all recipients a copy of this License along with the Program.

You may charge any price or no price for each copy that you convey, and you may offer support or warranty protection for a fee.

5. Conveying Modified Source Versions

You may convey a work based on the Program, or the modifications to produce it from the Program, in the form of source code under the terms of section 4, provided that you also meet all of these conditions:

a) The work must carry prominent notices stating that you modified it, and giving a relevant date.

b) The work must carry prominent notices stating that it is released under this License and any conditions added under section 7. This requirement modifies the requirement in section 4 to "keep intact all notices".

c) You must license the entire work, as a whole, under this License to anyone who comes into possession of a copy. This License will therefore apply, along with any applicable section 7 additional terms, to the whole of the work, and all its parts, regardless of how they are packaged. This License gives no permission to license the work in any other way, but it does not invalidate such permission if you have separately received it.

d) If the work has interactive user interfaces, each must display Appropriate Legal Notices; however, if the Program has interactive interfaces that do not display Appropriate Legal Notices, your work need not make them do so.

A compilation of a covered work with other separate and independent works, which are not by their nature extensions of the covered work, and which are not combined with it such as to form a larger program, in or on a volume of a storage or distribution medium, is called an "aggregate" if the compilation and its resulting copyright are not used to limit the access or legal rights of the compilation's users beyond what the individual works permit. Inclusion of a covered work in an aggregate does not cause this License to apply to the other parts of the aggregate.

6. Conveying Non-Source Forms

You may convey a covered work in object code form under the terms of sections 4 and 5, provided that you also convey the machine-readable Corresponding Source under the terms of this License, in one of these ways:

a) Convey the object code in, or embodied in, a physical product (including a physical distribution medium), accompanied by the Corresponding Source fixed on a durable physical medium customarily used for software interchange.

b) Convey the object code in, or embodied in, a physical product (including a physical distribution medium), accompanied by a written offer, valid for at least three years and valid for as long as you offer spare parts or customer support for that product model, to give anyone who possesses the object code either (1) a copy of the Corresponding Source for all the software in the product that is covered by this License, on a durable physical medium customarily used for software interchange, for a price no more than your reasonable cost of physically performing this conveying of source, or (2) access to copy the Corresponding Source from a network server at no charge.

c) Convey individual copies of the object code with a copy of the written offer to provide the Corresponding Source. This alternative is allowed only occasionally and noncommercially, and only if you received the object code with such an offer, in accord with subsection 6b.

d) Convey the object code by offering access from a designated place (gratis or for a charge), and offer equivalent access to the Corresponding Source in the same way through the same place at no further charge. You need not require recipients to copy the Corresponding Source along with the object code. If the place to copy the object code is a network server, the Corresponding Source may be on a different server (operated by you or a third party) that supports equivalent copying facilities, provided you maintain clear directions next to the object code saying where to find the Corresponding Source. Regardless of what server hosts the Corresponding Source, you remain obligated to ensure that it is available for as long as needed to satisfy these requirements.

e) Convey the object code using peer-to-peer transmission, provided you inform other peers where the object code and Corresponding Source of the work are being offered to the general public at no charge under subsection 6d.

A separable portion of the object code, whose source code is excluded from the Corresponding Source as a System Library, need not be included in conveying the object code work.

A "User Product" is either (1) a "consumer product", which means any tangible personal property which is normally used for personal, family, or household purposes, or (2) anything designed or sold for incorporation into a dwelling. In determining whether a product is a consumer product, doubtful cases shall be resolved in favor of coverage. For a particular product received by a particular user, "normally used" refers to a typical or common use of that class of product, regardless of the status of the particular user or of the way in which the particular user actually uses, or expects or is expected to use, the product. A product is a consumer product regardless of whether the product has substantial commercial, industrial or non-consumer uses, unless such uses represent the only significant mode of use of the product.

"Installation Information" for a User Product means any methods, procedures, authorization keys, or other information required to install and execute modified versions of a covered work in that User Product from a modified version of its Corresponding Source. The information must suffice to ensure that the continued functioning of the modified object code is in no case prevented or interfered with solely because modification has been made.

If you convey an object code work under this section in, or with, or specifically for use in, a User Product, and the conveying occurs as part of a transaction in which the right of possession and use of the User Product is transferred to the recipient in perpetuity or for a fixed term (regardless of how the transaction is characterized), the Corresponding Source conveyed under this section must be accompanied by the Installation Information. But this requirement does not apply if neither you nor any third party retains the ability to install modified object code on the User Product (for example, the work has been installed in ROM).

The requirement to provide Installation Information does not include a requirement to continue to provide support service, warranty, or updates for a work that has been modified or installed by the recipient, or for the User Product in which it has been modified or installed. Access to a network may be denied when the modification itself materially and adversely affects the operation of the network or violates the rules and protocols for communication across the network.

Corresponding Source conveyed, and Installation Information provided, in accord with this section must be in a format that is publicly documented (and with an implementation available to the public in source code form), and must require no special password or key for unpacking, reading or copying.

7. Additional Terms

"Additional permissions" are terms that supplement the terms of this License by making exceptions from one or more of its conditions. Additional permissions that are applicable to the entire Program shall be treated as though they were included in this License, to the extent that they are valid under applicable law. If additional permissions apply only to part of the Program, that part may be used separately under those permissions, but the entire Program remains governed by this License without regard to the additional permissions.

When you convey a copy of a covered work, you may at your option remove any additional permissions from that copy, or from any part of it. (Additional permissions may be written to require their own removal in certain cases when you modify the work.) You may place additional permissions on material, added by you to a covered work, for which you have or can give appropriate copyright permission.

Notwithstanding any other provision of this License, for material you add to a covered work, you may (if authorized by the copyright holders of that material) supplement the terms of this License with terms:

a) Disclaiming warranty or limiting liability differently from the terms of sections 15 and 16 of this License; or

b) Requiring preservation of specified reasonable legal notices or author attributions in that material or in the Appropriate Legal Notices displayed by works containing it; or

c) Prohibiting misrepresentation of the origin of that material, or requiring that modified versions of such material be marked in reasonable ways as different from the original version; or

d) Limiting the use for publicity purposes of names of licensors or authors of the material; or

e) Declining to grant rights under trademark law for use of some trade names, trademarks, or service marks; or

f) Requiring indemnification of licensors and authors of that material by anyone who conveys the material (or modified versions of it) with contractual assumptions of liability to the recipient, for any liability that these contractual assumptions directly impose on those licensors and authors.

All other non-permissive additional terms are considered "further restrictions" within the meaning of section 10. If the Program as you received it, or any part of it, contains a notice stating that it is governed by this License along with a term that is a further restriction, you may remove that term. If a license document contains a further restriction but permits relicensing or conveying under this License, you may add to a covered work material governed by the terms of that license document, provided that the further restriction does not survive such relicensing or conveying.

If you add terms to a covered work in accord with this section, you must place, in the relevant source files, a statement of the additional terms that apply to those files, or a notice indicating where to find the applicable terms.

Additional terms, permissive or non-permissive, may be stated in the form of a separately written license, or stated as exceptions; the above requirements apply either way.

8. Termination

You may not propagate or modify a covered work except as expressly provided under this License. Any attempt otherwise to propagate or modify it is void, and will automatically terminate your rights under this License (including any patent licenses granted under the third paragraph of section 11).

However, if you cease all violation of this License, then your license from a particular copyright holder is reinstated (a) provisionally, unless and until the copyright holder explicitly and finally terminates your license, and (b) permanently, if the copyright holder fails to notify you of the violation by some reasonable means prior to 60 days after the cessation.

Moreover, your license from a particular copyright holder is reinstated permanently if the copyright holder notifies you of the violation by some reasonable means, this is the first time you have received notice of violation of this License (for any work) from that copyright holder, and you cure the violation prior to 30 days after your receipt of the notice.

Termination of your rights under this section does not terminate the licenses of parties who have received copies or rights from you under this License. If your rights have been terminated and not permanently reinstated, you do not qualify to receive new licenses for the same material under section 10.

9. Acceptance Not Required for Having Copies

You are not required to accept this License in order to receive or run a copy of the Program. Ancillary propagation of a covered work occurring solely as a consequence of using peer-to-peer transmission to receive a copy likewise does not require acceptance. However, nothing other than this License grants you permission to propagate or modify any covered work. These actions infringe copyright if you do not accept this License. Therefore, by modifying or propagating a covered work, you indicate your acceptance of this License to do so.

10. Automatic Licensing of Downstream Recipients

Each time you convey a covered work, the recipient automatically receives a license from the original licensors, to run, modify and propagate that work, subject to this License. You are not responsible for enforcing compliance by third parties with this License.

An "entity transaction" is a transaction transferring control of an organization, or substantially all assets of one, or subdividing an organization, or merging organizations. If propagation of a covered work results from an entity transaction, each party to that transaction who receives a copy of the work also receives whatever licenses to the work the party's predecessor in interest had or could give under the previous paragraph, plus a right to possession of the Corresponding Source of the work from the predecessor in interest, if the predecessor has it or can get it with reasonable efforts.

You may not impose any further restrictions on the exercise of the rights granted or affirmed under this License. For example, you may not impose a license fee, royalty, or other charge for exercise of rights granted under this License, and you may not initiate litigation (including a cross-claim or counterclaim in a lawsuit) alleging that any patent claim is infringed by making, using, selling, offering for sale, or importing the Program or any portion of it.

11. Patents

A "contributor" is a copyright holder who authorizes use under this License of the Program or a work on which the Program is based. The work thus licensed is called the contributor's "contributor version".

A contributor's "essential patent claims" are all patent claims owned or controlled by the contributor, whether already acquired or hereafter acquired, that would be infringed by some manner, permitted by this License, of making, using, or selling its contributor version, but do not include claims that would be infringed only as a consequence of further modification of the contributor version. For purposes of this definition, "control" includes the right to grant patent sublicenses in a manner consistent with the requirements of this License.

Each contributor grants you a non-exclusive, worldwide, royalty-free patent license under the contributor's essential patent claims, to make, use, sell, offer for sale, import and otherwise run, modify and propagate the contents of its contributor version.

In the following three paragraphs, a "patent license" is any express agreement or commitment, however denominated, not to enforce a patent (such as an express permission to practice a patent or covenant not to sue for patent infringement). To "grant" such a patent license to a party means to make such an agreement or commitment not to enforce a patent against the party.

If you convey a covered work, knowingly relying on a patent license, and the Corresponding Source of the work is not available for anyone to copy, free of charge and under the terms of this License, through a publicly available network server or other readily accessible means, then you must either (1) cause the Corresponding Source to be so available, or (2) arrange to deprive yourself of the benefit of the patent license for this particular work, or (3) arrange, in a manner consistent with the requirements of this License, to extend the patent license to downstream recipients. "Knowingly relying" means you have actual knowledge that, but for the patent license, your conveying the covered work in a country, or your recipient's use of the covered work in a country, would infringe one or more identifiable patents in that country that you have reason to believe are valid.

If, pursuant to or in connection with a single transaction or arrangement, you convey, or propagate by procuring conveyance of, a covered work, and grant a patent license to some of the parties receiving the covered work authorizing them to use, propagate, modify or convey a specific copy of the covered work, then the patent license you grant is automatically extended to all recipients of the covered work and works based on it.

A patent license is "discriminatory" if it does not include within the scope of its coverage, prohibits the exercise of, or is conditioned on the non-exercise of one or more of the rights that are specifically granted under this License. You may not convey a covered work if you are a party to an arrangement with a third party that is in the business of distributing software, under which you make payment to the third party based on the extent of your activity of conveying the work, and under which the third party grants, to any of the parties who would receive the covered work from you, a discriminatory patent license (a) in connection with copies of the covered work conveyed by you (or copies made from those copies), or (b) primarily for and in connection with specific products or compilations that contain the covered work, unless you entered into that arrangement, or that patent license was granted, prior to 28 March 2007.

Nothing in this License shall be construed as excluding or limiting any implied license or other defenses to infringement that may otherwise be available to you under applicable patent law.

12. No Surrender of Others' Freedom

If conditions are imposed on you (whether by court order, agreement or otherwise) that contradict the conditions of this License, they do not excuse you from the conditions of this License. If you cannot convey a covered work so as to satisfy simultaneously your obligations under this License and any other pertinent obligations, then as a consequence you may not convey it at all. For example, if you agree to terms that obligate you to collect a royalty for further conveying from those to whom you convey the Program, the only way you could satisfy both those terms and this License would be to refrain entirely from conveying the Program.

13. Use with the GNU Affero General Public License

Notwithstanding any other provision of this License, you have permission to link or combine any covered work with a work licensed under version 3 of the GNU Affero General Public License into a single combined work, and to convey the resulting work. The terms of this License will continue to apply to the part which is the covered work, but the special requirements of the GNU Affero General Public License, section 13, concerning interaction through a network will apply to the combination as such.

14. Revised Versions of this License

The Free Software Foundation may publish revised and/or new versions of the GNU General Public License from time to time. Such new versions will be similar in spirit to the present version, but may differ in detail to address new problems or concerns.

Each version is given a distinguishing version number. If the Program specifies that a certain numbered version of the GNU General Public License "or any later version" applies to it, you have the option of following the terms and conditions either of that numbered version or of any later version published by the Free Software Foundation. If the Program does not specify a version number of the GNU General Public License, you may choose any version ever published by the Free Software Foundation.

If the Program specifies that a proxy can decide which future versions of the GNU General Public License can be used, that proxy's public statement of acceptance of a version permanently authorizes you to choose that version for the Program.

Later license versions may give you additional or different permissions. However, no additional obligations are imposed on any author or copyright holder as a result of your choosing to follow a later version.

15. Disclaimer of Warranty

THERE IS NO WARRANTY FOR THE PROGRAM, TO THE EXTENT PERMITTED BY APPLICABLE LAW. EXCEPT WHEN OTHERWISE STATED IN WRITING THE COPYRIGHT HOLDERS AND/OR OTHER PARTIES PROVIDE THE PROGRAM "AS IS" WITHOUT WARRANTY OF ANY KIND, EITHER EXPRESSED OR IMPLIED, INCLUDING, BUT NOT LIMITED TO, THE IMPLIED WARRANTIES OF MERCHANTABILITY AND FITNESS FOR A PARTICULAR PURPOSE. THE ENTIRE RISK AS TO THE QUALITY AND PERFORMANCE OF THE PROGRAM IS WITH YOU. SHOULD THE PROGRAM PROVE DEFECTIVE, YOU ASSUME THE COST OF ALL NECESSARY SERVICING, REPAIR OR CORRECTION.

16. Limitation of Liability

IN NO EVENT UNLESS REQUIRED BY APPLICABLE LAW OR AGREED TO IN WRITING WILL ANY COPYRIGHT HOLDER, OR ANY OTHER PARTY WHO

MODIFIES AND/OR CONVEYS THE PROGRAM AS PERMITTED ABOVE, BE LIABLE TO YOU FOR DAMAGES, INCLUDING ANY GENERAL, SPECIAL, INCIDENTAL OR CONSEQUENTIAL DAMAGES ARISING OUT OF THE USE OR INABILITY TO USE THE PROGRAM (INCLUDING BUT NOT LIMITED TO LOSS OF DATA OR DATA BEING RENDERED INACCURATE OR LOSSES SUSTAINED BY YOU OR THIRD PARTIES OR A FAILURE OF THE PROGRAM TO OPERATE WITH ANY OTHER PROGRAMS), EVEN IF SUCH HOLDER OR OTHER PARTY HAS BEEN ADVISED OF THE POSSIBILITY OF SUCH DAMAGES.

17. Interpretation of Sections 15 and 16

If the disclaimer of warranty and limitation of liability provided above cannot be given local legal effect according to their terms, reviewing courts shall apply local law that most closely approximates an absolute waiver of all civil liability in connection with the Program, unless a warranty or assumption of liability accompanies a copy of the Program in return for a fee.

END OF TERMS AND CONDITIONS

GPL Version II

GNU GENERAL PUBLIC LICENSE

Version 2, June 1991
Copyright (C) 1989, 1991 Free Software Foundation, Inc.
51 Franklin Street, Fifth Floor, Boston, MA 02110-1301, USA
Everyone is permitted to copy and distribute verbatim copies of this license document, but changing it is not allowed.

Preamble

The licenses for most software are designed to take away your freedom to share and change it. By contrast, the GNU General Public License is intended to guarantee your freedom to share and change free software–to make sure the software is free for all its users. This General Public License applies to most of the Free Software Foundation's software and to any other program whose authors commit to using it. (Some other Free Software Foundation software is covered by the GNU Lesser General Public License instead.) You can apply it to your programs, too.

When we speak of free software, we are referring to freedom, not price. Our General Public Licenses are designed to make sure that you have the freedom to distribute copies of free software (and charge for this service if you wish), that you receive source code or can get it if you want it, that you can change the software or use pieces of it in new free programs; and that you know you can do these things.

To protect your rights, we need to make restrictions that forbid anyone to deny you these rights or to ask you to surrender the rights. These restrictions translate to certain responsibilities for you if you distribute copies of the software, or if you modify it.

For example, if you distribute copies of such a program, whether gratis or for a fee, you must give the recipients all the rights that you have. You must make sure that they, too, receive or can get the source code. And you must show them these terms so they know their rights.

We protect your rights with two steps: (1) copyright the software, and (2) offer you this license which gives you legal permission to copy, distribute and/or modify the software.

Also, for each author's protection and ours, we want to make certain that everyone understands that there is no warranty for this free software. If the software is modified by someone else and passed on, we want its recipients to know that what they have is not the original, so that any problems introduced by others will not reflect on the original authors' reputations.

Finally, any free program is threatened constantly by software patents. We wish to avoid the danger that redistributors of a free program will individually obtain patent licenses, in effect making the program proprietary. To prevent this, we have made it clear that any patent must be licensed for everyone's free use or not licensed at all.

The precise terms and conditions for copying, distribution and modification follow.

TERMS AND CONDITIONS FOR COPYING, DISTRIBUTION AND MODIFICATION

0

This License applies to any program or other work which contains a notice placed by the copyright holder saying it may be distributed under the terms of this General Public License. The "Program", below, refers to any such program or work, and a "work based on the Program" means either the Program or any derivative work under copyright law: that is to say, a work containing the Program or a portion of it, either verbatim or with modifications and/or translated into another language. (Hereinafter, translation is included without limitation in the term "modification".) Each licensee is addressed as "you".

Activities other than copying, distribution and modification are not covered by this License; they are outside its scope. The act of running the Program is not restricted, and the output from the Program is covered only if its contents constitute a work based on the Program (independent of having been made by running the Program). Whether that is true depends on what the Program does.

1

You may copy and distribute verbatim copies of the Program's source code as you receive it, in any medium, provided that you conspicuously and appropriately publish on each copy an appropriate copyright notice and disclaimer of warranty; keep intact all the notices that refer to this License and to the absence of any warranty; and give any other recipients of the Program a copy of this License along with the Program.

You may charge a fee for the physical act of transferring a copy, and you may at your option offer warranty protection in exchange for a fee.

2

You may modify your copy or copies of the Program or any portion of it, thus forming a work based on the Program, and copy and distribute such modifications or work under the terms of Section 1 above, provided that you also meet all of these conditions:

a) You must cause the modified files to carry prominent notices stating that you changed the files and the date of any change.

b) You must cause any work that you distribute or publish, that in whole or in part contains or is derived from the Program or any part thereof, to be licensed as a whole at no charge to all third parties under the terms of this License.

c) If the modified program normally reads commands interactively when run, you must cause it, when started running for such interactive use in the most ordinary way, to print or display an announcement including an appropriate copyright notice and a notice that there is no warranty (or else, saying that you provide a warranty) and that users may redistribute the program under these conditions, and telling the user how to view a copy of this License. (Exception: if the Program itself is interactive but does not normally print such an announcement, your work based on the Program is not required to print an announcement.)

These requirements apply to the modified work as a whole. If identifiable sections of that work are not derived from the Program, and can be reasonably considered independent and separate works in themselves, then this License, and its terms, do not apply to those sections when you distribute them as separate works. But when you distribute the same sections as part of a whole which is a work based on the Program, the distribution of the whole must be on the terms of this License, whose permissions for other licensees extend to the entire whole, and thus to each and every part regardless of who wrote it.

Thus, it is not the intent of this section to claim rights or contest your rights to work written entirely by you; rather, the intent is to exercise the right to control the distribution of derivative or collective works based on the Program.

In addition, mere aggregation of another work not based on the Program with the Program (or with a work based on the Program) on a volume of a storage or distribution medium does not bring the other work under the scope of this License.

3

You may copy and distribute the Program (or a work based on it, under Section 2) in object code or executable form under the terms of Sections 1 and 2 above provided that you also do one of the following:

a) Accompany it with the complete corresponding machine-readable source code, which must be distributed under the terms of Sections 1 and 2 above on a medium customarily used for software interchange; or,

b) Accompany it with a written offer, valid for at least three years, to give any third party, for a charge no more than your cost of physically performing source distribution, a complete machine-readable copy of the corresponding source code, to be distributed under the terms of Sections 1 and 2 above on a medium customarily used for software interchange; or,

c) Accompany it with the information you received as to the offer to distribute corresponding source code. (This alternative is allowed only for noncommercial distribution and only if you received the program in object code or executable form with such an offer, in accord with Subsection b above.)

The source code for a work means the preferred form of the work for making modifications to it. For an executable work, complete source code means all the source code for all modules it contains, plus any associated interface definition files, plus the scripts used to control compilation and installation of the executable. However, as a special exception, the source code distributed need not include anything that is normally distributed (in either source or binary form) with the major components (compiler, kernel, and so on) of the operating system on which the executable runs, unless that component itself accompanies the executable.

If distribution of executable or object code is made by offering access to copy from a designated place, then offering equivalent access to copy the source code from the same place counts as distribution of the source code, even though third parties are not compelled to copy the source along with the object code.

4

You may not copy, modify, sublicense, or distribute the Program except as expressly provided under this License. Any attempt otherwise to copy, modify, sublicense or distribute the Program is void, and will automatically terminate your rights under this License. However, parties who have received copies, or rights, from you under this License will not have their licenses terminated so long as such parties remain in full compliance.

5

You are not required to accept this License, since you have not signed it. However, nothing else grants you permission to modify or distribute the Program or its derivative works.

These actions are prohibited by law if you do not accept this License. Therefore, by modifying or distributing the Program (or any work based on the Program), you indicate your acceptance of this License to do so, and all its terms and conditions for copying, distributing or modifying the Program or works based on it.

6

Each time you redistribute the Program (or any work based on the Program), the recipient automatically receives a license from the original licensor to copy, distribute or modify the Program subject to these terms and conditions. You may not impose any further restrictions on the recipients' exercise of the rights granted herein. You are not responsible for enforcing compliance by third parties to this License.

7

If, as a consequence of a court judgment or allegation of patent infringement or for any other reason (not limited to patent issues), conditions are imposed on you (whether by court order, agreement or otherwise) that contradict the conditions of this License, they do not excuse you from the conditions of this License. If you cannot distribute so as to satisfy simultaneously your obligations under this License and any other pertinent obligations, then as a consequence you may not distribute the Program at all. For example, if a patent license would not permit royalty-free redistribution of the Program by all those who receive copies directly or indirectly through you, then the only way you could satisfy both it and this License would be to refrain entirely from distribution of the Program.

If any portion of this section is held invalid or unenforceable under any particular circumstance, the balance of the section is intended to apply and the section as a whole is intended to apply in other circumstances.

It is not the purpose of this section to induce you to infringe any patents or other property right claims or to contest validity of any such claims; this section has the sole purpose of protecting the integrity of the free software distribution system, which is implemented by public license practices. Many people have made generous contributions to the wide range of software distributed through that system in reliance on consistent application of that system; it is up to the author/donor to decide if he or she is willing to distribute software through any other system and a licensee cannot impose that choice.

This section is intended to make thoroughly clear what is believed to be a consequence of the rest of this License.

8

If the distribution and/or use of the Program is restricted in certain countries either by patents or by copyrighted interfaces, the original copyright holder who places the Program under this License may add an explicit geographical distribution limitation excluding those countries, so that distribution is permitted only in or among countries not thus excluded. In such case, this License incorporates the limitation as if written in the body of this License.

9

The Free Software Foundation may publish revised and/or new versions of the General Public License from time to time. Such new versions will be similar in spirit to the present version, but may differ in detail to address new problems or concerns.

Each version is given a distinguishing version number. If the Program specifies a version number of this License which applies to it and "any later version", you have the option of following the terms and conditions either of that version or of any later version published by the Free Software Foundation. If the Program does not specify a version number of this License, you may choose any version ever published by the Free Software Foundation.

10

If you wish to incorporate parts of the Program into other free programs whose distribution conditions are different, write to the author to ask for permission. For software which is copyrighted by the Free Software Foundation, write to the Free Software Foundation; we sometimes make exceptions for this. Our decision will be guided by the two goals of preserving the free status of all derivatives of our free software and of promoting the sharing and reuse of software generally.

NO WARRANTY

11

BECAUSE THE PROGRAM IS LICENSED FREE OF CHARGE, THERE IS NO WARRANTY FOR THE PROGRAM, TO THE EXTENT PERMITTED BY APPLICABLE LAW. EXCEPT WHEN OTHERWISE STATED IN WRITING THE COPYRIGHT HOLDERS AND/OR OTHER PARTIES PROVIDE THE PROGRAM "AS IS" WITHOUT WARRANTY OF ANY KIND, EITHER EXPRESSED OR IMPLIED, INCLUDING, BUT NOT LIMITED TO, THE IMPLIED WARRANTIES OF MERCHANTABILITY AND FITNESS FOR A PARTICULAR PURPOSE. THE ENTIRE RISK AS TO THE QUALITY AND PERFORMANCE OF THE PROGRAM IS WITH YOU. SHOULD THE PROGRAM PROVE DEFECTIVE, YOU ASSUME THE COST OF ALL NECESSARY SERVICING, REPAIR OR CORRECTION.

12

IN NO EVENT UNLESS REQUIRED BY APPLICABLE LAW OR AGREED TO IN WRITING WILL ANY COPYRIGHT HOLDER, OR ANY OTHER PARTY WHO MAY MODIFY AND/OR REDISTRIBUTE THE PROGRAM AS PERMITTED

ABOVE, BE LIABLE TO YOU FOR DAMAGES, INCLUDING ANY GENERAL, SPECIAL, INCIDENTAL OR CONSEQUENTIAL DAMAGES ARISING OUT OF THE USE OR INABILITY TO USE THE PROGRAM (INCLUDING BUT NOT LIMITED TO LOSS OF DATA OR DATA BEING RENDERED INACCURATE OR LOSSES SUSTAINED BY YOU OR THIRD PARTIES OR A FAILURE OF THE PROGRAM TO OPERATE WITH ANY OTHER PROGRAMS), EVEN IF SUCH HOLDER OR OTHER PARTY HAS BEEN ADVISED OF THE POSSIBILITY OF SUCH DAMAGES.

END OF TERMS AND CONDITIONS

Index

D

T

U

V

W

X

Z